An Introduction to Soil Dynamics

Theory and Applications of Transport in Porous Media

Series Editor:
Jacob Bear, *Department of Civil and Environmental Engineering,
Technion – Israel Institute of Technology, Haifa, Israel*

Volume 24

For further volumes: http://www.springer.com/series/6612.

An Introduction to Soil Dynamics

Arnold Verruijt

Delft University of Technology, Delft,
The Netherlands

Arnold Verruijt
Delft University of Technology
2628 CN Delft
Netherlands
a.verruijt@verruijt.net

A CD-ROM accompanies this book containing programs for waves in piles, propagation of earthquakes in soils, waves in a half space generated by a line load, a point load, a strip load, or a moving load, and the propagation of a shock wave in a saturated elastic porous material.

Computer programs are also available from the website http://geo.verruijt.net

ISBN 978-94-007-3096-0   ISBN 978-90-481-3441-0 (eBook)
DOI 10.1007/978-90-481-3441-0
Springer Dordrecht Heidelberg London New York

Additional material to this book can be downloaded from http://extras.springer.com

Springer is part of Springer Science+Business Media (www.springer.com)

Preface

This book gives the material for an introductory course on Soil Dynamics, as given for about 10 years at the Delft University of Technology for students of civil engineering, and updated continuously since 1994.

The book presents the basic principles of elastodynamics and the major solutions of problems of interest for geotechnical engineering. For most problems the full analytical derivation of the solution is given, mainly using integral transform methods. These methods are presented briefly in Appendix A. The elastostatic solutions of many problems are also given, as an introduction to the elastodynamic solutions, and as possible limiting states of the corresponding dynamic problems. For a number of problems of elastodynamics of a half space exact solutions are given, in closed form, using methods developed by Pekeris and De Hoop. Some of these basic solutions are derived in full detail, to assist in understanding the beautiful techniques used in deriving them. For many problems the main functions for a computer program to produce numerical data and graphs are given, in C. Some approximations in which the horizontal displacements are disregarded, an approximation suggested by Westergaard and Barends, are also given, because they are much easier to derive, may give a first insight in the response of a foundation, and may be a stepping stone to solving the more difficult complete elastodynamic problems.

The book is directed towards students of engineering, and may be giving more details of the derivations of the solutions than strictly necessary, or than most other books on elastodynamics give, but this may be excused by my own difficulties in studying the subject, and by helping students with similar difficulties.

The book starts with a chapter on the behaviour of the simplest elementary system, a system consisting of a mass, supported by a linear spring and a linear damper. The main purpose of this chapter is to define the basic properties of dynamical systems, for future reference. In this chapter the major forms of damping of importance for soil dynamics problems, viscous damping and hysteretic damping, are defined and their properties are investigated.

Chapters 2 and 3 are devoted to one dimensional problems: wave propagation in piles, and wave propagation in layers due to earthquakes in the underlying layers, as first developed in the 1970s at the University of California, Berkeley. In these chapters the mathematical methods of Laplace and Fourier transforms, characteristics, and separation of variables, are used and compared. Some simple numerical models are also presented.

The next two chapters (Chaps. 4 and 5) deal with the important effect that soils are usually composed of two constituents: solid particles and a fluid, usually water, but perhaps oil, or a mixture of a liquid and gas. Chapter 4 presents the classical theory, due to Terzaghi, of semi-static consolidation, and some elementary solutions. In Chap. 5 the extension to the dynamical case is presented, mainly for the one dimensional case, as first presented by De Josselin de Jong and Biot, in 1956. The solution for the propagation of waves in a one dimensional column is presented, leading to the important conclusion that for most problems a practically saturated soil can be considered as a medium in which the solid particles and the fluid move and deform together, which in soil mechanics is usually denoted as a state of undrained deformations. For an elastic solid skeleton this means that the soil behaves as an elastic material with Poisson's ratio close to 0.5.

Chapters 6 and 7 deal with the solution of problems of cylindrical and spherical symmetry. In the chapter on cylindrically symmetric problems the propagation of waves in an infinite medium introduces Rayleigh's important principle of the radiation condition, which expresses that in an infinite medium no waves can be expected to travel from infinity towards the interior of the body.

Chapters 8 and 9 give the basic theory of the theory of elasticity for static and dynamic problems. Chapter 8 also gives the solution for some of the more difficult problems, involving mixed boundary value conditions. The corresponding dynamic problems still await solution, at least in analytic form. Chapter 9 presents the basics of dynamic problems in elastic continua, including the general properties of the most important types of waves: compression waves, shear waves, Rayleigh waves and Love waves, which appear in other chapters.

Chapter 10, on confined elastodynamics, presents an approximate theory of elastodynamics, in which the horizontal deformations are artificially assumed to vanish, an approximation due to Westergaard and generalized by Barends. This makes it possible to solve a variety of problems by simple means, and resulting in relatively simple solutions. It should be remembered that these are approximate solutions only, and that important features of the complete solutions, such as the generation of Rayleigh waves, are excluded. These approximate solutions are included in the present book because they are so much simpler to derive and to analyze than the full elastodynamic solutions. The full elastodynamic solutions of the problems considered in this chapter are given in Chaps. 11–13.

In soil mechanics the elastostatic solutions for a line load or a distributed load on a half plane are of great importance because they provide basic solutions for the stress distribution in soils due to loads on the surface. In Chaps. 11 and 12 the solution for two corresponding elastodynamic problems, a line load on a half plane and a strip load on a half plane, are derived. These chapters rely heavily on the theory developed by Cagniard and De Hoop. The solutions for impulse loads, which can be found in many publications, are first given, and then these are used as the basics for the solutions for the stresses in case of a line load constant in time. These solutions should tend towards the well known elastostatic limits, as they indeed do.

An important aspect of these solutions is that for large values of time the Rayleigh wave is clearly observed, in agreement with the general wave theory for a half plane. Approximate solutions valid for large values of time, including the Rayleigh waves, are derived for the line load and the strip load. These approximate solutions may be useful as the basis for the analysis of problems with a more general type of loading.

Chapter 13 presents the solution for a point load on an elastic half space, a problem first solved analytically by Pekeris. The solution is derived using integral transforms and an elegant transformation theorem due to Bateman and Pekeris. In this chapter numerical values are obtained using numerical integration of the final integrals.

In Chap. 14 some problems of moving loads are considered. Closed form solutions appear to be possible for a moving wave load, and for a moving strip load, assuming that the material possesses some hysteretic damping.

Chapter 15, finally, presents some practical considerations on foundation vibrations. On the basis of solutions derived in earlier chapters approximate solutions are expressed in the form of equivalent springs and dampings.

The text has been prepared using the LATEX version (Lamport, 1994) of the program TEX (Knuth, 1986). The PICTEX macros (Wichura, 1987) have been used to prepare the figures. Modern software provides a major impetus to the production of books and papers in facilitating the illustration of complex solutions by numerical and graphical examples. In this book many solutions are accompanied by parts of computer programs that have been used to produce the figures, so that readers can compose their own programs. It is all the more appropriate to acknowledge the effort that must have been made by earlier authors and their associates in producing their publications. A case in point is the paper by Lamb, more than a century ago, with many illustrative figures, for which the computations were made by Mr. Woodall.

The programs used to produce many of the illustrations in the book can be downloaded from the website http://geo.verruijt.net. Updates of these programs will be published on this website. Early versions of the book have been published on this website, leading to helpful comments by readers from all over the world.

Many thanks are due to Professor A.T. de Hoop for his many helpful and constructive ideas and comments, and to Dr. C. Cornejo Córdova for several years of joint research. Further comments will be greatly appreciated.

Delft   Arnold Verruijt

Contents

CD-ROM

A CD-ROM accompanies this book containing programs for waves in piles, propagation of earthquakes in soils, waves in a half space generated by a line load, a point load, a strip load, or a moving load, and the propagation of a shock wave in a saturated elastic porous material.

Chapter 1
Vibrating Systems

In this chapter a classical basic problem of dynamics will be considered, for the purpose of introducing various concepts and properties. The system to be considered is a single mass, supported by a linear spring and a viscous damper. The response of this simple system will be investigated, for various types of loading, such as a periodic load and a step load. In order to demonstrate some of the mathematical techniques the problems are solved by various methods, such as harmonic analysis using complex response functions, and the Laplace transform method.

1.1 Single Mass System

Consider the system of a single mass, supported by a spring and a dashpot, in which the damping is of a viscous character, see Fig. 1.1. The spring and the damper form a connection between the mass and an immovable base (for instance the earth).

According to Newton's second law the equation of motion of the mass is

$$m\frac{d^2u}{dt^2} = P(t),$$ (1.1)

where $P(t)$ is the total force acting upon the mass m, and u is the displacement of the mass.

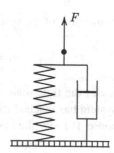

Fig. 1.1 Mass supported by spring and damper

A. Verruijt, *An Introduction to Soil Dynamics*,
Theory and Applications of Transport in Porous Media 24,
© Springer Science+Business Media B.V. 2010

It is now assumed that the total force P consists of an external force $F(t)$, and the reaction of a spring and a damper. In its simplest form a spring leads to a force linearly proportional to the displacement u, and a damper leads to a response linearly proportional to the velocity du/dt. If the spring constant is k and the viscosity of the damper is c, the total force acting upon the mass is

$$P(t) = F(t) - ku - c\frac{du}{dt}. \qquad (1.2)$$

Thus the equation of motion for the system is

$$m\frac{d^2u}{dt^2} + c\frac{du}{dt} + ku = F(t). \qquad (1.3)$$

The response of this simple system will be analyzed by various methods, in order to be able to compare the solutions with various problems from soil dynamics. In many cases a problem from soil dynamics can be reduced to an equivalent single mass system, with an equivalent mass, an equivalent spring constant, and an equivalent viscosity (or damping). The main purpose of many studies is to derive expressions for these quantities. Therefore it is essential that the response of a single mass system under various types of loading is fully understood. For this purpose both free vibrations and forced vibrations of the system will be considered in some detail.

1.2 Characterization of Viscosity

The damper has been characterized in the previous section by its viscosity c. Alternatively this element can be characterized by a response time of the spring-damper combination. The response of a system of a parallel spring and damper to a unit step load of magnitude F_0 is

$$u = \frac{F_0}{k}[1 - \exp(-t/t_r)], \qquad (1.4)$$

where t_r is the response time of the system, defined by

$$t_r = c/k. \qquad (1.5)$$

This quantity expresses the time scale of the response of the system. After a time of say $t \approx 4t_r$ the system has reached its final equilibrium state, in which the spring dominates the response. If $t < t_r$ the system is very stiff, with the damper dominating its behaviour.

1.3 Free Vibrations

When the system is unloaded, i.e. $F(t) = 0$, the possible vibrations of the system are called free vibrations. They are described by the homogeneous equation

$$m\frac{d^2u}{dt^2} + c\frac{du}{dt} + ku = 0. \tag{1.6}$$

An obvious solution of this equation is $u = 0$, which means that the system is at rest. If it is at rest initially, say at time $t = 0$, then it remains at rest. It is interesting to investigate, however, the response of the system when it has been brought out of equilibrium by some external influence. For convenience of the future discussions we write

$$\omega_0 = \sqrt{k/m}, \tag{1.7}$$

and

$$2\zeta = \omega_0 t_r = \frac{c}{m\omega_0} = \frac{c\omega_0}{k} = \frac{c}{\sqrt{km}}. \tag{1.8}$$

The quantity ω_0 will turn out to be the resonance frequency of the undamped system, and ζ will be found to be a measure for the damping in the system.

With (1.7) and (1.8) the differential equation can be written as

$$\frac{d^2u}{dt^2} + 2\zeta\omega_0\frac{du}{dt} + \omega_0^2 u = 0. \tag{1.9}$$

This is an ordinary linear differential equation, with constant coefficients. According to the standard approach in the theory of linear differential equations the solution of the differential equation is sought in the form

$$u = A\exp(\alpha t), \tag{1.10}$$

where A is a constant, probably related to the initial value of the displacement u, and α is as yet unknown. Substitution into (1.9) gives

$$\alpha^2 + 2\zeta\omega_0\alpha + \omega_0^2 = 0. \tag{1.11}$$

This is called the characteristic equation of the problem. The assumption that the solution is an exponential function, see (1.10), appears to be justified, if (1.11) can be solved for the unknown parameter α. The possible values of α are determined by the roots of the quadratic equation (1.11). These roots are, in general,

$$\alpha_{1,2} = -\zeta\omega_0 \pm \omega_0\sqrt{\zeta^2 - 1}. \tag{1.12}$$

These solutions may be real, or they may be complex, depending upon the sign of the quantity $\zeta^2 - 1$. Thus, the character of the response of the system depends upon the value of the damping ratio ζ, because this determines whether the roots are real or complex. The various possibilities will be considered separately below. Because many systems are only slightly damped, it is most convenient to first consider the case of small values of the damping ratio ζ.

Small Damping

When the damping ratio is smaller than 1, $\zeta < 1$, the roots of the characteristic equation (1.11) are both complex,

$$\alpha_{1,2} = -\zeta\omega_0 \pm i\omega_0\sqrt{1 - \zeta^2}, \tag{1.13}$$

where i is the imaginary unit, $i = \sqrt{-1}$. In this case the solution can be written as

$$u = A_1\exp(i\omega_1 t)\exp(-\zeta\omega_0 t) + A_2\exp(-i\omega_1 t)\exp(-\zeta\omega_0 t), \tag{1.14}$$

where

$$\omega_1 = \omega_0\sqrt{1 - \zeta^2}. \tag{1.15}$$

The complex exponential function $\exp(i\omega_1 t)$ may be expressed as

$$\exp(i\omega_1 t) = \cos(\omega_1 t) + i\sin(\omega_1 t). \tag{1.16}$$

Therefore the solution (1.14) may also be written in terms of trigonometric functions, which is often more convenient,

$$u = C_1\cos(\omega_1 t)\exp(-\zeta\omega_0 t) + C_2\sin(\omega_1 t)\exp(-\zeta\omega_0 t). \tag{1.17}$$

The constants C_1 and C_2 depend upon the initial conditions. When these initial conditions are that at time $t = 0$ the displacement is given to be u_0 and the velocity is zero, it follows that the final solution is

$$\frac{u}{u_0} = \frac{\cos(\omega_1 t - \psi)}{\cos(\psi)}\exp(-\zeta\omega_0 t), \tag{1.18}$$

where ψ is a phase angle, defined by

$$\tan(\psi) = \frac{\omega_0\zeta}{\omega_1} = \frac{\zeta}{\sqrt{1 - \zeta^2}}. \tag{1.19}$$

The solution (1.18) is a damped sinusoidal vibration. It is a fluctuating function, with its zeroes determined by the zeroes of the function $\cos(\omega_1 t - \psi)$, and its amplitude gradually diminishing, according to the exponential function $\exp(-\zeta\omega_0 t)$.

The solution is shown graphically in Fig. 1.2 for various values of the damping ratio ζ. If the damping is small, the frequency of the vibrations is practically equal to that of the undamped system, ω_0, see also (1.15). For larger values of the

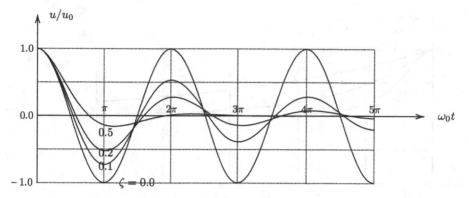

Fig. 1.2 Free vibrations of a weakly damped system

damping ratio the frequency is slightly smaller. The influence of the frequency on the amplitude of the response then appears to be very large. For large frequencies the amplitude becomes very small. If the frequency is so large that the damping ratio ζ approaches 1 the character of the solution may even change from that of a damped fluctuation to the non-fluctuating response of a strongly damped system. These conditions are investigated below.

Critical Damping

When the damping ratio is equal to 1, $\zeta = 1$, the characteristic equation (1.11) has two equal roots,

$$\alpha_{1,2} = -\omega_0. \tag{1.20}$$

In this case the damping is said to be *critical*. The solution of the problem in this case is, taking into account that there is a double root,

$$u = (A + Bt)\exp(-\omega_0 t), \tag{1.21}$$

where the constants A and B must be determined from the initial conditions. When these are again that at time $t = 0$ the displacement is u_0 and the velocity is zero, it follows that the final solution is

$$u = u_0(1 + \omega_0 t)\exp(-\omega_0 t). \tag{1.22}$$

This solution is shown in Fig. 1.3, together with some results for large damping ratios.

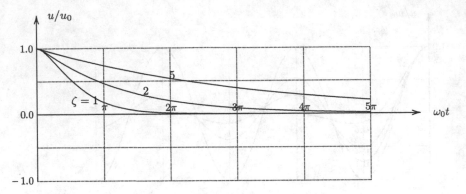

Fig. 1.3 Free vibrations of a strongly damped system

Large Damping

When the damping ratio is greater than 1 ($\zeta > 1$) the characteristic equation (1.11) has two real roots,

$$\alpha_{1,2} = -\zeta\omega_0 \pm \omega_0\sqrt{\zeta^2 - 1}. \tag{1.23}$$

The solution for the case of a mass point with an initial displacement u_0 and an initial velocity zero now is

$$\frac{u}{u_0} = \frac{\omega_2}{\omega_2 - \omega_1}\exp(-\omega_1 t) - \frac{\omega_1}{\omega_2 - \omega_1}\exp(-\omega_2 t), \tag{1.24}$$

where

$$\omega_1 = \omega_0(\zeta - \sqrt{\zeta^2 - 1}), \tag{1.25}$$

and

$$\omega_2 = \omega_0(\zeta + \sqrt{\zeta^2 - 1}). \tag{1.26}$$

This solution is also shown graphically in Fig. 1.3, for $\zeta = 2$ and $\zeta = 5$. It appears that in these cases, with large damping, the system will not oscillate, but will monotonously tend towards the equilibrium state $u = 0$.

1.4 Forced Vibrations

In the previous section the possible free vibrations of the system have been investigated, assuming that there was no load on the system. When there is a certain load, periodic or not, the response of the system also depends upon the characteristics of this load. This case of *forced vibrations* is studied in this section and the next. In the present section the load is assumed to be periodic.

For a periodic load the force $F(t)$ can be written, in its simplest form, as

$$F = F_0 \cos(\omega t), \tag{1.27}$$

where ω is the given circular frequency of the load. In engineering practice the frequency is sometimes expressed by the frequency of oscillation f, defined as the number of cycles per unit time (cps, cycles per second),

$$f = \omega/2\pi. \tag{1.28}$$

In order to study the response of the system to such a periodic load it is most convenient to write the force as

$$F = \Re\{F_0 \exp(i\omega t)\}, \tag{1.29}$$

where the symbol $\Re$ indicates the real value of the term between brackets. If it is assumed that F_0 is real the two expressions (1.27) and (1.29) are equivalent.

The solution for the displacement u is now also written in terms of a complex variable,

$$u = \Re\{U \exp(i\omega t)\}, \tag{1.30}$$

where U in general will appear to be complex. Substitution of (1.30) and (1.29) into the differential equation (1.3) gives

$$(k + ic\omega - m\omega^2)U = F_0. \tag{1.31}$$

Actually, only the real part of this equation is obtained, but it is convenient to add the (irrelevant) imaginary part of the equation, so that a fully complex equation is obtained. After all the calculations have been completed the real part should be considered only, in accordance with (1.30).

The solution of the problem defined by (1.31) is

$$U = \frac{F_0/k}{1 + 2i\zeta\omega/\omega_0 - \omega^2/\omega_0^2}, \tag{1.32}$$

where, as before,

$$\omega_0 = \sqrt{k/m}, \quad 2\zeta = \frac{c}{m\omega_0} = \frac{c\omega_0}{k} = \frac{c}{\sqrt{km}}. \tag{1.33}$$

The quantity ω_0 is the resonance frequency of the undamped system, and ζ is a measure for the damping in the system.

With (1.30) and (1.32) the displacement is now found to be

$$u = u_0 \cos(\omega t - \psi), \tag{1.34}$$

where the amplitude u_0 is given by

$$u_0 = \frac{F_0/k}{\sqrt{(1 - \omega^2/\omega_0^2)^2 + (2\zeta\omega/\omega_0)^2}}, \tag{1.35}$$

and the phase angle ψ is given by

$$\tan\psi = \frac{2\zeta\,\omega/\omega_0}{1 - \omega^2/\omega_0^2}. \tag{1.36}$$

In terms of the original parameters the amplitude can be written as

$$u_0 = \frac{F_0/k}{\sqrt{(1 - m\omega^2/k)^2 + (c\omega/k)^2}}, \tag{1.37}$$

and in terms of these parameters the phase angle ψ is given by

$$\tan\psi = \frac{c\omega/k}{1 - m\omega^2/k}. \tag{1.38}$$

It is interesting to note that for the case of a system of zero mass these expressions tend towards simple limits,

$$m = 0 \;:\; u_0 = \frac{F_0/k}{\sqrt{1 + (c\omega/k)^2}}, \tag{1.39}$$

and

$$m = 0 \;:\; \tan\psi = \frac{c\omega}{k}. \tag{1.40}$$

The amplitude of the system, as described by (1.35), is shown graphically in Fig. 1.4, as a function of the frequency, and for various values of the damping ratio ζ. It appears that for small values of the damping ratio there is a definite maximum of the response curve, which even becomes infinitely large if $\zeta \to 0$. This is called *resonance* of the system. If the system is undamped resonance occurs if $\omega = \omega_0 = \sqrt{k/m}$. This is sometimes called the *eigen frequency* of the free vibrating system.

One of the most interesting aspects of the solution is the behaviour near resonance. Actually the maximum response occurs when the slope of the curve in

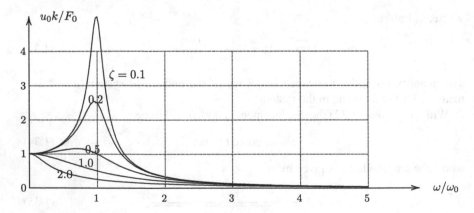

Fig. 1.4 Amplitude of forced vibration

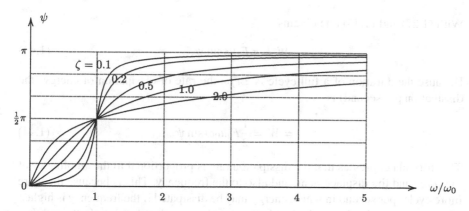

Fig. 1.5 Phase angle of forced vibration

Fig. 1.4 is horizontal. This is the case when $du_0/d\omega = 0$, or, with (1.35),

$$\frac{du_0}{d\omega} = 0 : \frac{\omega}{\omega_0} = \sqrt{1 - 2\zeta^2}. \tag{1.41}$$

For small values of the damping ratio ζ this means that the maximum amplitude occurs if the frequency ω is very close to ω_0, the resonance frequency of the un-damped system. For large values of the damping ratio the resonance frequency may be somewhat smaller, even approaching 0 when $2\zeta^2$ approaches 1. When the damping ratio is very large, the system will never show any sign of resonance. Of course the price to be paid for this very stable behaviour is the installation of a damping element with a very high viscosity.

The phase angle ψ is shown in a similar way in Fig. 1.5. For small frequencies, that is for quasi-static loading, the amplitude of the system approaches the static response F_0/k, and the phase angle is practically 0. In the neighbourhood of the resonance frequency of the undamped system (i.e. if $\omega/\omega_0 \approx 1$) the phase angle is about $\pi/2$, which means that the amplitude is maximal when the force is zero, and vice versa. For very rapid fluctuations the inertia of the system may prevent practically all vibrations (as indicated by the very small amplitude, see Fig. 1.4), but the system moves out of phase, as indicated by the phase angle approaching π, see Fig. 1.5.

Dissipation of Work

An interesting quantity is the dissipation of work during a full cycle. This can be derived by calculating the work done by the force during a full cycle,

$$W = \int_{\omega t = 0}^{2\pi} F \frac{du}{dt} dt. \tag{1.42}$$

With (1.27) and (1.34) one obtains

$$W = \pi F_0 u_0 \sin \psi. \tag{1.43}$$

Because the duration of a full cycle is $2\pi/\omega$ the rate of dissipation of energy (the dissipation per second) is

$$D = \dot{W} = \frac{1}{2} F_0 u_0 \omega \sin \psi. \tag{1.44}$$

This formula expresses that the dissipation rate is proportional to the amplitudes of the force and the displacement, and also to the frequency. This is because there are more cycles per second in which energy may be dissipated if the frequency is higher. The proportionality factor $\sin \psi$, which depends upon the phase angle ψ, and thus upon the viscosity c, see (1.8), finally expresses the relative part of the energy that is dissipated. The maximum of this factor is 1, if the displacement and the force are out of phase. Its minimum is 0, when the viscosity of the damper is zero.

Using the expressions for $\tan \psi$ and F_0/u_0 given in (1.35) and (1.36) the formula for the energy dissipation per cycle can also be written in various other forms. One of the simplest expressions appears to be

$$W = \pi c \omega u_0^2. \tag{1.45}$$

This shows that the energy dissipation is zero for static loading (when the frequency is zero), or when the viscosity vanishes. It may be noted that the formula suggests that the energy dissipation may increase indefinitely when the frequency is very large, but this is not true. For very high frequencies the displacement u_0 becomes very small. In this respect the original formula, (1.43), is a more useful general expression.

1.5 Equivalent Spring and Damping

The analysis of the response of a system to a periodic load, characterized by a time function $\exp(i\omega t)$, often leads to a relation of the form

$$F = (K + iC\omega)U, \tag{1.46}$$

where U is the amplitude of a characteristic displacement, F is the amplitude of the force, and K and C may be complicated functions of the parameters representing the properties of the system, and perhaps also of the frequency ω. Comparison of this relation with (1.31) shows that this response function is of the same character as that of a combination of a spring and a damper. This means that the system can be considered as equivalent with such a spring-damper system, with equivalent stiffness K and equivalent damping C. The response of the system can then be analyzed using the properties of a spring-damper system. This type of equivalence will be used in Chap. 15 to study the response of a vibrating mass on an elastic half plane. The method can also be used to study the response of a foundation pile in an elastic layer. Actually, it is often very convenient and useful to try to represent the response of a complicated system to a harmonic load in the form of an equivalent spring stiffness K and an equivalent damping C.

In the special case of a sinusoidal displacement one may write

$$u = \Im\{U \exp(i\omega t)\} = U \sin(\omega t), \tag{1.47}$$

if U is real. The corresponding force now is, with (1.46),

$$F = \Im\{(K + iC\omega)U \exp(i\omega t)\}, \tag{1.48}$$

or,

$$F = \{K \sin(\omega t) + C\omega \cos(\omega t)\}U. \tag{1.49}$$

This is another useful form of the general relation between force and displacement in case of a spring K and damping C.

1.6 Solution by Laplace Transform Method

It may be interesting to present also the method of solution of the original differential equation (1.3),

$$m\frac{d^2u}{dt^2} + c\frac{du}{dt} + ku = F(t), \tag{1.50}$$

by the Laplace transform method. This is a general technique, that enables to solve the problem for any given load $F(t)$ (Churchill, 1972). As an example the problem will be solved for a step load, applied at time $t = 0$,

$$F(t) = \begin{cases} 0, & \text{if } t < 0, \\ F_0, & \text{if } t > 0. \end{cases} \tag{1.51}$$

It is assumed that at time $t = 0$ the system is at rest, so that both the displacement u and the velocity du/dt are zero at time $t = 0$.

The Laplace transform of the displacement u is defined as

$$\bar{u} = \int_0^\infty u \exp(-st)\, dt, \tag{1.52}$$

where s is the Laplace transform variable. The most characteristic property of the Laplace transform is that differentiation with respect to time t is transformed into multiplication by the transform parameter s. Thus the differential equation (1.50) becomes

$$(ms^2 + cs + k)\bar{u} = \int_0^\infty F(t) \exp(-st)\, dt = \frac{F_0}{s}. \tag{1.53}$$

Again it is convenient to introduce the characteristic frequency ω_0 and the damping ratio ζ, see (1.7) and (1.8), such that

$$k = \omega_0^2 m, \tag{1.54}$$

and

$$c = 2\zeta m \omega_0. \tag{1.55}$$

The solution of the algebraic equation (1.53) is

$$\bar{u} = \frac{F_0/m}{s(s + \omega_1)(s + \omega_2)}, \tag{1.56}$$

where

$$\omega_1 = \omega_0\left(\zeta - i\sqrt{1 - \zeta^2}\right), \tag{1.57}$$

and

$$\omega_2 = \omega_0\left(\zeta + i\sqrt{1 - \zeta^2}\right). \tag{1.58}$$

These definitions are in agreement with (1.25) and (1.26) given above.

The solution (1.56) can also be written as

$$\bar{u} = \frac{F_0}{m}\left\{\frac{1}{\omega_1\omega_2 s} - \frac{1}{\omega_1(\omega_2 - \omega_1)(s + \omega_1)} + \frac{1}{\omega_2(\omega_2 - \omega_1)(s + \omega_2)}\right\}. \tag{1.59}$$

In this form the solution is suitable for inverse Laplace transformation. The result is

$$u = \frac{F_0}{m}\left\{\frac{1}{\omega_1\omega_2} - \frac{\exp(-\omega_1 t)}{\omega_1(\omega_2 - \omega_1)} + \frac{\exp(-\omega_2 t)}{\omega_2(\omega_2 - \omega_1)}\right\}. \tag{1.60}$$

Using the definitions (1.57) and (1.58) and some elementary mathematical operations this expression can also be written as

$$u = \frac{F_0}{k}\left\{1 - \left[\cos\left(\omega_0 t\sqrt{1 - \zeta^2}\right) + \frac{\zeta}{\sqrt{1 - \zeta^2}}\sin\left(\omega_0 t\sqrt{1 - \zeta^2}\right)\right]\exp(-\zeta\omega_0 t)\right\}. \tag{1.61}$$

This formula applies for all values of the damping ratio ζ. For values larger than 1, however, the formula is inconvenient because then the factor $\sqrt{1 - \zeta^2}$ is imaginary. For such cases the formula can better be written in the equivalent form

$$u = \frac{F_0}{k}\left\{1 - \left[\cosh\left(\omega_0 t\sqrt{\zeta^2 - 1}\right) + \frac{\zeta}{\sqrt{\zeta^2 - 1}}\sinh\left(\omega_0 t\sqrt{\zeta^2 - 1}\right)\right]\exp(-\zeta\omega_0 t)\right\}. \tag{1.62}$$

For the case of critical damping, $\zeta = 1$, both formulas contain a factor $0/0$, and the solution seems to degenerate. For that case a simple expansion of the functions near

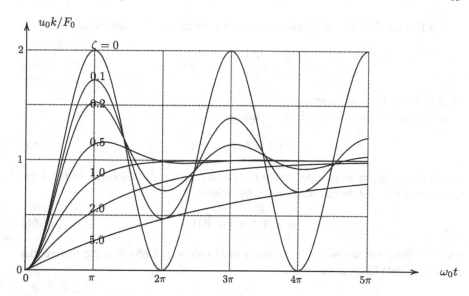

Fig. 1.6 Response to step load

$\zeta = 1$ gives, however,

$$\zeta = 1 \; : \; u = \frac{F_0}{k}\{1 - (1 + \omega_0 t)\exp(-\omega_0 t)\}. \tag{1.63}$$

Figure 1.6 shows the response of the system as a function of time, for various values of the damping ratio. It appears that an oscillating response occurs if the damping is smaller than critical. When there is absolutely no damping these oscillations will continue forever, but damping results in the oscillations gradually vanishing. The system will ultimately approach its new equilibrium state, with a displacement F_0/k. When the damping is sufficiently large, such that $\zeta > 1$, the oscillations are suppressed, and the system will approach its equilibrium state by a monotonously increasing function.

It has been shown in this section that the Laplace transform method can be used to solve the dynamic problem in a straightforward way. For a step load this solution method leads to a relatively simple closed form solution, which can be obtained by elementary means. For other types of loading the analysis may be more complicated, however, depending upon the characteristics of the load function.

1.7 Hysteretic Damping

In this section an alternative form of damping is introduced, *hysteretic damping*, which may be better suited to describe the damping in soils.

It is first recalled that the basic equation of a single mass system is, see (1.3),

$$m\frac{d^2u}{dt^2} + c\frac{du}{dt} + ku = F(t),$$ (1.64)

where c is the viscous damping.

In the case of forced vibrations the load is

$$F(t) = F_0\cos(\omega t),$$ (1.65)

where F_0 is a given amplitude, and ω is a given frequency. As seen in Sect. 1.4 the response of the system can be obtained by writing

$$u = \Re\{U\exp(i\omega t)\},$$ (1.66)

where U may be complex. Substitution of (1.66) and (1.65) into the differential equation (1.64) leads to the equation

$$(k + ic\omega - m\omega^2)U = F_0.$$ (1.67)

In Sect. 1.4 it was assumed that the viscosity c is a constant. In that case the damping ratio ζ was defined as

$$2\zeta = \frac{c}{m\omega_0} = \frac{c\omega_0}{k} = \frac{c}{\sqrt{km}},$$ (1.68)

where

$$\omega_0 = \sqrt{k/m},$$ (1.69)

the resonance frequency (or *eigen frequency*) of the undamped system. All this means that the influence of the damping depends upon the frequency, see for instance Fig. 1.4, which shows that the amplitude of the vibrations tends towards zero when $\omega/\omega_0 \to \infty$.

A different type of damping is *hysteretic damping*, which may be used to represent the damping caused in a vibrating system by dry friction. In this case it is assumed that the factor $c\omega/k$ is constant. The damping ratio ζ_h is now defined as

$$2\zeta_h = \omega t_r = \frac{c\omega}{k}.$$ (1.70)

It is often considered that hysteretic damping is a more realistic representation of the behaviour of soils than viscous damping. The main reason is that the irreversible (plastic) deformations that occur in soils under cyclic loading are independent of the frequency of the loading. This can be expressed by a constant damping ratio ζ_h as defined here.

Equation (1.67) now can be written as

$$k(1 + 2i\zeta_h - \omega^2/\omega_0^2)U = F_0,$$ (1.71)

with the solution

$$U = \frac{F_0/k}{1 + 2i\zeta_h - \omega^2/\omega_0^2}. \tag{1.72}$$

The displacement u now is

$$u = u_0 \cos(\omega t - \psi_h), \tag{1.73}$$

where the amplitude u_0 is given by

$$u_0 = \frac{F_0/k}{\sqrt{(1 - \omega^2/\omega_0^2)^2 + 4\zeta_h^2}}, \tag{1.74}$$

and the phase angle ψ_h is given by

$$\tan\psi = \frac{2\zeta_h}{1 - \omega^2/\omega_0^2}. \tag{1.75}$$

For a system of zero mass these expressions tend towards simple limits,

$$m = 0 \ : \ u_0 = \frac{F_0/k}{\sqrt{1 + 4\zeta_h^2}}, \tag{1.76}$$

and

$$m = 0 \ : \ \tan\psi_h = 2\zeta_h. \tag{1.77}$$

These formulas express that in this case both the amplitude and the phase shift are constant, independent of the frequency ω. This means that the response of the system is independent of the speed of loading and unloading. This is a familiar characteristic of materials such as soft soils (especially granular materials) under cyclic loading. For this reason hysteretic damping seems to be a more realistic form of damping in soils than viscous damping (Hardin, 1965; Verruijt, 1999).

The amplitude of the system, as described by (1.74), is shown graphically in Fig. 1.7, as a function of the frequency, and for various values of the hysteretic damping ratio ζ_h. The behaviour is very similar to that of a system with viscous damping, see Fig. 1.4, except for small values of the frequency. However, in this system the influence of the mass dominates the response, especially for high frequencies.

The phase angle is shown in Fig. 1.8. Again it appears that the main difference with the system having viscous damping occurs for small values of the frequency. For large values of the frequency the influence of the mass appears to dominate the response of the system.

It should be noted that in the absence of mass the response of a system with hysteretic damping is quite different from that of a system with viscous damping, as demonstrated by the difference between (1.39) and (1.76). In a system with viscous damping the amplitude tends towards zero for high frequencies, see (1.39), whereas

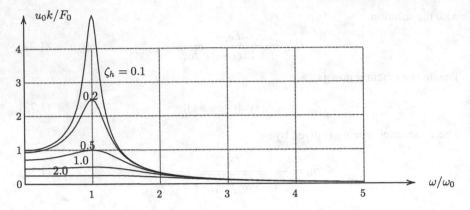

Fig. 1.7 Amplitude of forced vibration, hysteretic damping

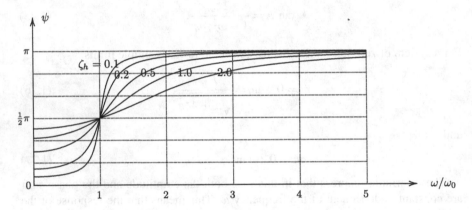

Fig. 1.8 Phase angle of forced vibration, hysteretic damping

in a system with hysteretic damping (and zero mass) the amplitude is independent of the frequency, see (1.76).

Chapter 2
Waves in Piles

In this chapter the problem of the propagation of compression waves in piles is studied. This problem is of importance when considering the behaviour of a foundation pile and the soil during pile driving, and under dynamic loading, such as the behaviour of a pile in the foundation of a railway bridge. Because of the one-dimensional character of the problem, and the simple shape of the pile, usually having a constant cross section and a long length, this is one of the simplest problems of wave propagation in a mathematical sense, and therefore it may be used to illustrate some of the main characteristics of engineering dynamics. Several methods of analysis will be used: the Laplace transform method, separation of variables, the method of characteristics, and numerical solution methods.

2.1 One-Dimensional Wave Equation

First, the case of a free standing pile will be considered, ignoring the interaction with the soil. In later sections the friction interaction with the surrounding soil, and the interaction with the soil at the base will be considered.

Consider a pile of constant cross sectional area A, consisting of a linear elastic material, with modulus of elasticity E. If there is no friction along the shaft of the pile the equation of motion of an element is

$$\frac{\partial N}{\partial z} = \rho A \frac{\partial^2 w}{\partial t^2}, \qquad (2.1)$$

where ρ is the mass density of the material, and w is the displacement in axial direction. The normal force N is related to the stress by

$$N = \sigma A,$$

and the stress is related to the strain by Hooke's law for the pile material

$$\sigma = E\varepsilon.$$

A. Verruijt, *An Introduction to Soil Dynamics*,
Theory and Applications of Transport in Porous Media 24,
© Springer Science+Business Media B.V. 2010

Fig. 2.1 Element of pile

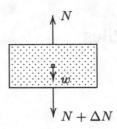

Finally, the strain is related to the vertical displacement w by the relation

$$\varepsilon = \partial w / \partial z.$$

Thus the normal force N is related to the vertical displacement w by the relation

$$N = EA \frac{\partial w}{\partial z}. \tag{2.2}$$

Substitution of (2.2) into (2.1) gives

$$E \frac{\partial^2 w}{\partial z^2} = \rho \frac{\partial^2 w}{\partial t^2}. \tag{2.3}$$

This is the *wave equation*. It can be solved analytically, for instance by the Laplace transform method, separation of variables, or by the method of characteristics, or it can be solved numerically. All these techniques are presented in this chapter. The analytical solution will give insight into the behaviour of the solution. A numerical model is particularly useful for more complicated problems, involving friction along the shaft of the pile, and non-uniform properties of the pile and the soil.

2.2 Solution by Laplace Transform Method

Many problems of one-dimensional wave propagation can be solved conveniently by the Laplace transform method (Churchill, 1972), see also Appendix A. Some examples of this technique are given in this section.

2.2.1 Pile of Infinite Length

The Laplace transform of the displacement w is defined by

$$\overline{w}(z, s) = \int_0^{\infty} w(z, t) \exp(-st) \, dt, \tag{2.4}$$

where s is the Laplace transform parameter, which can be assumed to have a positive real part. Now consider the problem of a pile of infinite length, which is initially at rest, and on the top of which a constant pressure is applied, starting at time $t = 0$. The Laplace transform of the differential equation (2.3) now is

$$\frac{d^2\overline{w}}{dz^2} = \frac{s^2}{c^2}\overline{w},$$

(2.5)

where c is the wave velocity,

$$c = \sqrt{E/\rho}.$$

(2.6)

The solution of the ordinary differential equation (2.5) that vanishes at infinity is

$$\overline{w} = A\exp(-sz/c).$$

(2.7)

The integration constant A, which may depend upon the transformation parameter s, can be obtained from the boundary condition. For a constant pressure p_0 applied at the top of the pile this boundary condition is

$$z = 0, \ t > 0 : \ E\frac{\partial w}{\partial z} = -p_0.$$

(2.8)

The Laplace transform of this boundary condition is

$$z = 0 : \ E\frac{d\overline{w}}{dz} = -\frac{p_0}{s}.$$

(2.9)

With (2.7) the value of the constant A can now be determined. The result is

$$A = \frac{pc}{Es^2},$$

(2.10)

so that the final solution of the transformed problem is

$$\overline{w} = \frac{pc}{Es^2}\exp(-sz/c).$$

(2.11)

The inverse transform of this function can be found in elementary tables of Laplace transforms, see for instance Abramowitz and Stegun (1964) or Churchill (1972). The final solution now is

$$w = \frac{pc(t - z/c)}{E}\,H(t - z/c),$$

(2.12)

where $H(t - t_0)$ is Heaviside's unit step function, defined as

$$H(t - t_0) = \begin{cases} 0, & \text{if } t < t_0, \\ 1, & \text{if } t > t_0. \end{cases}$$

(2.13)

The solution (2.12) indicates that a point in the pile remains at rest as long as $t < z/c$. From that moment on (this is the moment of arrival of the wave) the point starts to move, with a linearly increasing displacement, which represents a constant velocity.

It may seem that this solution is in disagreement with Newton's second law, which states that the velocity of a mass point will linearly increase in time when a constant force is applied. In the present case the velocity is constant. The moving mass gradually increases, however, so that the results are really in agreement with Newton's second law: the momentum (mass times velocity) linearly increases with time. Actually, Newton's second law is the basic principle involved in deriving the basic differential equation (2.3), so that no disagreement is possible, of course.

2.2.2 Pile of Finite Length

The Laplace transform method can also be used for the analysis of waves in piles of finite length. Many solutions can be found in the literature (Churchill, 1972; Carslaw and Jaeger, 1948). An example will be given below.

Consider the case of a pile of finite length, say h, see Fig. 2.2. The boundary $z = 0$ is free of stress, and the boundary $z = h$ undergoes a sudden displacement, at time $t = 0$. Thus the boundary conditions are

$$z = 0, \ t > 0 \ : \ \frac{\partial w}{\partial z} = 0, \tag{2.14}$$

and

$$z = h, \ t > 0 \ : \ w = w_0. \tag{2.15}$$

The general solution of the transformed differential equation

$$\frac{d^2\overline{w}}{dz^2} = \frac{s^2}{c^2}\,\overline{w}, \tag{2.16}$$

is

$$\overline{w} = A\exp(sz/c) + B\exp(-sz/c). \tag{2.17}$$

The constants A and B (which may depend upon the Laplace transform parameter s) can be determined from the transforms of the boundary conditions (2.14) and (2.15). The result is

$$\overline{w} = \frac{w_0}{s}\,\frac{\cosh(sz/c)}{\cosh(sh/c)}. \tag{2.18}$$

The mathematical problem now remaining is to find the inverse transform of this expression. This can be accomplished by using the complex inversion integral

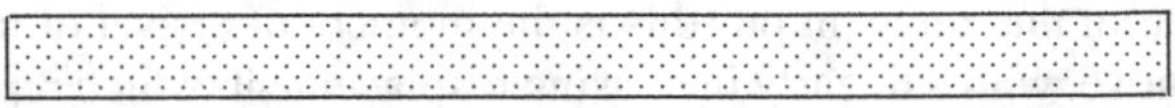

Fig. 2.2 Pile of finite length

Fig. 2.3 Displacement of
free end

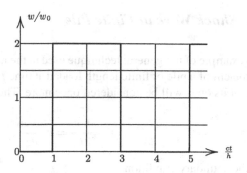

(Churchill, 1972), or its simplified form, the Heaviside expansion theorem, see Appendix A. This gives, after some elementary mathematical analysis,

$$\frac{w}{w_0} = 1 - \frac{4}{\pi} \sum_{k=0}^{\infty} \frac{(-1)^k}{(2k+1)} \cos\left[(2k+1)\frac{\pi z}{2h}\right] \cos\left[(2k+1)\frac{\pi ct}{2h}\right]. \qquad (2.19)$$

As a special case one may consider the displacement of the free end $z = 0$. This is found to be

$$\frac{w}{w_0} = 1 - \frac{4}{\pi} \sum_{k=0}^{\infty} \frac{(-1)^k}{(2k+1)} \cos\left[(2k+1)\frac{\pi ct}{2h}\right]. \qquad (2.20)$$

This expression is of the form of a Fourier series. Actually, it is the same series as the one given in the example in Appendix A, except for a constant factor and some changes in notation. The summation of the series is shown in Fig. 2.3.

It appears that the free end remains at rest for a time h/c, then suddenly shows a displacement $2w_0$ for a time span $2h/c$, and then switches continuously between zero displacement and $2w_0$. The physical interpretation, which may become more clear after considering the solution of the problem by the method of characteristics in a later section, is that a compression wave starts to travel at time $t = 0$ towards the free end, and then is reflected as a tension wave in order that the end remains free. The time h/c is the time needed for a wave to travel through the entire length of the pile.

2.3 Separation of Variables

For certain problems, especially problems of continuous vibrations, the differential equation (2.3) can be solved conveniently by a method known as *separation of variables*. Two examples will be considered in this section.

2.3.1 Shock Wave in Finite Pile

As an example of the general technique used in the method of separation of variables the problem of a pile of finite length loaded at time $t = 0$ by a constant displacement at one of its ends will be considered once more. The differential equation is

$$\frac{\partial^2 w}{\partial t^2} = c^2 \frac{\partial^2 w}{\partial z^2}, \tag{2.21}$$

with the boundary conditions

$$z = 0, \ t > 0 \ : \ \frac{\partial w}{\partial z} = 0, \tag{2.22}$$

and

$$z = h, \ t > 0 \ : \ w = w_0. \tag{2.23}$$

The first condition expresses that the boundary $z = 0$ is a free end, and the second condition expresses that the boundary $z = h$ is displaced by an amount w_0 at time $t = 0$. The initial conditions are supposed to be that the pile is at rest at $t = 0$.

The solution of the problem is now sought in the form

$$w = w_0 + Z(z)T(t). \tag{2.24}$$

The basic assumption here is that solutions can be written as a product of two functions, a function $Z(z)$, which depends upon z only, and another function $T(t)$, which depends only on t. Substitution of (2.24) into the differential equation (2.21) gives

$$\frac{1}{c^2} \frac{1}{T} \frac{d^2 T}{dt^2} = \frac{1}{Z} \frac{d^2 Z}{dz^2}. \tag{2.25}$$

The left hand side of this equation depends upon t only, the right hand side depends upon z only. Therefore the equation can be satisfied only if both sides are equal to a certain constant. This constant may be assumed to be negative or positive. If it is assumed that this constant is negative one may write

$$\frac{1}{Z} \frac{d^2 Z}{dz^2} = -\lambda^2, \tag{2.26}$$

where λ is an unknown constant. The general solution of (2.26) is

$$Z = C_1 \cos(\lambda z) + C_2 \sin(\lambda z), \tag{2.27}$$

where C_1 and C_2 are constants. They can be determined from the boundary conditions. Because dZ/dz must be 0 for $z = 0$ it follows that $C_2 = 0$. If now it is required that $Z = 0$ for $z = h$, in order to satisfy the boundary condition (2.23), it follows that a non-zero solution can be obtained only if $\cos(\lambda h) = 0$, which can be satisfied if

$$\lambda = \lambda_k = (2k+1)\frac{\pi}{2h}, \quad k = 0, 1, 2, \ldots. \tag{2.28}$$

On the other hand, one obtains for the function T

$$\frac{1}{T}\frac{d^2 T}{dt^2} = -c^2 \lambda^2, \tag{2.29}$$

with the general solution

$$T = A\cos(\lambda ct) + B\sin(\lambda ct). \tag{2.30}$$

The solution for the displacement w can now be written as

$$w = w_0 + \sum_{k=0}^{\infty}\left[A_k\cos(\lambda_k ct) + B_k\sin(\lambda_k ct)\right]\cos(\lambda_k z). \tag{2.31}$$

The velocity now is

$$\frac{\partial w}{\partial t} = \sum_{k=0}^{\infty}\left[-A_k\lambda_k c\sin(\lambda_k ct) + B_k\lambda_k c\cos(\lambda_k ct)\right]\cos(\lambda_k z). \tag{2.32}$$

Because this must be zero for $t = 0$ and all values of z, to satisfy the initial condition of rest, it follows that $B_k = 0$. Furthermore, the initial condition that the displacement must also be zero for $t = 0$, now leads to the equation

$$\sum_{k=0}^{\infty} A_k\cos(\lambda_k z) = -w_0, \tag{2.33}$$

which must be satisfied for all values of z in the range $0 < z < h$. This is the standard problem from Fourier series analysis, see Appendix A. It can be solved by multiplication of both sides by $\cos(\lambda_j z)$, and then integrating both sides over z from $z = 0$ to $z = h$. The result is

$$A_k = \frac{4}{\pi}\frac{w_0}{(2k+1)}(-1)^k. \tag{2.34}$$

Substitution of this result into the solution (2.31) now gives finally, with $B_k = 0$,

$$\frac{w}{w_0} = 1 + \frac{4}{\pi}\sum_{k=0}^{\infty}\frac{(-1)^k}{(2k+1)}\cos\left[(2k+1)\frac{\pi z}{2h}\right]\cos\left[(2k+1)\frac{\pi ct}{2h}\right]. \tag{2.35}$$

This is exactly the same result as found earlier by using the Laplace transform method, see (2.19). It may give some confidence that both methods lead to the same result.

The solution (2.35) can be seen as a summation of periodic solutions, each combined with a particular shape function. Usually a periodic function is written as $\cos(\omega t)$. In this case it appears that the possible frequencies are

$$\omega = \omega_k = (2k+1)\frac{\pi c}{2h}, \quad k = 0, 1, 2, \ldots. \tag{2.36}$$

These are usually called the *characteristic frequencies*, or *eigen frequencies* of the system. The corresponding shape functions

$$\psi_k(z) = \cos\left[(2k+1)\frac{\pi z}{2h}\right], \quad k = 0, 1, 2, \ldots, \tag{2.37}$$

are the *eigen functions* of the system.

2.3.2 Periodic Load

The solution is much simpler if the load is periodic, because then it can be assumed that all displacements are periodic. As an example the problem of a pile of finite length, loaded by a periodic load at one end, and rigidly supported at its other end, will be considered, see Fig. 2.4. In this case the boundary conditions at the left side boundary, where the pile is supported by a rigid wall or foundation, is

$$z = 0 : w = 0. \tag{2.38}$$

The boundary condition at the other end is

$$z = h : \sigma = E\frac{\partial w}{\partial z} = -p_0 \sin(\omega t), \tag{2.39}$$

where h is the length of the pile, and ω is the frequency of the periodic load.

It is again assumed that the solution of the partial differential equation (2.3) can be written as the product of a function of z and a function of t. In particular, because the load is periodic, it is now assumed that

$$w = W(z)\sin(\omega t). \tag{2.40}$$

Substitution into the differential equation (2.3) shows that this equation can indeed be satisfied, provided that the function $W(z)$ satisfies the ordinary differential equation

$$\frac{d^2 W}{dz^2} = -\frac{\omega^2}{c^2}W, \tag{2.41}$$

where $c = \sqrt{E/\rho}$, the wave velocity.

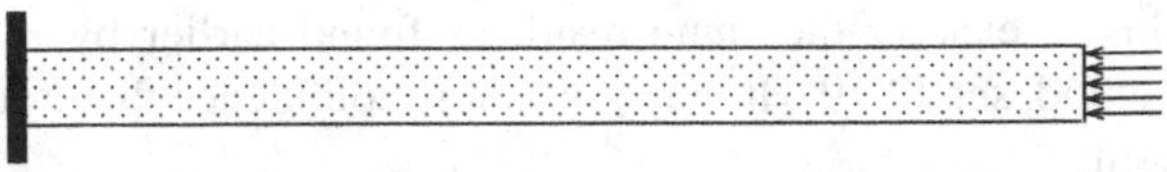

Fig. 2.4 Pile loaded by periodic pressure

The solution of the differential equation (2.41) that also satisfies the two boundary conditions (2.38) and (2.39) is

$$W(z) = -\frac{p_0 c}{E\omega} \frac{\sin(\omega z/c)}{\cos(\omega h/c)}. \tag{2.42}$$

This means that the final solution of the problem is, with (2.42) and (2.40),

$$w(z, t) = -\frac{p_0 c}{E\omega} \frac{\sin(\omega z/c)}{\cos(\omega h/c)} \sin(\omega t). \tag{2.43}$$

It can easily be verified that this solution satisfies all requirements, because it satisfies the differential equation, and both boundary conditions. Thus a complete solution has been obtained by elementary procedures. Of special interest is the motion of the free end of the pile. This is found to be

$$w(h, t) = w_0 \sin(\omega t), \tag{2.44}$$

where

$$w_0 = -\frac{p_0 c}{E\omega} \tan(\omega h/c). \tag{2.45}$$

The amplitude of the total force, $F_0 = -p_0 A$, can be written as

$$F_0 = \frac{EA}{c} \frac{\omega}{\tan(\omega h/c)} w_0. \tag{2.46}$$

Resonance

It may be interesting to consider the case that the frequency ω is equal to one of the eigen frequencies of the system,

$$\omega = \omega_k = (2k+1)\frac{\pi c}{2h}, \quad k = 0, 1, 2, \ldots. \tag{2.47}$$

In that case $\cos(\omega h/c) = 0$, and the amplitude of the displacement, as given by (2.45), becomes infinitely large. This phenomenon is called *resonance* of the system. If the frequency of the load equals one of the eigen frequencies of the system, this may lead to very large displacements, indicating resonance.

In engineering practice the pile may be a concrete foundation pile, for which the order of magnitude of the wave velocity c is about 3000 m/s, and for which a normal length h is 20 m. In civil engineering practice the frequency ω is usually not very large, at least during normal loading. A relatively high frequency is say $\omega = 20 \, \text{s}^{-1}$. In that case the value of the parameter $\omega h/c$ is about 0.13, which is rather small, much smaller than all eigen frequencies (the smallest of which occurs for $\omega h/c = \pi/2$). The function $\tan(\omega h/c)$ in (2.46) may now be approximated by its argument, so that this expression reduces to

$$\omega h/c \ll 1 : \quad F_0 \approx \frac{EA}{h} w_0. \tag{2.48}$$

This means that the pile can be considered to behave, as a first approximation, as a spring, without mass, and without damping. In many situations in civil engineering practice the loading is so slow, and the elements are so stiff (especially when they consist of concrete or steel), that the dynamic analysis can be restricted to the motion of a single spring.

It must be noted that the approximation presented above is not always justified. When the material is soft (e.g. soil) the velocity of wave propagation may not be that high. And loading conditions with very high frequencies may also be of importance, for instance during installation (pile driving). In general one may say that in order for dynamic effects to be negligible, the loading must be so slow that the frequency is considerably smaller than the smallest eigen frequency.

2.4 Solution by Characteristics

A powerful method of solution for problems of wave propagation in one dimension is provided by the method of characteristics. This method is presented in this section.

The wave equation (2.3) has solutions of the form

$$w = f_1(z - ct) + f_2(z + ct), \tag{2.49}$$

where f_1 and f_2 are arbitrary functions, and c is the velocity of propagation of waves,

$$c = \sqrt{E/\rho}. \tag{2.50}$$

In mathematics the directions $z = ct$ and $z = -ct$ are called the *characteristics*. The solution of a particular problem can be obtained from the general solution (2.49) by using the initial conditions and the boundary conditions.

A convenient way of constructing solutions is by writing the basic equations in the following form

$$\frac{\partial \sigma}{\partial z} = \rho \frac{\partial v}{\partial t}, \tag{2.51}$$

$$\frac{\partial \sigma}{\partial t} = E \frac{\partial v}{\partial z}, \tag{2.52}$$

where v is the velocity, $v = \partial w/\partial t$, and σ is the stress in the pile.

In order to simplify the basic equations two new variables ξ and η are introduced, defined by

$$\xi = z - ct, \qquad \eta = z + ct. \tag{2.53}$$

Equations (2.51) and (2.52) can now be transformed into

$$\frac{\partial \sigma}{\partial \xi} + \frac{\partial \sigma}{\partial \eta} = \rho c \left(-\frac{\partial v}{\partial \xi} + \frac{\partial v}{\partial \eta} \right),$$ (2.54)

$$\frac{\partial \sigma}{\partial \xi} - \frac{\partial \sigma}{\partial \eta} = \rho c \left(\frac{\partial v}{\partial \xi} + \frac{\partial v}{\partial \eta} \right),$$ (2.55)

from which it follows, by addition or subtraction of the two equations, that

$$\frac{\partial (\sigma - Jv)}{\partial \eta} = 0,$$ (2.56)

$$\frac{\partial (\sigma + Jv)}{\partial \xi} = 0,$$ (2.57)

where J is the impedance,

$$J = \rho c = \sqrt{E\rho}.$$ (2.58)

In terms of the original variables z and t the equations are

$$\frac{\partial (\sigma - Jv)}{\partial (z + ct)} = 0,$$ (2.59)

$$\frac{\partial (\sigma + Jv)}{\partial (z - ct)} = 0.$$ (2.60)

These equations mean that the quantity $\sigma - Jv$ is independent of $z + ct$, and $\sigma + Jv$ is independent of $z - ct$. This means that

$$\sigma - Jv = f_1(z - ct),$$ (2.61)

$$\sigma + Jv = f_2(z + ct).$$ (2.62)

These equations express that the quantity $\sigma - Jv$ is a function of $z - ct$ only, and that $\sigma + Jv$ is a function of $z + ct$ only. This means that $\sigma - Jv$ is constant when $z - ct$ is constant, and that $\sigma + Jv$ is constant when $z + ct$ is constant. These properties enable to construct solutions, either in a formal analytical way, or graphically, by mapping the solution, as represented by the variables σ and Jv, onto the plane of the independent variables z and ct.

As an example let there be considered the case of a free pile, which is hit at its upper end $z = 0$ at time $t = 0$ such that the stress at that end is $-p$. The other end, $z = h$, is free, so that the stress is zero there. The initial state is such that all velocities are zero. The solution is illustrated in Fig. 2.5. In the upper figure, the diagram of z and ct has been drawn, with lines of constant $z - ct$ and lines of constant $z + ct$. Because initially the velocity v and the stress σ are zero throughout the pile, the condition in each point of the pile is represented by the point 1 in the lower figure, the diagram of σ and Jv. The points in the lower left corner of the upper diagram (this region is marked 1) can all be reached from points on the axis

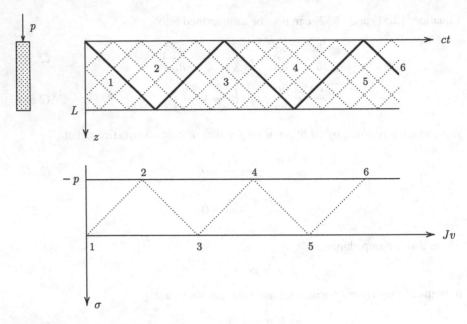

Fig. 2.5 The method of characteristics

$ct = 0$ (for which $\sigma = 0$ and $Jv = 0$) by a downward going characteristic, i.e. lines $z - ct = $ constant. Thus in all these points $\sigma - Jv = 0$. At the bottom of the pile the stress is always zero, $\sigma = 0$. Thus in the points in region 1 for which $z = 0$ the velocity is also zero, $Jv = 0$. Actually, in the entire region $1 : \sigma = Jv = 0$, because all these points can be reached by an upward going characteristic and a downward going characteristic from points where $\sigma = Jv = 0$. The point 1 in the lower diagram thus is representative for all points in region 1 in the upper diagram.

For $t > 0$ the value of the stress σ at the upper boundary $z = 0$ is $-p$, for all values of t. The velocity is unknown, however. The axis $z = 0$ in the upper diagram can be reached from points in the region 1 along lines for which $z + ct = $ constant. Therefore the corresponding point in the diagram of σ and Jv must be located on the line for which $\sigma + Jv = $ constant, starting from point 1. Because the stress σ at the top of the pile must be $-p$ the point in the lower diagram must be point 2. This means that the velocity is $Jv = p$, or $v = p/J$. This is the velocity of the top of the pile for a certain time, at least for $ct = 2h$, if h is the length of the pile, because all points for which $z = 0$ and $ct < 2h$ can be reached from region 1 along characteristics $z + ct = $ constant.

At the lower end of the pile the stress σ must always be zero, because the pile was assumed to be not supported. Points in the upper diagram on the line $z = h$ can be reached from region 2 along lines of constant $x - ct$. Therefore they must be located on a line of constant $N - Jv$ in the lower diagram, starting from point 2. This gives point 3, which means that the velocity at the lower end of the pile is now $v = 2p/J$. This velocity applies to all points in the region 3 in the upper diagram.

Fig. 2.6 Velocity of the
bottom of the pile

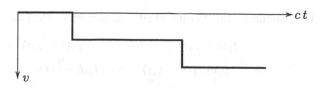

In this way the velocity and the stress in the pile can be analyzed in successive steps. The thick lines in the upper diagram are the boundaries of the various regions. If the force at the top continues to be applied, as is assumed in Fig. 2.5, the velocity of the pile increases continuously. Figure 2.6 shows the velocity of the bottom of the pile as a function of time. The velocity gradually increases with time, because the pressure p at the top of the pile continues to act. This is in agreement with Newton's second law, which states that the velocity will increase linearly under the influence of a constant force.

2.5 Reflection and Transmission of Waves

An interesting aspect of wave propagation in continuous media is the behaviour of waves at surfaces of discontinuity of the material properties. In order to study this phenomenon let us consider the propagation of a short shock wave in a pile consisting of two materials, see Fig. 2.7. A compression wave is generated in the pile by a pressure of short duration at the left end of the pile. The pile consists of two materials: first a stiff section, and then a very long section of smaller stiffness.

The solution of the basic equations in the first section can be written as

$$v = v_1 = f_1(z - c_1 t) + f_2(z + c_1 t), \tag{2.63}$$

$$\sigma = \sigma_1 = -\rho_1 c_1 f_1(z - c_1 t) + \rho_1 c_1 f_2(z + c_1 t), \tag{2.64}$$

where ρ_1 is the density of the material in that section, and c_1 is the wave velocity, $c_1 = \sqrt{E_1/\rho_1}$. It can easily be verified that this solution satisfies the two basic differential equations (2.51) and (2.52).

In the second part of the pile the solution is

$$v = v_2 = g_1(z - c_2 t) + g_2(z + c_2 t), \tag{2.65}$$

$$\sigma = \sigma_2 = -\rho_2 c_2 g_1(z - c_2 t) + \rho_2 c_2 g_2(z + c_2 t), \tag{2.66}$$

where ρ_2 and c_2 are the density and the wave velocity in that part of the pile.

At the interface of the two materials the value of z is the same in both solutions, say $z = h$, and the condition is that both the velocity v and the normal stress σ must

Fig. 2.7 Non-homogeneous
pile

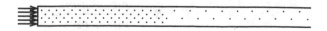

be continuous at that point, at all values of time. Thus one obtains

$$f_1(h - c_1 t) + f_2(h + c_1 t) = g_1(h - c_2 t) + g_2(h + c_2 t), \tag{2.67}$$

$$-\rho_1 c_1 f_1(h - c_1 t) + \rho_1 c_1 f_2(h + c_1 t)$$
$$= -\rho_2 c_2 g_1(h - c_2 t) + \rho_2 c_2 g_2(h + c_2 t). \tag{2.68}$$

If we write

$$f_1(h - c_1 t) = F_1(t), \tag{2.69}$$

$$f_2(h + c_1 t) = F_2(t), \tag{2.70}$$

$$g_1(h - c_2 t) = G_1(t), \tag{2.71}$$

$$g_2(h + c_2 t) = G_2(t), \tag{2.72}$$

then the continuity conditions are

$$F_1(t) + F_2(t) = G_1(t) + G_2(t), \tag{2.73}$$

$$-\rho_1 c_1 F_1(t) + \rho_1 c_1 F_2(t) = -\rho_2 c_2 G_1(t) + \rho_2 c_2 G_2(t). \tag{2.74}$$

In general these equations are, of course, insufficient to solve for the four functions. However, if it is assumed that the pile is very long (or, more generally speaking, when the value of time is so short that the wave reflected from the end of the pile has not yet arrived), it may be assumed that the solution representing the wave coming from the end of the pile is zero, $G_2(t) = 0$. In that case the solutions F_2 and G_1 can be expressed in the first wave, F_1, which is the wave coming from the top of the pile. The result is

$$F_2(t) = \frac{\rho_1 c_1 - \rho_2 c_2}{\rho_1 c_1 + \rho_2 c_2} F_1(t), \tag{2.75}$$

$$G_1(t) = \frac{2\rho_1 c_1}{\rho_1 c_1 + \rho_2 c_2} F_1(t), \tag{2.76}$$

This means, for instance, that whenever the first wave $F_1(t) = 0$ at the interface, then there is no reflected wave, $F_2(t) = 0$, and there is no transmitted wave either, $G_1(t) = 0$. On the other hand, when the first wave has a certain value at the interface, then the values of the reflected wave and the transmitted wave at that point may be calculated from the relations (2.75) and (2.76). If the values are known the values at later times may be calculated using the relations (2.69)–(2.72).

The procedure may be illustrated by an example. Therefore let it be assumed that the two parts of the pile have the same density, $\rho_1 = \rho_2$, but the stiffness in the first section is 9 times the stiffness in the rest of the pile, $E_1 = 9E_2$. This means that the wave velocities differ by a factor 3, $c_1 = 3c_2$. The reflection coefficient and the transmission coefficient now are, with (2.75) and (2.76),

$$R_v = \frac{\rho_1 c_1 - \rho_2 c_2}{\rho_1 c_1 + \rho_2 c_2} = 0.5, \tag{2.77}$$

Fig. 2.8 Reflection and transmission (velocity)

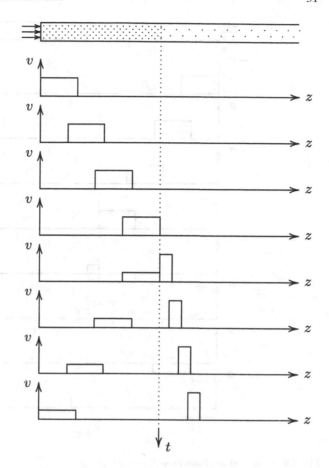

$$T_v = \frac{2\rho_1 c_1}{\rho_1 c_1 + \rho_2 c_2} = 1.5. \tag{2.78}$$

The behaviour of the solution is illustrated graphically in Fig. 2.8, which shows the velocity profile at various times. In the first four diagrams the incident wave travels toward the interface. During this period there is no reflected wave, and no transmitted wave in the second part of the pile. As soon as the incident wave hits the interface a reflected wave is generated, and a wave is transmitted into the second part of the pile. The magnitude of the velocities in this transmitted wave is 1.5 times the original wave, and it travels a factor 3 slower. The magnitude of the velocities in the reflected wave is 0.5 times those in the original wave.

The stresses in the two parts of the pile are shown in graphical form in Fig. 2.9. The reflection coefficient and the transmission coefficient for the stresses can be obtained using (2.64) and (2.66). The result is

$$R_\sigma = -\frac{\rho_1 c_1 - \rho_2 c_2}{\rho_1 c_1 + \rho_2 c_2} = -0.5, \tag{2.79}$$

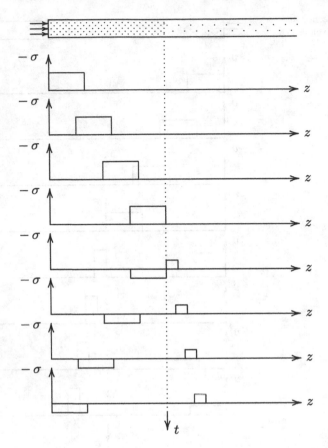

Fig. 2.9 Reflection and transmission (stress)

$$T_\sigma = \frac{2\rho_2 c_2}{\rho_1 c_1 + \rho_2 c_2} = 0.5, \tag{2.80}$$

where it has been taken into account that the form of the solution for the stresses, see (2.64) and (2.66), involves factors ρc, and signs of the terms different from those in the expressions for the velocity. In the case considered here, where the first part of the pile is 9 times stiffer than the rest of the pile, it appears that the reflected wave leads to stresses of the opposite sign in the first part. Thus a compression wave in the pile is reflected in the first part by tension.

It may be interesting to note the two extreme cases of reflection. When the second part of the pile is so soft that it can be entirely disregarded (or, when the pile consists only of the first part, which is free to move at its end), the reflection coefficient for the velocity is $R_v = 1$, and for the stress it is $R_\sigma = -1$. This means that in this case a compression wave is reflected as a tension wave of equal magnitude. The velocity in the reflected wave is in the same direction as in the incident wave.

Fig. 2.10 Graphical solution using characteristics

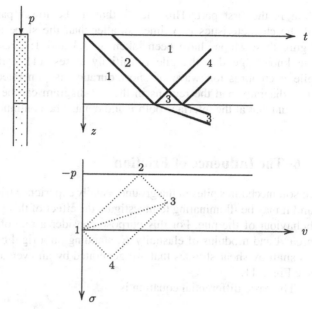

If the second part of the pile is infinitely stiff (or, if the pile meets a rigid foundation after the first part) the reflection coefficient for the velocity is $R_v = -1$, and for the stresses it is $R_\sigma = 1$. Thus, in this case a compression wave is reflected as a compressive wave of equal magnitude. These results are of great importance in pile driving. When a pile hits a very soft layer, a tension wave may be reflected from the end of the pile, and a concrete pile may not be able to withstand these tensile stresses. Thus, the energy supplied to the pile must be reduced in this case, for instance by reducing the height of fall of the hammer. When the pile hits a very stiff layer the energy of the driving equipment may be increased without the risk of generating tensile stresses in the pile, and this may help to drive the pile through this stiff layer. Of course, great care must be taken when the pile tip suddenly passes from the very stiff layer into a soft layer. Experienced pile driving operators use these basic principles intuitively.

It may be noted that tensile stresses may also be generated in a pile when an upward traveling (reflected) wave reaches the top of the pile, which by that time may be free of stress. This phenomenon has caused severe damage to concrete piles, in which cracks developed near the top of the pile, because concrete cannot withstand large tensile stresses. In order to prevent this problem, driving equipment has been developed that continues to apply a compressive force at the top of the pile for a relatively long time. Also, the use of prestressed concrete results in a considerable tensile strength of the material.

The problem considered in this section can also be analyzed graphically, by using the method of characteristics, see Fig. 2.10. The data given above imply that the wave velocity in the second part of the pile is 3 times smaller than in the first part, and that the impedance in the second part is also 3 times smaller

than in the first part. This means that in the lower part of the pile the slope
of the characteristics is 3 times smaller than the slope in the upper part. In the
figure these slopes have been taken as 1:3 and 1:1, respectively. Starting from
the knowledge that the pile is initially at rest (1), and that at the top of the
pile a compression wave of short duration is generated (2), the points in the
v, σ-diagram, and the regions in the z, t-diagram can be constructed, taking into
account that at the interface both v and σ must be continuous.

2.6 The Influence of Friction

In soil mechanics piles in the ground usually experience friction along the pile shaft,
and it may be illuminating to investigate the effect of this friction on the mechanical
behaviour of the pile. For this purpose consider a pile of constant cross sectional
area A and modulus of elasticity E, standing on a rigid base, and supported along
its shaft by shear stresses that are generated by an eventual movement of the pile,
see Fig. 2.11.

The basic differential equation is

$$EA\frac{\partial^2 w}{\partial z^2} - C\tau = \rho A\frac{\partial^2 w}{\partial t^2},\tag{2.81}$$

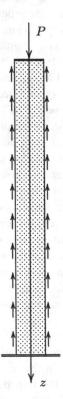

Fig. 2.11 Pile in soil, with
friction

where C is the circumference of the pile shaft, and τ is the shear stress. It is assumed, as a first approximation, that the shear stress is linearly proportional to the vertical displacement of the pile,

$$\tau = kw, \tag{2.82}$$

where the constant k has the character of a subgrade modulus. The differential equation (2.81) can now be written as

$$\frac{\partial^2 w}{\partial z^2} - \frac{w}{H^2} = \frac{1}{c^2} \frac{\partial^2 w}{\partial t^2}, \tag{2.83}$$

where H is a length parameter characterizing the ratio of the axial pile stiffness to the friction constant,

$$H^2 = \frac{EA}{kC}, \tag{2.84}$$

and c is the usual wave velocity, defined by

$$c^2 = E/\rho. \tag{2.85}$$

The boundary conditions are supposed to be

$$z = 0 : N = EA\frac{\partial w}{\partial z} = -P\sin(\omega t), \tag{2.86}$$

$$z = L : w = 0. \tag{2.87}$$

The first boundary condition expresses that at the top of the pile it is loaded by a periodic force, of amplitude P and circular frequency ω. The second boundary condition expresses that at the bottom of the pile no displacement is possible, indicating that the pile is resting upon solid rock.

The problem defined by the differential equation (2.83) and the boundary conditions (2.86) and (2.87) can easily be solved by the method of separation of variables. In this method it is assumed that the solution can be written as the product of a function of z and a factor $\sin(\omega t)$. It turns out that all the conditions are met by the solution

$$w = \frac{PH}{EA\alpha} \frac{\sinh[\alpha(L-z)/H]}{\cosh(\alpha L/H)} \sin(\omega t), \tag{2.88}$$

where α is given by

$$\alpha = \sqrt{1 - \omega^2 H^2/c^2}. \tag{2.89}$$

The displacement at the top of the pile, w_t, is of particular interest. If this is written as

$$w_t = \frac{P}{K} \sin(\omega t), \tag{2.90}$$

the spring constant K appears to be

$$K = \frac{EA}{L} \frac{\alpha L/H}{\tanh(\alpha L/H)}. \tag{2.91}$$

The first term in the right hand side is the spring constant in the absence of friction, when the elasticity is derived from the deformation of the pile only.

The behaviour of the second term in (2.91) depends upon the frequency ω through the value of the parameter α, see (2.89). It should be noted that for values of $\omega H/c > 1$ the parameter α becomes imaginary, say $\alpha = i\beta$, where now

$$\beta = \sqrt{\omega^2 H^2/c^2 - 1}. \tag{2.92}$$

The spring constant can then be written more conveniently as

$$\omega H/c > 1 \; : \; K = \frac{EA}{L} \frac{\beta L/H}{\tan(\beta L/H)}. \tag{2.93}$$

This formula implies that for certain values of $\omega H/c$ the spring constant will be zero, indicating resonance. These values correspond to the eigen values of the system. For certain other values the spring constant is infinitely large. For these values of the frequency the system appears to be very stiff. In such a case part of the pile is in compression and another part is in tension, such that the total strains from bottom to top just cancel.

The value of the spring constant is shown, as a function of the frequency, in Fig. 2.12, for $H/L = 1$. This figure contains data for both ranges of the parameters.

It is interesting to consider the probable order of magnitude of the parameters in engineering practice. For this purpose the value of the subgrade modulus k must first be evaluated. This parameter can be estimated to be related to the soil stiffness by a formula of the type $k = E_s/D$, where E_s is the modulus of elasticity of the soil (assuming that the deformations are small enough to justify the definition of such a quantity), and D is the width of the pile. For a circular concrete pile of diameter D the value of the characteristic length H now is, with (2.84),

$$H^2 = \frac{EA}{kC} = \frac{E_c D^2}{2E_s}. \tag{2.94}$$

Under normal conditions, with a pile being used in soft soil, the ratio of the elastic moduli of concrete and soil is about 1000, and most piles have diameters of about 0.40 m. This means that $H \approx 10$ m. Furthermore the order of magnitude of the wave propagation velocity c in concrete is about 3000 m/s. This means that the parameter $\omega H/c$ will usually be small compared to 1, except for phenomena of very high frequency, such as may occur during pile driving. In many civil engineering problems, where the fluctuations originate from wind or wave loading, the frequency is usually about 1 s^{-1} or smaller, so that the order of magnitude of the parameter $\omega H/c$ is about 0.01. In such cases the value of α will be very close to 1, see (2.89). This indicates that the response of the pile is practically static.

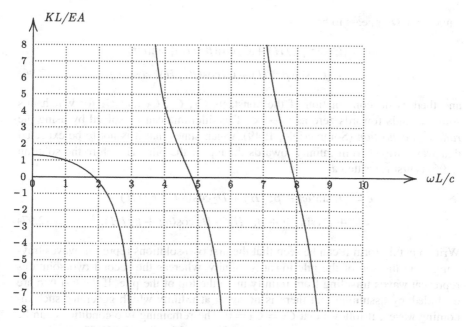

Fig. 2.12 Spring constant ($H/L = 1$)

If the loading is due to the passage of a heavy train, at a velocity of 100 km/h, and with a distance of the wheels of 5 m, the period of the loading is about 1/6 s, and thus the frequency is about 30 s^{-1}. In such cases the parameter $\omega H/c$ may not be so small, indicating that dynamic effects may indeed be relevant.

Infinitely Long Pile

A case of theoretical interest is that of an infinitely long pile, $L \to \infty$. If the frequency is low this limiting case can immediately be obtained from the general solution (2.91), because then the function $\tanh(\alpha L/H)$ can be approximated by its asymptotic value 1. The result is

$$L \to \infty, \ \omega H/c < 1 : \ K = \frac{EA\alpha}{H}. \tag{2.95}$$

This solution degenerates when the dimensionless frequency $\omega H/c = 1$, because then $\alpha = 0$, see (2.89). Such a zero spring constant indicates resonance of the system.

For frequencies larger than this resonance frequency the solution (2.93) can not be used, because the function $\tan(\beta L/H)$ continues to fluctuate when its argument tends towards infinity. Therefore the problem must be studied again from the beginning, but now for an infinitely long pile. The general solution of the differential equation now appears to be

$$w = [C_1 \sin(\beta z/H) + C_2 \cos(\beta z/H)] \sin(\omega t)$$
$$+ [C_3 \sin(\beta z/H) + C_4 \cos(\beta z/H)] \cos(\omega t), \qquad (2.96)$$

and there is no combination of the constants C_1, C_2, C_3 and C_4 for which this solution tends towards zero as $z \to \infty$. This dilemma can be solved by using the *radiation condition* (Sommerfeld, 1949), which states that it is not to be expected that waves travel from infinity towards the top of the pile. Therefore the solution (2.96) is first rewritten as

$$w = D_1 \sin(\omega t - \beta z/H) + D_2 \cos(\omega t - \beta z/H)$$
$$+ D_3 \sin(\omega t + \beta z/H) + D_4 \cos(\omega t + \beta z/H). \qquad (2.97)$$

Written in this form it can be seen that the first two solutions represent waves traveling from the top of the pile towards infinity, whereas the second two solutions represent waves traveling from infinity up to the top of the pile. If the last two are excluded, by assuming that there is no agent at infinity which generates such incoming waves, it follows that $C_3 = C_4 = 0$. The remaining two conditions can be determined from the boundary condition at the top of the pile, (2.86). The final result is

$$L \to \infty, \ \omega H/c > 1 : w = \frac{PH}{EA\beta} \sin(\omega t - \beta z/H). \qquad (2.98)$$

This solution applies only if the frequency is larger than the eigen frequency of the system, which is defined by $\omega H/c = 1$. It may be noted that the solution (2.98) also degenerates for $\omega H/c = 1$ because then $\beta = 0$, see (2.92).

2.7 Numerical Solution

In order to construct a numerical model for the solution of wave propagation problems the basic equations are written in a numerical form. For this purpose the pile is subdivided into n elements, all of the same length Δz. The displacement w_i and the velocity v_i of an element are defined in the centroid of element i, and the normal forces N_i are defined at the boundary between elements i and $i + 1$, see Fig. 2.13. The friction force acting on element i is denoted by F_i. This particular choice for

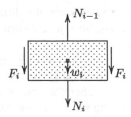

Fig. 2.13 Element of pile

the definition of the various quantities either at the centroid of the elements or at their boundaries, has a physical background. The velocity derives its meaning from a certain mass, whereas the normal force is an interaction between the material on both sides of a section. It is interesting to note, however, that this way of modeling, sometimes denoted as *leap frog* modeling, also has distinct mathematical advantages, with respect to accuracy and stability.

The equation of motion of an element is

$$N_i - N_{i-1} + F_i = \rho A \Delta z \frac{v_i(t + \Delta t) - v_i(t)}{\Delta t} \qquad (i = 1, \ldots, n). \qquad (2.99)$$

It should be noted that there are $n + 1$ normal forces, from N_0 to N_n. The force N_0 can be considered to be the force at the top of the pile, and N_n is the force at the bottom end of the pile.

The displacement w_i is related to the velocity v_i by the equation

$$v_i = \frac{w_i(t + \Delta t) - w_i(t)}{\Delta t} \qquad (i = 1, \ldots, n). \qquad (2.100)$$

The deformation is related to the normal force by Hooke's law, which can be formulated as

$$N_i = EA \frac{w_{i+1} - w_i}{\Delta z} \qquad (i = 1, \ldots, n - 1). \qquad (2.101)$$

Here EA is the product of the modulus of elasticity E and the area A of the cross section.

The values of the normal force at the top and at the bottom of the pile, N_0 and N_n are supposed to be given by the boundary conditions.

Example

A simple example may serve to illustrate the numerical algorithm. Suppose that the pile is initially at rest, and let a constant force P be applied at the top of the pile, with the bottom end being free. In this case the boundary conditions are

$$N_0 = -P, \qquad (2.102)$$

and

$$N_n = 0. \qquad (2.103)$$

The friction forces are supposed to be zero.

At time $t = 0$ all quantities are zero, except N_0. A new set of velocities can now be calculated from (2.99). Actually, this will make only one velocity non-zero, namely v_1, which will then be

$$v_1 = \frac{P \Delta t}{\rho A \Delta z}. \qquad (2.104)$$

Next, a new set of values for the displacements can be calculated from (2.100). Again, in the first time step, only one value will be non-zero, namely

$$w_1 = v_1 \Delta t = \frac{P(\Delta t)^2}{\rho A \Delta z}. \tag{2.105}$$

Finally, a new set of values for the normal force can be calculated from (2.101). This will result in N_1 getting a value, namely

$$N_1 = -EA \frac{w_1}{\Delta z} = -P \frac{c^2(\Delta t)^2}{(\Delta z)^2}. \tag{2.106}$$

This process can now be repeated, using the equations in the same order.

An important part of the numerical process is the value of the time step used. The description of the process given above indicates that in each time step the non-zero values of the displacements, velocities and normal forces increase by 1 in downward direction. This suggests that in each time step a wave travels into the pile over a distance Δz. In the previous section, when considering the analytical solution of a similar problem (actually, the same problem), it was found that waves travel in the pile at a velocity

$$c = \sqrt{E/\rho}. \tag{2.107}$$

Combining these findings suggests that the ratio of spatial step and time step should be

$$\Delta z = c \Delta t. \tag{2.108}$$

It may be noted that this means that (2.104) reduces to

$$v_1 = \frac{P}{\rho A c}. \tag{2.109}$$

The expression in the denominator is precisely what was defined as the impedance in the previous section, see (2.58), and the value P/J corresponds exactly to what was found in the analytical solution. Equation (2.105) now gives

$$w_1 = \frac{P \Delta t}{\rho A c}, \tag{2.110}$$

and the value of N_1 after one time step is found to be, from (2.106),

$$N_1 = -P. \tag{2.111}$$

Again this corresponds exactly with the analytical solution. If the time step is chosen different from the critical time step the numerical solution will show considerable deviations from the correct analytical solution. This is usually denoted as *numerical diffusion*.

All this confirms the propriety of the choice (2.108) for the relation between time step and spatial step. In a particular problem the spatial step is usually chosen first, by subdividing the pile length into a certain number of elements. Then the time step may be determined from (2.108).

It should be noted that the choice of the time step is related to the algorithm proposed here. When using a different algorithm it may be more appropriate to use a different (usually smaller) time step than the critical time step used here (Bowles, 1974).

The calculations described above can be performed by the program IMPACT, for the case of a pile loaded at its top by a constant force, for a short time. The main function in this program is given below, with the quantities S, V and W denoting the stress, the velocity and the displacement.

```
void Calculate(void)
  {
  int j;
  if (T>TT) S[0]=0;else S[0]=1;
  for (j=1;j<=N;j++) V[j]+=(S[j]-S[j-1])/(RHO*C);
  for (j=1;j<=N;j++) W[j]+=V[j]*DT;
  for (j=1;j<N;j++) S[j]=E*(W[j+1]-W[j])/DX;
  }
```

The main function of the program IMPACT.

The program uses interactive input, in which the user may edit the input data before the calculations are started. The program will show the stresses in the pile on the screen, in graphical form. An example is shown in Fig. 2.14. In this case the pile has been subdivided into 500 elements, and the figure shows the stresses in the pile after 200, 400, 600, 800 and 1000 time steps. It appears that the block wave is traveling through the pile without any deformation, and it is reflected at the free bottom as a tensile wave of the same magnitude. All this is in agreement with the general theory presented in earlier sections of this chapter.

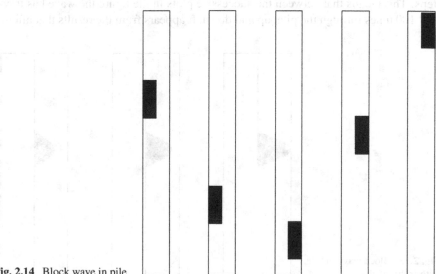

Fig. 2.14 Block wave in pile

2.8 A Simple Model for a Pile with Friction

When there is friction along the shaft of the pile, this can be introduced through the variables F_i, see (2.99). It should then be known how the friction force depends upon variables such as the local displacement and the local velocity. A simple model is to assume that the friction is proportional to the velocity, always acting in the direction opposite to the velocity. The program FRICTION can perform these calculations. The main function of this program is reproduced below, for the case of a single sinusoidal wave applied at the top of the pile.

```
void Calculate(void)
{
  int j;
  if (T>TT) S[0]=0;else S[0]=(F/AREA)*sin(PI*T/TT);
  for (j=1;j<=N;j++) V[j]+=(S[j]-S[j-1]-FR*DX*CIRC*V[j])/(RHO*AREA*C);
  for (j=1;j<=N;j++) W[j]+=V[j]*DT;
  for (j=1;j<N;j++)  S[j]=E*AREA*(W[j+1]-W[j])/DX;
}
```

The main function of the program FRICTION.

The variable FR in this program is the shear stress generated along the shaft of the pile in case of a unit velocity (1 m/s). In professional programs a more sophisticated formula for the friction may be used, in which the friction not only depends upon the velocity but also on the displacement, in a non-linear way. Also a model for the resistance at the point of the pile may be introduced, and the possibility of a layered soil, see for instance Bowles (1974).

Output of the program is shown in Fig. 2.15. The pile has been divided into 200 elements, its length is 20 m, and its cross section is a square of 0.40 m × 0.40 m. The maximum applied force is 100 kN, and the shear stress by friction is 1 kN/m^2 if the local velocity is 1 m/s.

Results for the stresses in the pile are shown after 100, 20100 and 40100 time steps. This means that between the successive plots in the figure the wave has traveled 100 times through the pile, up and down. It appears from the results that after a

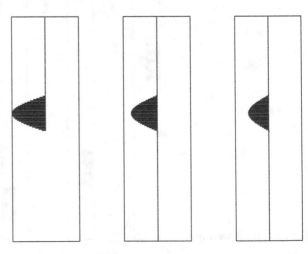

Fig. 2.15 Block wave in pile, with friction

large number of time steps the magnitude of the stresses is indeed decreased by the effect of the friction.

It may be mentioned that the program becomes unstable if the friction constant is taken too large, or if the initial wave is discontinuous, as in the case of a block wave. These unwanted effects can be eliminated by using a more sophisticated numerical method, such as the finite element method, see for instance Brinkgreve and Vermeer (2002).

Problems

2.1 A free pile is hit by a normal force of short duration. Analyze the motion of the pile by the method of characteristics, using a diagram as in Fig. 2.5.

2.2 Extend the diagram shown in Fig. 2.10 towards the right, so that the reflected wave hits the top of the pile, and is again reflected there.

2.3 As a first order approximation of (2.46) the response of a pile may be considered to be equivalent to a spring, see (2.48). Show, by using an approximation of the function $\tan(\omega h/c)$ by its first two terms, that a second order approximation is by a spring and a mass. Show, by comparison with (1.37), that the equivalent mass is $\frac{2}{3}$ of the total mass of the pile.

2.4 Verify some of the characteristic data shown in Fig. 2.12. For instance, check the values for $\omega L/c = 0$ and $\omega L/c = 1$, and check the zeroes of the spring constant.

Chapter 3
Earthquakes in Soft Layers

In this chapter the response of a soft soil layer to an earthquake in the base rock underlying the soft soil layer is considered, see Fig. 3.1. An earthquake generates various waves in the rock, resulting in waves of vertical displacements and horizontal displacements along the rock surface. These will generate compression waves and shear waves in the overlying soil. It is generally assumed that the most important component is the wave of horizontal displacements at the rock surface, which generates shear waves in the soil. Waves of vertical displacements at the rock surface will generate compression waves in the soil, and these may lead to vertical displacements of considerable magnitude, and damage of the structure on top of the soil, but usually structural damage due to such vertical compression waves remains limited. It is usually considered assumed that most structural damage is caused by shear waves in the soil, for instance collapse of the columns in the structure. For this reason the considerations in this chapter will be restricted to shear waves in the soil. Some solutions of this *ground response* problem will be presented, mainly for a homogeneous linear elastic layer, carrying a certain mass, representing the structure. The effect of hysteretic damping in the soft soil will also be considered.

   The type of model considered in this chapter is a typical example of an engineering approximation, using certain assumptions (a thin layer of soft elastic soil on a

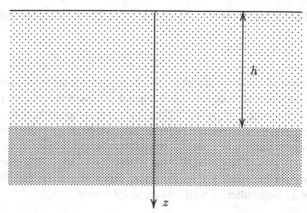

Fig. 3.1 Soft soil on hard base rock

A. Verruijt, *An Introduction to Soil Dynamics*,
Theory and Applications of Transport in Porous Media 24,
© Springer Science+Business Media B.V. 2010

hard base rock of large depth, and a periodic wave in the rock, of relatively large wave length) that are supposed to be applicable in a large class of field situations. The model has been developed at the University of California by Idriss and Seed (1968), and has later been generalized to a soil consisting of several layers with non-linear properties, see for instance Kramer (1996). The model can be considered as a simplified case of a Love wave, see Chap. 9.

It should be noted that it is assumed that the soil is strong and stiff enough to accommodate the shear stresses produced by the shaking of the base rock without failure of the soil. In particular, the possibility of liquefaction of a loose sandy soil is not considered. In areas where earthquakes may be expected great care should be taken to avoid the risk of soil liquefaction, preferably by not building on loose soils, or compacting such soils before any structure is built upon it. For a practical approach to the analysis of the liquefaction risk during earthquakes see Seed and Idriss (1982).

3.1 Earthquake Parameters

It is assumed that the earthquake generates traveling waves in the base rock, which can be described by the following equation for the horizontal displacements at the upper surface of the rock,

$$u = u_0 \sin[\omega(t - x/c_2)] = u_0 \sin(\omega t - \lambda_2 x) = u_0 \sin\left(\frac{2\pi t}{T} - \frac{2\pi x}{L_2}\right), \quad (3.1)$$

where u is the lateral displacement at the surface of the rock, u_0 its amplitude, ω is the dominant frequency of the wave, and c_2 is its propagation velocity in the rock. The parameter λ_2 is the wave number. The wave period T and the wave length L_2 of the wave are related to the frequency ω and the wave number λ_2 by the relations

$$\omega = \frac{2\pi}{T}, \qquad \lambda_2 = \frac{2\pi}{L_2}. \quad (3.2)$$

The propagation velocity can be related to the shear modulus μ and the mass density ρ of the rock by the equation

$$c_2 = \sqrt{\mu_2/\rho_2}. \quad (3.3)$$

Normal values of the shear modulus of rock are of the order of magnitude of $\mu_2 \approx 10\,\text{GPa} = 10^{10}\,\text{kg/ms}^2$, and normal values of the density of the rock are of the order of magnitude of $\rho_2 \approx 2500\,\text{kg/m}^3$. This means that normal values of the velocity of propagation are of the order of magnitude of $c_2 \approx 2000\,\text{m/s}$. These values are also representative of concrete, which can be considered as an artificial rock.

It should be noted that the stiffness of rock in engineering practice may well be somewhat smaller than the value given above, so that the velocity of propagation may be smaller as well, say $c_2 = 1500\,\text{m/s}$.

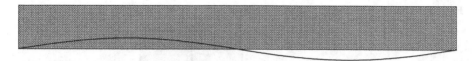

Fig. 3.2 Wave length compared to thickness of layer, $L_2/h = 10$

The dominant period of the waves generated by an earthquake usually is in the range

$$T = 0.1 \text{ s–}0.5 \text{ s}. \tag{3.4}$$

In this chapter an average value of $T = 0.2$ s will normally be used. In that case the dominant frequency is, with (3.2),

$$\omega \approx 30 \text{ s}^{-1}. \tag{3.5}$$

Because $\lambda = \omega/c$ it now follows that

$$\lambda_2 \approx 0.0150 \text{ m}^{-1}, \tag{3.6}$$

so that the wave length is, with (3.2),

$$L_2 \approx 400 \text{ m}. \tag{3.7}$$

The thickness of the layers of soft soil above the base rock often is in the range of $h = 10$ m–40 m. This means that the wave length L_2 is an order of magnitude larger than the thickness h, see for instance Fig. 3.2, in which the wave length is 10 times the thickness of the layer. This means that over a reasonably large horizontal distance the displacement at the bottom of the soil is the same. This justifies the assumption that in the soil the wave is one-dimensional, in vertical direction. Or, to be more precise, it can be assumed that throughout the soil the horizontal displacements will be of the form

$$u = f(z) \sin[\omega(t - x/c_2)], \tag{3.8}$$

where $f(z)$ is a function of z only. The factor x/c_2 in (3.8) indicates that the horizontal coordinate x results in a phase shift of magnitude x/c_2, which is constant if x is constant.

3.2 Horizontal Vibrations

In this section the propagation of horizontal vibrations in a column of elastic soil, generated by the horizontal motion of the base rock, is considered. As mentioned above, it is assumed that in each column the problem is one-dimensional, with the displacement being a function of the vertical coordinate z and time only.

Fig. 3.3 Element of column

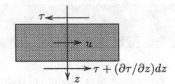

The basic differential equation is the one-dimensional wave equation, which can be derived as follows. The first basic equation is the equation of motion of an element of the column, see Fig. 3.3,

$$\frac{\partial \tau}{\partial z} = \rho \frac{\partial^2 u}{\partial t^2}, \tag{3.9}$$

where ρ is the density of the soil. The second equation is the equation of elasticity,

$$\tau = \mu \gamma = \mu \left(\frac{\partial u}{\partial z} + \frac{\partial w}{\partial x} \right), \tag{3.10}$$

where μ is the shear modulus of the soil, and γ is the shear deformation. It is now assumed, on the basis of the observation from the previous section that the wave length L_2 of the waves in x-direction is very large compared to the layer thickness h, that the derivative $\partial w / \partial x$ is small compared to $\partial u / \partial z$. Thus (3.11) reduces to

$$\tau = \mu \gamma = \mu \frac{\partial u}{\partial z}. \tag{3.11}$$

It now follows from (3.9) and (3.11) that

$$\frac{\partial^2 u}{\partial t^2} = c^2 \frac{\partial^2 u}{\partial z^2}, \tag{3.12}$$

where c is the propagation velocity of shear waves in the soil,

$$c = \sqrt{\mu/\rho}. \tag{3.13}$$

Equation (3.12) is the *wave equation*.

It may be noted that usually the soil is a two phase medium, consisting of particles and water, but for shear deformations this has no effect.

3.2.1 Unloaded Soil Layer

For the simplest case, namely that of a homogeneous layer with no surface load, the boundary conditions may be supposed to be

$$z = h : u = u_0 \sin[\omega(t - x/c_2)], \tag{3.14}$$

and

$$z = 0 : \quad \frac{\partial u}{\partial z} = 0. \tag{3.15}$$

The first boundary condition expresses that at the lower boundary of the soil layer a sinusoidal wave is acting, and the second boundary condition expresses that the top of the soil layer (the soil surface) is free of stress. The vertical displacement w has been disregarded, or, to be more precise, it has been assumed that the derivative $\partial w / \partial x$ is small, compared to $\partial u / \partial z$.

The solution of the problem defined by (3.12), (3.14) and (3.15) is

$$u = u_0 \frac{\cos(\omega z / c)}{\cos(\omega h / c)} \sin[\omega(t - x / c_2)]. \tag{3.16}$$

It can easily be verified that this solution satisfies all necessary conditions. The displacements are all in phase with the vibration of the base rock. This is caused by the simplicity of the problem considered, without damping, for instance. If the amplitude of the displacements at the top of the layer is denoted by u_t, it follows that

$$u_t = u_0 \frac{1}{\cos(\omega h / c)}. \tag{3.17}$$

This is always larger than the value at the base, because the function $\cos(\omega h / c)$ is always smaller than 1. For certain values of the frequency the amplitude at the surface of the layer may become infinitely large, indicating resonance. The smallest frequency for which this occurs is when $\omega h / c = \pi / 2$, or

$$\omega = \omega_1 = \frac{\pi c}{2h} = 1.571 \frac{c}{h}. \tag{3.18}$$

When the frequency ω is expressed as $2\pi / T$, where T is the period of the vibration, then the first occurrence of resonance is for a period

$$T = T_1 = \frac{4h}{c}. \tag{3.19}$$

Because h / c is the travel time of a single wave through the layer, upward or downward, this means that resonance occurs if the period of the vibration is such that a wave travels 4 times through the layer. This can be understood by noting the effect of a quarter wave during which a periodic shear stress is acting at the base of the layer. A shear wave will travel through the column, and will be reflected at the free top as a shear wave of opposite sign. When this shear wave reaches the bottom of the column again, after having travelled over a length h, it will be reflected at the rigid bottom as another shear wave of opposite sign. This in its turn will be reflected at the top of the column as a shear wave of the original sign, and this wave has to travel over another column length h to arrive at the bottom of the column. Interference may take place if the wave has travelled over a distance of $4h$, and meets another wave of the same sign. This will be the case if the period of the wave $T = 4h/c$. In this case an ever stronger wave will be generated in the soil layer, indicating resonance.

In dry soils the elastic modulus may be approximated by the expression

$$\mu \approx \tfrac{1}{2}C\sigma_v, \tag{3.20}$$

where, for dynamic loading, C is the compression coefficient of the soil (which is about 250–2500 for sand, and 100–1000 for clay), and σ_v is the vertical (effective) stress. The average stress in the layer is

$$\sigma_v \approx \tfrac{1}{2}\rho gh, \tag{3.21}$$

where g is the gravity constant ($g \approx 10$ m/s^2). It now follows that the wave velocity c can be approximated by

$$c \approx \tfrac{1}{2}\sqrt{Cgh}. \tag{3.22}$$

For a layer of sand of 10 m thickness, assuming $C = 1000$, the value of this wave velocity will be about 150 m/s, which is an order of magnitude smaller than the wave velocity in rock-like materials. The value of the first eigen frequency now is, with (3.18), $\omega_1 \approx 25$ s^{-1}. As this may be very close to the dominant frequency of earthquake motion, which was given as approximately 30 s^{-1}, see (3.5), it follows that an earthquake may lead to large displacements at the soil surface, if the conditions are unfavorable.

It should be noted that in this section damping, which is an essential property of soft soils, has not been taken into account. Damping will be considered in a later section, and will be found to have a moderating effect, but first the case of a shear wave in a layer with a certain surface load will be considered.

3.2.2 Soil Layer with Surface Load

As a second example consider the case of a soil layer loaded by a mass at its surface, see Fig. 3.4. In this case the boundary conditions are

$$z = h \ : \ u = u_0 \sin[\omega(t - x/c_2)], \tag{3.23}$$

and

$$z = 0 \ : \ \rho d\frac{\partial^2 u}{\partial t^2} = \tau = \mu\frac{\partial u}{\partial z}, \tag{3.24}$$

where d is a measure for the surface load, with the mass of the surface load expressed as the thickness of an equivalent soil layer, and τ is the shear stress transmitted between the surface load and the foundation soil. Equation (3.24) can also be written as

$$z = 0 \ : \ d\frac{\partial^2 u}{\partial t^2} = c^2\frac{\partial u}{\partial z}. \tag{3.25}$$

Fig. 3.4 Soil layer with
surface load

The solution of the problem defined by (3.12), (3.23) and (3.25) is

$$\frac{u}{u_0} = \frac{\cos(\omega z/c) - (\omega d/c)\sin(\omega z/c)}{\cos(\omega h/c) - (\omega d/c)\sin(\omega h/c)} \sin[\omega(t - x/c_2)]. \qquad (3.26)$$

It can easily be verified that this solution satisfies all necessary conditions, and that
it reduces to the solution of the previous case if the mass of the surface load tends
towards zero ($d \to 0$). As in the previous example the displacements are all in phase
with the vibration of the base rock, as a result of the simplicity of the problem
considered, with damping being disregarded, for instance.

The amplitude of the vertical displacement at the top of the layer is

$$\frac{u_t}{u_0} = \frac{1}{\cos(\omega h/c) - (\omega d/c)\sin(\omega h/c)}. \qquad (3.27)$$

As an example one may consider the case of a sandy soil, with $c = 300$ m/s and
a wave of frequency $\omega = 30$ s^{-1}. If the thickness of the soil layer is 20 m, and the
equivalent thickness of the surface load is 2 m (indicating a small house), the value
of the parameters $\omega h/c$ and $\omega d/c$ is 2, respectively 0.2. In that case the amplitude
at the top of the soil layer (and of the surface load) is found to be 1.67 times the amplitude at the base, indicating a certain amplification of the effect of the earthquake.
The amplification depends very much on the values of the various parameters of the
soil and the earthquake, and may be considerably larger than the value obtained in
this example.

Actually, even resonance may occur, as indicated by an infinitely large amplitude, if the denominator of the fraction in (3.27) vanishes. If the smallest resonance
frequency is again denoted by ω_1, its value can be determined from the condition

$$\left(\frac{\omega_1 d}{c}\right) = \cot\left(\frac{\omega_1 h}{c}\right). \qquad (3.28)$$

This will be considered in some more detail later, in Sect. 5.4.3, with damping also
taken into account.

From a point of view of theoretical verification it is interesting to consider in particular the case of rather slow vibrations, when the parameter $\omega h/c$ is small compared to 1. In that case the resonance frequency ω_1, as defined by (3.28), can be obtained from the relation

$$\omega_1^2 = \frac{c^2}{hd} = \frac{\mu}{\rho hd} = \frac{\mu A/h}{\rho Ad},$$ (3.29)

where A is the area of the column considered. The quantity $\mu A/h$ can be considered as the spring stiffness k of the column, and ρAd is the total mass m of the surface load. Thus the resonance frequency can also be written as

$$\omega_1^2 = \frac{k}{m}.$$ (3.30)

This result is in perfect agreement with the result obtained in Chap. 1 for the resonance frequency of a system of a discrete spring and mass. It appears that the result obtained in this section is in agreement with the result for a discrete spring and mass if the dimensionless frequency parameter is small enough ($\omega h/c \ll 1$). This is the case, for instance, if the soil is sufficiently stiff, or if the frequency is very small, or if the soil layer is very thin.

3.3 Shear Waves in a Gibson Material

The stiffness of a soil usually increases with the effective stress, and thus with depth. A simple relation is obtained if it is assumed that the shear modulus increases linearly with depth,

$$\mu = \mu_0 z/h.$$ (3.31)

Stresses and deformations of materials of this type have been investigated extensively by Gibson (1967). For this reason the material is often denoted as a *Gibson material*.

The basic differential equation is, for a non-homogeneous material,

$$\rho \frac{\partial^2 u}{\partial t^2} = \frac{\partial \tau}{\partial z} = \frac{\partial}{\partial z}\left(\mu \frac{\partial u}{\partial z}\right),$$ (3.32)

where τ is the shear stress. With (3.31) the differential equation now is found to be

$$\frac{1}{c^2} \frac{\partial^2 u}{\partial t^2} = \frac{z}{h} \frac{\partial^2 u}{\partial z^2} + \frac{1}{h} \frac{\partial u}{\partial z},$$ (3.33)

where now

$$c = \sqrt{\mu_0/\rho}.$$ (3.34)

Again restriction is made to sinusoidal fluctuations,

$$u(z, t) = f(z) \sin[\omega(t - x/c_2)]. \tag{3.35}$$

Substitution of this expression into (3.33) leads to the following ordinary differential equation for the function f,

$$\frac{d^2 f}{dz^2} + \frac{1}{z}\frac{df}{dz} + \frac{\omega^2 h}{c^2 z} f = 0. \tag{3.36}$$

The general solution of this equation is

$$f = A J_0(2\omega\sqrt{zh}/c) + B Y_0(2\omega\sqrt{zh}/c), \tag{3.37}$$

where $J_0(x)$ and $Y_0(x)$ are Bessel functions of order zero, and of the first and second kind, respectively (Abramowitz and Stegun, 1964).

Let the boundary condition at the surface be that the shear stress is zero,

$$z = 0 : \tau = 0. \tag{3.38}$$

It then follows that the coefficient B must be zero. If the other boundary condition is that at a depth h the amplitude of the displacements is u_0,

$$z = h : u = u_0 \sin[\omega(t - x/c_2)], \tag{3.39}$$

then the coefficient A is found to be

$$A = \frac{u_0}{J_0(2\omega h/c)}. \tag{3.40}$$

The final solution now is

$$u = u_0 \frac{J_0(2\omega\sqrt{zh}/c)}{J_0(2\omega h/c)} \sin[\omega(t - x/c_2)]. \tag{3.41}$$

The amplitude at the surface is

$$u_t = \frac{u_0}{J_0(2\omega h/c)}. \tag{3.42}$$

This is always larger than the amplitude at the base. For certain values of the frequency the amplitude even becomes infinitely large, again indicating resonance. The smallest value of the frequency for which this occurs (to be denoted by ω_1) is determined by the first zero of the Bessel function $J_0(x)$, which occurs for $x = 2.405$ (Abramowitz and Stegun, 1964; p. 409). Hence

$$\omega_1 = 1.202 \frac{c}{h}. \tag{3.43}$$

This is about 23% smaller than in the case of a homogeneous layer with its constant shear modulus equal to the value obtained here at a depth $z = h$, see (3.18).

3.4 Hysteretic Damping

In this section the influence of damping is investigated, for a homogenous linear elastic layer. This will appear to have a considerable effect, reducing the displacements at the surface if the damping coefficient is large enough. The effect of damping on the surface vibrations of soft soil layers produced by an earthquake in the underlying rock has been investigated by Idriss and Seed (1968), for a class of non-homogeneous layers, with a shear modulus increasing with depth. The damping was introduced in that model by a friction force on each element proportional to its velocity, simulating the resistance due to some viscous resistance, see also Das (1993) and Kramer (1996). In this chapter damping will be introduced by a hysteretic effect in the stress-strain relation of the soil, simulating irreversible (plastic) deformations in each complete cycle.

3.4.1 Basic Equations

The basic partial differential equation can be established by considering the equation of motion and the constitutive relation of the material. The equation of motion is, as before, see (3.9),

$$\frac{\partial \tau}{\partial z} = \rho \frac{\partial^2 u}{\partial t^2}. \tag{3.44}$$

The constitutive relation is assumed to be

$$\tau = \mu \gamma + \mu t_r \frac{\partial \gamma}{\partial t} = \mu \frac{\partial u}{\partial z} + \mu t_r \frac{\partial^2 u}{\partial t \partial z}, \tag{3.45}$$

where t_r is the response time of the material, see Chap. 1, which may be used to characterize the damping of the material. For a viscous material this can be considered to be a given constant. In such cases the effect of damping depends upon the frequency of the loading, with the material becoming very stiff for very high frequencies. For soils this is not realistic, as the damping is considered to be produced by irreversible plastic deformations of the material. In order to describe hysteretic damping it is assumed that the product ωt_r is constant. This can be taken into account by introducing a dimensionless damping parameter ζ such that

$$2\zeta = \omega t_r. \tag{3.46}$$

The constitutive relation (3.45) can now be written as

$$\tau = \mu \frac{\partial u}{\partial z} + 2\mu(\zeta/\omega) \frac{\partial^2 u}{\partial t \partial z}. \tag{3.47}$$

It follows from (3.44) and (3.47) that

$$\frac{1}{c^2} \frac{\partial^2 u}{\partial t^2} = \frac{\partial^2 u}{\partial z^2} + \frac{2\zeta}{\omega} \frac{\partial^3 u}{\partial t \partial z^2}, \tag{3.48}$$

which is the basic differential equation to be considered.

For a harmonic vibration, with frequency ω, the solution can be assumed to be of the form

$$u = f(z) \sin[\omega(t - x/c_2)] + g(z) \cos[\omega(t - x/c_2)]. \tag{3.49}$$

Substitution into the differential equation shows that the functions $f(z)$ and $g(z)$ must satisfy the differential equations

$$\frac{d^2 f}{dz^2} + \frac{\omega^2}{c^2} f - 2\zeta \frac{d^2 g}{dz^2} = 0, \tag{3.50}$$

and

$$\frac{d^2 g}{dz^2} + \frac{\omega^2}{c^2} g + 2\zeta \frac{d^2 f}{dz^2} = 0. \tag{3.51}$$

The general solution of the system of (3.50) and (3.51) is

$$f = A_1 \exp[(p+iq)z] + A_2 \exp[(p-iq)z]$$
$$+ A_3 \exp[-(p+iq)z] + A_4 \exp[-(p-iq)z], \tag{3.52}$$

$$g = -i A_1 \exp[(p+iq)z] + i A_2 \exp[(p-iq)z]$$
$$- i A_3 \exp[-(p+iq)z] + i A_4 \exp[-(p-iq)z], \tag{3.53}$$

where p and q must be determined from the equations

$$p^2 - q^2 = -\frac{\omega^2/c^2}{1 + 4\zeta^2}, \tag{3.54}$$

$$2pq = 2\zeta \frac{\omega^2/c^2}{1 + 4\zeta^2}. \tag{3.55}$$

The values of the parameters p and q can most easily be determined by introducing the complex variable

$$p + iq = r \sin(\phi) + i r \cos(\phi), \tag{3.56}$$

so that

$$p = r \sin(\phi), \qquad q = r \cos(\phi). \tag{3.57}$$

The angle ϕ can then be determined from the condition

$$2\phi = \arctan(2\zeta), \tag{3.58}$$

and the radius r can be determined from the condition

$$r^4 = \frac{\omega^4/c^4}{1 + 4\zeta^2}. \tag{3.59}$$

These parameters have been chosen such that they reduce to a simple form in the absence of damping. Actually, when $\zeta = 0$ the parameters are $p = 0$ and $q = \omega/c$.

The integration constants A_1, A_2, A_3 and A_4 in the general solution given in (3.52) and (3.53) must be determined from the boundary conditions at the top and the bottom of the layer. Two cases will be considered: an unloaded soil layer and a layer with a given surface load.

3.4.2 Unloaded Soil Layer

For an unloaded soil layer the boundary conditions are

$$z = 0 \; : \; \tau = 0. \tag{3.60}$$

$$z = h \; : \; u = u_0 \sin[\omega(t - x/c_2)]. \tag{3.61}$$

The four integration constants can easily be determined from these conditions. The final solution then is

$$
\begin{aligned}
Au/u_0 = {} & \cosh(ph)\cos(qh)\cosh(pz)\cos(qz)\sin[\omega(t - x/c_2)] \\
& + \sinh(ph)\sin(qh)\sinh(pz)\sin(qz)\sin[\omega(t - x/c_2)] \\
& + \cosh(ph)\cos(qh)\sinh(pz)\sin(qz)\cos[\omega(t - x/c_2)] \\
& - \sinh(ph)\sin(qh)\cosh(pz)\cos(qz)\cos[\omega(t - x/c_2)], \quad (3.62)
\end{aligned}
$$

where

$$A = \cosh^2(ph) - \sin^2(qh). \tag{3.63}$$

If the amplitude of the vibration at the top is denoted by u_t its value is found to be

$$\frac{u_t}{u_0} = \frac{1}{\sqrt{A}}. \tag{3.64}$$

When there is no damping this solution reduces to the result obtained before, in (3.17). Actually, for $\zeta = 0$ the quantity $\sqrt{A}$ reduces to $\cos(\omega h/c)$, so that then the ratio of the two amplitudes becomes $1/\cos(\omega h/c)$, which is in agreement with (3.17).

The amplitude of the displacements at the top of the layer is shown, as a function of the dimensionless frequency $\omega h/c$, and for three values of the damping ratio ζ, in Fig. 3.5. For very small frequencies ($\omega h/c \to 0$), i.e. for the static case, the displacement at the top is equal to the displacement at the bottom, $u_t/u_0 = 1$. Furthermore, it appears that for very small values of the damping ratio very large values of the displacements may be obtained for certain frequencies, near $\omega h/c = \pi/2, 3\pi/2, \ldots$ etc. This is the resonance effect observed before, in Sect. 3.2.1. Finally, it follows from Fig. 3.5 that the amplification of the displacements is considerably reduced if the damping ratio ζ increases.

Fig. 3.5 Amplitude of wave
at the top of the layer

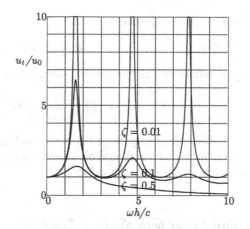

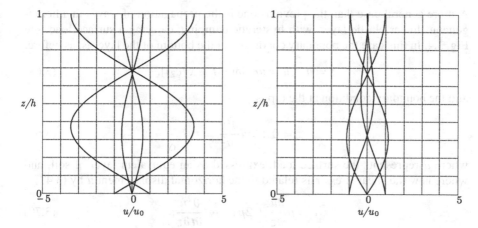

Fig. 3.6 Displacements as a function of depth, $\omega h/c = 5$, $\zeta = 0.01$ and $\zeta = 0.2$

The displacements are shown as a function of the vertical coordinate z in Fig. 3.6, for $\omega h/c = 5$, $\zeta = 0.01$ and $\zeta = 0.2$, and for four values of time, namely $\omega t = 0$, $\pi/2$, π, $3\pi/2$. For the case of small damping ($\zeta = 0.01$) the amplification factor for the amplitude of the displacements at the top of the layer should be about $1/\cos(\omega h/c) = 3.525$, see (3.17). The results shown in the left part of Fig. 3.6 appear to confirm this value. The right part of the figure shows the influence of damping on the displacements. If the damping ratio is taken as $\zeta = 0.2$ the amplitude of the displacements at the top is considerably smaller, as expected.

If the damping ratio is small, $\zeta \ll 1$, the amplitude of the wave at the top can be shown to be

$$\zeta \ll 1 : \quad \frac{u_t}{u_0} \approx \frac{1}{\sqrt{W^2\zeta^2 + \cos^2(W)}}, \tag{3.65}$$

where W is the dimensionless frequency,

$$W = \omega h / c. \tag{3.66}$$

The largest value of the amplitude ratio occurs if $W = \pi/2$,

$$\zeta \ll 1 \; : \; \left(\frac{u_t}{u_0}\right)_{max} \approx \frac{2}{\pi\zeta}. \tag{3.67}$$

For $\zeta = 0$ this is infinitely large, as before indicating resonance. If $\zeta = 0.1$ the maximum amplitude ratio should be approximately 6.366, using (3.67). This value is confirmed by the data shown in Fig. 3.5.

3.4.3 Soil Layer with Surface Load

A more general case than the previous one is the propagation of waves in a homogeneous linear elastic layer, with hysteretic damping, carrying a surface load, see Fig. 3.4. In this case the boundary conditions at the bottom of the layer is, as before,

$$z = h \; : \; u = u_0 \sin[\omega(t - x/c_2)], \tag{3.68}$$

and the boundary condition at the top is

$$z = 0 \; : \; \rho d \frac{\partial^2 u}{\partial t^2} = \tau, \tag{3.69}$$

where d represents the surface load, expressed as an equivalent layer of soil, and where now the shear stress τ is related to the horizontal displacement u by (3.47),

$$\tau = \mu \frac{\partial u}{\partial z} + 2\mu(\zeta/\omega) \frac{\partial^2 u}{\partial t \partial z}. \tag{3.70}$$

It now follows from (3.69) and (3.70) that the boundary condition at the top can be written as

$$z = 0 \; : \; \frac{d}{c^2} \frac{\partial^2 u}{\partial t^2} = \frac{\partial u}{\partial z} + \frac{2\zeta}{\omega} \frac{\partial^2 u}{\partial t \partial z}. \tag{3.71}$$

The general solution of the problem for a material with hysteretic damping can been written in the form of (3.49),

$$u = f(z) \sin[\omega(t - x/c_2)] + g(z) \cos[\omega(t - x/c_2)], \tag{3.72}$$

where the functions $f(z)$ and $g(z)$ are given by (3.52) and (3.53). This solution can also be written as

$$u = C_1 \exp(pz) \cos[\omega(t - x/c_2 + qz)] + C_2 \exp(pz) \sin[\omega(t - x/c_2 + qz)]$$
$$+ C_3 \exp(-pz) \cos[\omega(t - x/c_2 - qz)]$$
$$+ C_4 \exp(-pz) \sin[\omega(t - x/c_2 - qz)]. \tag{3.73}$$

This form is more convenient for the formulation of the boundary conditions.

Substitution of the general solution (3.73) into the two boundary conditions leads, after some elementary algebra, to the following four equations

$$\left(ph - 2\zeta qh + \frac{d}{h}\frac{\omega^2 h^2}{c^2}\right)C_1 + (qh + 2\zeta ph)C_2$$

$$-\left(ph - 2\zeta qh - \frac{d}{h}\frac{\omega^2 h^2}{c^2}\right)C_3 - (qh + 2\zeta ph)C_4 = 0, \qquad (3.74)$$

$$-(qh + 2\zeta ph)C_1 + \left(ph - 2\zeta qh + \frac{d}{h}\frac{\omega^2 h^2}{c^2}\right)C_2$$

$$+ (qh + 2\zeta ph)C_3 - \left(ph - 2\zeta qh - \frac{d}{h}\frac{\omega^2 h^2}{c^2}\right)C_4 = 0, \qquad (3.75)$$

$$\exp(ph)\cos(qh)C_1 + \exp(ph)\sin(qh)C_2$$

$$+ \exp(-ph)\cos(qh)C_3 - \exp(-ph)\sin(qh)C_4 = 0, \qquad (3.76)$$

$$-\exp(ph)\sin(qh)C_1 + \exp(ph)\cos(qh)C_2$$

$$+ \exp(-ph)\sin(qh)C_3 + \exp(-ph)\cos(qh)C_4 = u_0. \qquad (3.77)$$

The constants C_1, C_2, C_3 and C_4 can be determined from these equations. A numerical solution of the system of four linear equations is probably most convenient, especially because the data will be calculated by a simple computer program anyway. The parameters of the problem are the dimensionless frequency $\omega h/c$, the damping ratio ζ, and the dimensionless mass of the load, expressed as the ratio d/h.

The amplitude of the vibrations at the top of the soil layer are shown in Fig. 3.7, for $\zeta = 0.1$ and $\zeta = 0.5$, and for three values of the load, $d/h = 0$, 1, 10. The results for $d/h = 0$ are in agreement with those shown in Fig. 3.5 for $\zeta = 0.1$. The right part of the figure shows the influence of damping on the displacements. If the

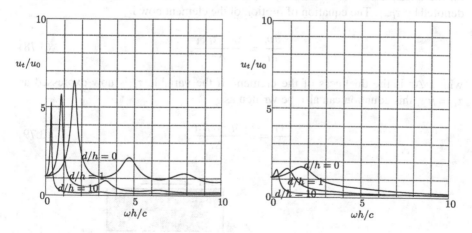

Fig. 3.7 Amplitude of wave at the top of the layer, $\zeta = 0.1$ and $\zeta = 0.5$

damping ratio is taken as $\zeta = 0.5$ the amplitude of the displacements at the top is considerably smaller, as can be expected.

3.5 Numerical Solution

All the analytical solutions presented above suffer from the defect that the stress-strain-relationship must be of rather simple form (linear elastic, with perhaps linear hysteretic damping), and that the soil properties must be homogeneous. Real soils are often composed of several layers of variable properties, and often they exhibit non-linear properties. Therefore a numerical solution may be considered, because this can more easily be generalized to non-linear and non-homogeneous properties. In this section a simple numerical solution method is presented, again with hysteretic damping.

The considerations will be restricted to one-dimensional problems, such as wave propagation in a soft layer, from a stiff deep layer to the surface. For this relatively simple class of problems there is little difference between the various existing numerical techniques, such as finite elements and finite differences. Therefore the simplest of these methods, an explicit finite difference method, will be used.

Basic Equations

It is most convenient to base the numerical model upon a description of the basic equations in terms of the lateral displacement u, the lateral velocity v, and the shear stress s. Let the soil layer be subdivided into a certain number (n) of elements, and let the velocity of a typical element be denoted by v_i, see Fig. 3.8. The shear stress on the lower surface is denoted by τ_i, and the shear stress at the upper surface is denoted by τ_{i-1}. The equation of motion of the element now is,

$$\rho \frac{\partial v_i}{\partial t} = \frac{\tau_i - \tau_{i-1}}{\Delta z}, \tag{3.78}$$

where Δz is the thickness of the element. If the variable τ_i is now expressed as $\tau_i = \mu s_i$ this equation can also be written as

$$\frac{\partial v_i}{\partial t} = c^2 \frac{s_i - s_{i-1}}{\Delta z}, \tag{3.79}$$

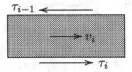

Fig. 3.8 Shear wave

where, as usual, c is the shear wave velocity, $c = \sqrt{\mu/\rho}$. The finite difference form of (3.79) is

$$v_i' = v_i + c^2 \frac{\Delta t}{\Delta z}(s_i - s_{i-1}), \tag{3.80}$$

where v_i' represents the velocity after a time interval Δt. The velocity is the time derivative of the displacement,

$$v_i = \frac{\partial u_i}{\partial t}, \tag{3.81}$$

or, in finite difference form,

$$u_i' = u_i + v_i \Delta t. \tag{3.82}$$

The shear stress can be related to the shear strain by the equation

$$\tau_i = \mu \frac{\partial u}{\partial z} + \mu t_r \frac{\partial v}{\partial z}, \tag{3.83}$$

where t_r is the characteristic time of the damping effect. As before we write

$$2\zeta = \omega t_r, \tag{3.84}$$

where ζ is the dimensionless damping ratio, and where ω is the frequency of the load, assuming that the load is periodic. This means that (3.83) can also be written as

$$s_i = \frac{\partial u}{\partial z} + \frac{2\zeta}{\omega} \frac{\partial v}{\partial z}. \tag{3.85}$$

The finite difference form of this equation is

$$s_i = (u_{i+1} - u_i)/\Delta z + (2\zeta/\omega)(v_{i+1} - v_i)/\Delta z. \tag{3.86}$$

A numerical model can now be developed as follows. If the problem is again that of the propagation of a shear wave from a certain depth to the surface of the soil, the boundary condition at the lower boundary of the layer can be considered to be

$$u_n = d\sin(\omega t), \qquad v_n = d\omega\sin(\omega t), \tag{3.87}$$

where d is the amplitude of the sinusoidal fluctuation, with frequency ω. Using (3.86) the shear stresses at every level (from $i = 1$ to $i = n - 1$) can now be calculated, assuming that the displacements in the layer itself are initially zero. Using (3.80) the velocities at the end of the time interval can then be calculated, and finally the displacements at the end of the time interval can be calculated using (3.82), from $i = 1$ to $i = n - 1$. This process can then be repeated for as many steps as desired.

The calculations can be executed by a computer program, with the main computation algorithm being reproduced below.

```
for (i=n-1;i>0;i--)
{
s[i]=(u[i+1]-u[i])/dz+(2*zeta/omega)*(v[i+1]-v[i])/dz;
v[i]=v[i]+c*c*dt*(s[i]-s[i-1])/dz;
u[i]=u[i]+v[i]*dt;
}
```

In a computer program the time step should be so small that instabilities are avoided. The magnitude of these time steps can most simply be investigated by considering the basic equation (3.48),

$$\frac{1}{c^2}\frac{\partial^2 u}{\partial t^2} = \frac{\partial^2 u}{\partial z^2} + \frac{2\zeta}{\omega}\frac{\partial^3 u}{\partial t \partial z^2}. \tag{3.88}$$

For $\zeta = 0$ this equation reduces to the standard wave equation

$$\frac{1}{c^2}\frac{\partial^2 u}{\partial t^2} = \frac{\partial^2 u}{\partial z^2}. \tag{3.89}$$

Numerical approximations of this equation, using the simplest finite difference approximations, usually are stable if in each time step the wave travels not more than a single spatial step. This leads to the following condition for the time step,

$$\Delta t \leq \Delta z/c. \tag{3.90}$$

This is the *Courant condition*, see Press et al. (1988).

For large values of the damping ratio ζ the basic equation (3.88) reduces to a diffusion equation for the velocity,

$$\frac{1}{c^2}\frac{\partial v}{\partial t} = \frac{2\zeta}{\omega}\frac{\partial^2 v}{\partial z^2}. \tag{3.91}$$

This can be solved numerically by a stable process if the following stability criterion is satisfied, see e.g. Press et al. (1988),

$$\Delta t \leq \frac{\omega \Delta z^2}{4\zeta c^2}. \tag{3.92}$$

It is suggested that in a computer program the time steps are taken small enough for both criteria to be satisfied.

A computer program using the method described here, QUAKE, may be used as an alternative to the analytical solutions presented in this chapter, and may be used as a basis for more general problems, of non-homogeneous layers, and perhaps involving non-linear soil properties. When comparing the results of a simple computer program with the analytical results it will be observed that there may be considerable deviations, especially for small values of time. This is a result of the initial condition in the numerical solution. It may take many cycles of vibrations before the numerical solution has reached the steady state that has been assumed in the analytical solutions. Actually, during a real earthquake the soil may not reach the steady state, and the results of a non-steady computation may be more realistic.

Problems

3.1 Investigate the influence of the frequency ω and the damping ratio ζ on the ratio of the displacements at the top and the bottom of a soft soil layer.

3.2 Using the computer program QUAKE, verify that the results of the program are in agreement with the analytical results given earlier, at least after many cycles of vibration.

Chapter 4
Theory of Consolidation

4.1 Consolidation

Soft soils such as sand and clay consist of small particles, and often the pore space between the particles is filled with water. In soil mechanics this is denoted as a saturated or a partially saturated porous medium. The deformation of such porous media depends upon the stiffness of the porous material, but also upon the behaviour of the fluid in the pores. If the permeability of the material is small, the deformations may be considerably hindered, or at least retarded, by the pore fluid. The simultaneous deformation of the porous material and flow of pore fluid is the subject of the theory of consolidation, often denoted as *poroelasticity*.

The theory was developed originally by Terzaghi (1925) for the one-dimensional case, and extended to three dimensions by Biot (1941), and it has been studied extensively since. In Terzaghi's original theory the pore fluid and the solid particles were assumed to be completely incompressible. This means that deformations of the porous medium are possible only by a rearrangement of the particles, and that volume changes must be accompanied by the expulsion of pore water. This is a good approximation of the real behaviour of soft soils, especially clay, and also soft sands. Such soils are highly compressible (deformations may be as large as several percents), whereas the constituents, particles and fluid are very stiff.

In later presentations of the theory, starting with those of Biot, compression of the pore fluid and compression of the particles has been taken into account. This generalization made it possible to also consider the deformations of materials such as sandstone and other porous rocks, which are very important in the engineering of deep reservoirs of oil or gas. The linear theory of poroelasticity (or consolidation) has now reached a stage where there is practically general consensus on the basic equations, see e.g. De Boer (2000), Wang (2000), Coussy (2004), Verruijt (2008b).

In this chapter the basic equations of the general theory of linear consolidation are derived, for the case of a linear material, and for pseudo-static deformations (in which inertial forces are disregarded). A simplified version of the theory, in which the soil deformation is assumed to be strictly vertical, is also presented in

A. Verruijt, *An Introduction to Soil Dynamics*,
Theory and Applications of Transport in Porous Media 24,
© Springer Science+Business Media B.V. 2010

this chapter. The analytical solutions for two simple examples are given. In the next chapter the generalization to dynamics is presented.

Before deriving the basic equations of consolidation it is convenient to consider some of the basic principles underlying the theory, especially the influence of the compressibilities of the two constituents (solid particles and pore fluid) on the behaviour of a porous medium in the absence of drainage.

4.1.1 Undrained Compression of a Porous Medium

Consider an element of porous soil or rock, of porosity n, saturated with a fluid. The element is loaded, in undrained condition, by an isotropic total stress $\Delta\sigma$. The resulting pore pressure is denoted by Δp. In order to determine the relation between Δp and $\Delta\sigma$ the load is considered to be applied in two stages: an increment of pressure both in the fluid and in the soil particles of magnitude Δp, and a load on the soil, without any pore pressures, of magnitude $\Delta\sigma$. Compatibility of the two stages, requiring that the total volume change is the sum of the volume changes of the fluid and the solid particles, will be required only for the combination of the two stages.

In the first stage, in which the stress in both fluid and particles is increased by Δp, the volume change of the pore fluid is

$$\Delta V_f = -nC_f\Delta p V,\tag{4.1}$$

where C_f is the compressibility of the pore fluid (which may include the compression of small amounts of isolated gas bubbles), and V is the total volume of the element considered. The volume change of the particles is

$$\Delta V_s = -(1-n)C_s\Delta p V,\tag{4.2}$$

where C_s is the compressibility of the solid material. Assuming that the solid particles all have the same compressibility, it follows that their uniform compression leads to a volume change of the pore space as well (at this stage compatibility of the deformations of fluid and particles is ignored) of the same magnitude. Thus the total volume change of the porous medium is

$$\Delta V = -nC_s V.\tag{4.3}$$

In the second stage the pressure in the fluid remains unchanged, so that there is no volume change of the fluid,

$$\Delta V_f = 0,\tag{4.4}$$

The stress increment $\Delta\sigma - \Delta p$ on the soil, at constant pore pressure, leads to an average stress increment in the solid particles of magnitude $(\Delta\sigma - \Delta p)/(1-n)$. The resulting volume change of the particles is

$$\Delta V_s = -C_s(\Delta\sigma - \Delta p)V.\tag{4.5}$$

The volume change of the porous medium as a whole in this stage also involves the deformations due to sliding and rolling at the contacts of the particles. Assuming that this is also a linear process, in a first approximation, it follows that in this stage of loading

$$\Delta V = -C_m(\Delta\sigma - \Delta p)V, \tag{4.6}$$

where C_m is the compressibility of the porous medium. It is to be expected that this is considerably larger than the compressibilities of the two constituents: fluid and solid particles, because the main mechanism of soil deformation is not so much the compression of the fluid or the particles, but rather the deformation due to a rearrangement of the particles, including sliding and rolling.

Due to both these two loadings the volume changes are

$$\Delta V_f = -nC_f\Delta pV, \tag{4.7}$$

$$\Delta V_s = -(1-n)C_s\Delta pV - C_s(\Delta\sigma - \Delta p)V, \tag{4.8}$$

$$\Delta V = -C_s\Delta pV - C_m(\Delta\sigma - \Delta p)V. \tag{4.9}$$

Because there is no drainage in the combined loading situation, by assumption, the total volume change must be equal to the sum of the volume changes of the fluid and the particles, $\Delta V = \Delta V_f + \Delta V_s$. This gives, with (4.7)–(4.8),

$$\frac{\Delta p}{\Delta\sigma} = B = \frac{1}{1 + n(C_f - C_s)/(C_m - C_s)}. \tag{4.10}$$

The derivation leading to this equation is due to Bishop (1973), but similar equations were given earlier by Gassmann (1951) and Geertsma (1957). The ratio $\Delta p/\Delta\sigma$ under isotropic loading is often denoted by B in soil mechanics (Skempton, 1954). In early developments, such as in Terzaghi's publications, the compressibilities of the fluid and of the solid particles were disregarded, $C_f = C_s = 0$. In that case $B = 1$, which is often used as a first approximation.

4.1.2 The Principle of Effective Stress

The effective stress, introduced by Terzaghi (1925), is defined as that part of the total stresses that governs the deformation of the soil or rock. It is assumed that the total stresses can be decomposed into the sum of the effective stresses and the pore pressure by writing

$$\sigma_{ij} = \sigma'_{ij} + \alpha p\delta_{ij}, \tag{4.11}$$

where σ_{ij} are the components of total stress, σ'_{ij} are the components of effective stress, p is the pore pressure (the pressure in the fluid in the pores), δ_{ij} are the Kronecker delta symbols ($\delta_{ij} = 1$ if $i = j$ and $\delta_{ij} = 0$ otherwise), and α is Biot's coefficient, which is unknown at this stage. For the isotropic parts of the stresses it follows from (4.11) that

$$\sigma = \sigma' + \alpha p. \tag{4.12}$$

In the case of an isotropic linear elastic porous material the relation between the volumetric strain ε and the isotropic effective stress is of the form

$$\varepsilon = \frac{\Delta V}{V} = -C_m \Delta \sigma' = -C_m \Delta \sigma - C_m \alpha p, \tag{4.13}$$

where, as before, C_m denotes the compressibility of the porous material, the inverse of its compression modulus, $C_m = 1/K$. Equation (4.13) should be in agreement with (4.9), which is the case only if

$$\alpha = 1 - C_s/C_m. \tag{4.14}$$

This expression for Biot's coefficient is generally accepted in rock mechanics (Biot and Willis, 1957) and in the mechanics of other porous materials, such as bone or skin (Coussy, 2004). For soft soils the value of α is close to 1.

If the coefficient α is taken as 1, the effective stress principle reduces to

$$\sigma_{ij} = \sigma'_{ij} + p\delta_{ij}. \tag{4.15}$$

This is the form in which the effective stress principle is often expressed in soil mechanics, on the basis of Terzaghi's original work (1925, 1943). This is often justified because soil mechanics practice usually deals with highly compressible clays or sands, in which the compressibility of the solid particles is very small compared to the compressibility of the porous material as a whole. In this case the effective stress is also the average of the forces transmitted in the isolated contact points between the particles. This is sometimes denoted as the *intergranular stress*.

4.2 Conservation of Mass

One of the major principles in the theory of consolidation is that the mass of the two components, water and solid particles, must be conserved. This will be formulated in this section.

Consider a porous material, consisting of a solid matrix or an assembly of particles, with a continuous pore space. The pore space is filled with a fluid, usually water, but possibly some other fluid, or a mixture of fluids. The average velocity of the fluid is denoted by $\mathbf{v}$ and the average velocity of the solids is denoted by $\mathbf{w}$. The densities are denoted by ρ_f and ρ_s, respectively, and the porosity by n.

The equations of conservation of mass of the solids and the fluid can be established by considering the flow into and out of an elementary volume, fixed in space, see Fig. 4.1. The mass of fluid in an elementary volume V is $n\rho_f V$. The increment of this mass per unit time is determined by the net inward flux across the surfaces

Fig. 4.1 Conservation of mass of the fluid

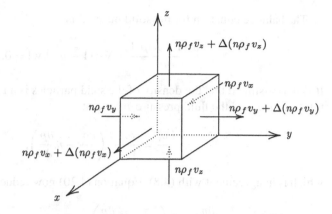

of the element. In y-direction the flow through the left and the right faces of the element shown in Fig. 4.1 (both having an area $\Delta x \Delta z$), leads to a net outward flux of magnitude

$$\Delta(n\rho_f v_y)\Delta x\Delta z = \frac{\Delta(n\rho_f v_y)}{\Delta y}V,$$

where V denotes the volume $\Delta x \Delta y \Delta z$. This leads to the following mass balance equation

$$\frac{\partial(n\rho_f)}{\partial t} + \frac{\partial(n\rho_f v_x)}{\partial x} + \frac{\partial(n\rho_f v_y)}{\partial y} + \frac{\partial(n\rho_f v_z)}{\partial z} = 0. \tag{4.16}$$

Using vector notation this can also be written as

$$\frac{\partial(n\rho_f)}{\partial t} + \nabla \cdot (n\rho_f \mathbf{v}) = 0. \tag{4.17}$$

The compressibility of the fluid can be expressed by assuming that the constitutive equation of the fluid is

$$\frac{d\rho_f}{dp} = \rho_f C_f, \tag{4.18}$$

which is in agreement with the definition of the fluid compressibility C_f in (4.7). For pure water the compressibility is $C_f \approx 0.5 \times 10^{-9}$ m^2/kN. For a fluid containing small amounts of a gas the compressibility may be considerably larger, however. It now follows from (4.17) that

$$\frac{\partial n}{\partial t} + nC_f\frac{\partial p}{\partial t} + \nabla \cdot (n\mathbf{v}) = 0, \tag{4.19}$$

where a term expressing the product of the fluid velocity and the pressure gradient has been disregarded, assuming that both are small quantities, so that the product is of second order.

The balance equation for the solid material is

$$\frac{\partial[(1-n)\rho_s]}{\partial t} + \nabla\cdot[(1-n)\rho_s\mathbf{w}] = 0. \tag{4.20}$$

It is now assumed that the density of the solid particles is a function of the isotropic total stress σ and the fluid pressure p, so that

$$\frac{\partial\rho_s}{\partial t} = \frac{\rho_s C_s}{1-n}\left(\frac{\partial\sigma}{\partial t} - n\frac{\partial p}{\partial t}\right), \tag{4.21}$$

which is in agreement with (4.8). Equation (4.20) now reduces to

$$-\frac{\partial n}{\partial t} + C_s\left(\frac{\partial\sigma}{\partial t} - n\frac{\partial p}{\partial t}\right) + \nabla\cdot[(1-n)\mathbf{w}] = 0, \tag{4.22}$$

where again a term expressing the product of a velocity and a gradient of stress or pressure has been disregarded.

The time derivative of the porosity n can easily be eliminated from (4.22) and (4.19) by adding these two equations. This gives

$$\nabla\cdot\mathbf{w} + \nabla\cdot[n(\mathbf{v}-\mathbf{w})] + n(C_f - C_s)\frac{\partial p}{\partial t} + C_s\frac{\partial\sigma}{\partial t} = 0. \tag{4.23}$$

The quantity $n(\mathbf{v}-\mathbf{w})$ is the porosity multiplied by the relative velocity of the fluid with respect to the solids. This is precisely what is intended by the *specific discharge*, which is the quantity that appears in Darcy's law for the flow of a fluid through a porous medium. It will be denoted by $\mathbf{q}$,

$$\mathbf{q} = n(\mathbf{v}-\mathbf{w}). \tag{4.24}$$

If the displacement vector of the solids is denoted by $\mathbf{u}$, the term $\nabla\cdot\mathbf{w}$ can also be written as $\partial\varepsilon/\partial t$, where ε is the volume strain,

$$\varepsilon = \nabla\cdot\mathbf{u}. \tag{4.25}$$

Equation (4.23) can now be written as

$$\frac{\partial\varepsilon}{\partial t} + n(C_f - C_s)\frac{\partial p}{\partial t} + C_s\frac{\partial\sigma}{\partial t} = -\nabla\cdot\mathbf{q}. \tag{4.26}$$

Because the isotropic total stress can be expressed as $\sigma = \sigma' + \alpha p$, see (4.12), and the isotropic effective stress can be related to the volume strain by $\sigma' = -\varepsilon/C_m$, where C_m is the compressibility of the porous medium, see (4.13), it follows that (4.26) can also be written as

$$\alpha\frac{\partial\varepsilon}{\partial t} + S_p\frac{\partial p}{\partial t} = -\nabla\cdot\mathbf{q}, \tag{4.27}$$

where S_p is the *storativity*, of the pore space,

$$S_p = nC_f + (\alpha - n)C_s. \qquad (4.28)$$

Equation (4.27) will be denoted here as the *fluid conservation equation*. It has been denoted as the *storage equation* in earlier literature. It is an important basic equation of the theory of consolidation. In its form (4.26) it admits a simple heuristic interpretation: the compression of the soil consists of the compression of the pore fluid and the particles plus the amount of fluid expelled from an element by flow. The equation actually expresses conservation of mass of fluids and solids, together with some notions about the compressibilities.

It may be noted that in deriving (4.27) a number of assumptions have been made, but these are all relatively realistic. Thus, it has been assumed that the solid particles and the fluid are linearly compressible, and some second order terms, consisting of the products of small quantities, have been disregarded. The fluid conservation equation (4.27) can be considered as a reasonably accurate description of physical reality.

4.3 Darcy's Law

In 1857 Darcy found, from experiments, that the specific discharge of a fluid in a porous material is proportional to the head loss. In terms of the quantities used in this chapter Darcy's law can be written as

$$\mathbf{q} = -\frac{\kappa}{\mu}(\nabla p - \rho_f \mathbf{g}), \qquad (4.29)$$

where κ is the (intrinsic) permeability of the porous material, μ is the viscosity of the fluid, and $\mathbf{g}$ is the gravity vector. The permeability depends upon the size of the pores. As a first approximation one may consider that the permeability κ is proportional to the square of the particle size.

If the coordinate system is such that the z-axis is pointing in upward vertical direction the components of the gravity vector are $g_x = 0$, $g_y = 0$, $g_z = -g$, and then Darcy's law may also be written as

$$q_x = -\frac{\kappa}{\mu}\frac{\partial p}{\partial x},$$

$$q_y = -\frac{\kappa}{\mu}\frac{\partial p}{\partial y}, \qquad (4.30)$$

$$q_z = -\frac{\kappa}{\mu}\left(\frac{\partial p}{\partial z} + \rho_f g\right).$$

The product $\rho_f g$ may also be written as γ_w, the volumetric weight of the fluid.

In soil mechanics practice the coefficient in Darcy's law is often expressed in terms of the hydraulic conductivity k rather than the permeability κ. This hydraulic conductivity is defined as

$$k = \frac{\kappa \rho_f g}{\mu}. \tag{4.31}$$

This means that Darcy's law can also be written as

$$q_x = -\frac{k}{\gamma_w}\frac{\partial p}{\partial x},$$
$$q_y = -\frac{k}{\gamma_w}\frac{\partial p}{\partial y}, \tag{4.32}$$
$$q_z = -\frac{k}{\gamma_w}\left(\frac{\partial p}{\partial z} + \gamma_w\right).$$

From these equations it follows that

$$\nabla \cdot \mathbf{q} = \frac{\partial q_x}{\partial x} + \frac{\partial q_y}{\partial y} + \frac{\partial q_z}{\partial z} = -\nabla \cdot \left(\frac{k}{\gamma_w}\nabla p\right), \tag{4.33}$$

if again a small second order term (involving the spatial derivative of the hydraulic conductivity) is disregarded.

Substitution of (4.33) into (4.27) gives

$$\alpha\frac{\partial \varepsilon}{\partial t} + S_p\frac{\partial p}{\partial t} = \nabla \cdot \left(\frac{k}{\gamma_w}\nabla p\right). \tag{4.34}$$

Compared to (4.27) the only additional assumption is the validity of Darcy's law. As Darcy's law usually gives a good description of flow in a porous medium, (4.34) can be considered as reasonably accurate.

4.4 Equilibrium Equations

The complete formulation of a fully three-dimensional problem requires a consideration of the principles of solid mechanics, including equilibrium, compatibility and the stress-strain-relations. In addition to these equations the initial conditions and the boundary conditions must be formulated. These equations are presented here, for a linear elastic material.

The equations of equilibrium can be established by considering the stresses acting upon the six faces of an elementary volume, see Fig. 4.2. In this figure only the six stress components in the y-direction are shown. The equilibrium equations in the three coordinate directions are

Fig. 4.2 Equilibrium of element

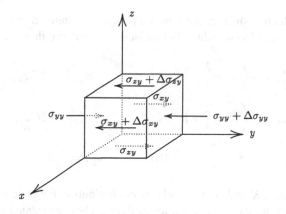

$$\frac{\partial \sigma_{xx}}{\partial x} + \frac{\partial \sigma_{yx}}{\partial y} + \frac{\partial \sigma_{zx}}{\partial z} - f_x = 0,$$

$$\frac{\partial \sigma_{xy}}{\partial x} + \frac{\partial \sigma_{yy}}{\partial y} + \frac{\partial \sigma_{zy}}{\partial z} - f_y = 0, \tag{4.35}$$

$$\frac{\partial \sigma_{xz}}{\partial x} + \frac{\partial \sigma_{yz}}{\partial y} + \frac{\partial \sigma_{zz}}{\partial z} - f_z = 0,$$

where f_x, f_y and f_z denote the components of a possible body force. In addition to these equilibrium conditions there are three equations of equilibrium of moments. These can be taken into account most conveniently by noting that they result in the symmetry of the stress tensor,

$$\sigma_{xy} = \sigma_{yx},$$

$$\sigma_{yz} = \sigma_{zy}, \tag{4.36}$$

$$\sigma_{zx} = \sigma_{xz}.$$

The stresses in these equations are total stresses. They are considered positive for compression, in agreement with common soil mechanics practice, but in contrast with the usual sign convention in solid mechanics.

The total stresses are related to the effective stresses by the generalized Terzaghi principle, see (4.11),

$$\sigma_{xx} = \sigma'_{xx} + \alpha p, \quad \sigma_{xy} = \sigma'_{xy}, \quad \sigma_{xz} = \sigma'_{xz},$$

$$\sigma_{yy} = \sigma'_{yy} + \alpha p, \quad \sigma_{yz} = \sigma'_{yz}, \quad \sigma_{yx} = \sigma'_{yx}, \tag{4.37}$$

$$\sigma_{zz} = \sigma'_{zz} + \alpha p, \quad \sigma_{zx} = \sigma'_{zx}, \quad \sigma_{zy} = \sigma'_{zy},$$

where α is Biot's coefficient, $\alpha = 1 - C_s/C_m$.

The effective stresses determine the deformations of the soil. The shear stresses can of course only be transmitted by the soil skeleton. As a first approximation the effective stresses are now supposed to be related to the strains by the generalized form of Hooke's law. For an isotropic material these relations are

$$\sigma'_{xx} = -\left(K - \frac{2}{3}G\right)\varepsilon - 2G\varepsilon_{xx}, \qquad \sigma'_{xy} = -2G\varepsilon_{xy}, \qquad \sigma'_{xz} = -2G\varepsilon_{xz},$$

$$\sigma'_{yy} = -\left(K - \frac{2}{3}G\right)\varepsilon - 2G\varepsilon_{yy}, \qquad \sigma'_{yz} = -2G\varepsilon_{yz}, \qquad \sigma'_{yx} = -2G\varepsilon_{yx}, \quad (4.38)$$

$$\sigma'_{zz} = -(K - \frac{2}{3}G)\varepsilon - 2G\varepsilon_{zz}, \qquad \sigma'_{zx} = -2G\varepsilon_{zx}, \qquad \sigma'_{zy} = -2G\varepsilon_{zy},$$

where K and G are the elastic coefficients of the material, the compression modulus and the shear modulus, respectively. They are related to the Lamé constants λ and μ by the relations

$$\lambda = K - \frac{2}{3}G, \quad G = \mu. \tag{4.39}$$

The compression modulus (or bulk modulus) K is the inverse of the compressibility C_m of the porous medium $K = 1/C_m$. In soil mechanics the compression modulus K and shear modulus G are often used as the two basic elastic coefficients because they so well describe the two different modes of deformation: compression and shear.

The volume strain ε in (4.38) is the sum of the three linear strains,

$$\varepsilon = \varepsilon_{xx} + \varepsilon_{yy} + \varepsilon_{zz}. \tag{4.40}$$

The strain components are related to the displacement components by the compatibility equations

$$\varepsilon_{xx} = \frac{\partial u_x}{\partial x}, \qquad \varepsilon_{xy} = \frac{1}{2}\left(\frac{\partial u_x}{\partial y} + \frac{\partial u_y}{\partial x}\right), \qquad \varepsilon_{xz} = \frac{1}{2}\left(\frac{\partial u_x}{\partial z} + \frac{\partial u_z}{\partial x}\right),$$

$$\varepsilon_{yy} = \frac{\partial u_y}{\partial y}, \qquad \varepsilon_{yz} = \frac{1}{2}\left(\frac{\partial u_y}{\partial z} + \frac{\partial u_z}{\partial y}\right), \qquad \varepsilon_{yx} = \frac{1}{2}\left(\frac{\partial u_y}{\partial x} + \frac{\partial u_x}{\partial y}\right), \quad (4.41)$$

$$\varepsilon_{zz} = \frac{\partial u_z}{\partial z}, \qquad \varepsilon_{zx} = \frac{1}{2}\left(\frac{\partial u_z}{\partial x} + \frac{\partial u_x}{\partial z}\right), \qquad \varepsilon_{zy} = \frac{1}{2}\left(\frac{\partial u_z}{\partial y} + \frac{\partial u_y}{\partial z}\right).$$

This completes the system of basic field equations. The total number of unknowns is 22 (9 stresses, 9 strains, 3 displacements and the pore pressure), and the total number of equations is also 22 (6 equilibrium equations, 9 compatibility equations, 6 independent stress-strain-relations, and the storage equation).

The system of equations can be simplified considerably by eliminating the stresses and the strains, finally expressing the equilibrium equations in the displacements. For a homogeneous material (when K and G are constant) these equations are

$$\left(K + \frac{1}{3}G\right)\frac{\partial \varepsilon}{\partial x} + G\nabla^2 u_x - \alpha\frac{\partial p}{\partial x} + f_x = 0,$$

$$\left(K + \frac{1}{3}G\right)\frac{\partial \varepsilon}{\partial y} + G\nabla^2 u_y - \alpha\frac{\partial p}{\partial y} + f_y = 0, \qquad (4.42)$$

$$\left(K + \frac{1}{3}G\right)\frac{\partial \varepsilon}{\partial z} + G\nabla^2 u_z - \alpha\frac{\partial p}{\partial z} + f_z = 0,$$

where the volume strain ε should now be expressed as

$$\varepsilon = \frac{\partial u_x}{\partial x} + \frac{\partial u_y}{\partial y} + \frac{\partial u_z}{\partial z}, \qquad (4.43)$$

and the operator ∇^2 is defined as

$$\nabla^2 = \frac{\partial^2}{\partial x^2} + \frac{\partial^2}{\partial y^2} + \frac{\partial^2}{\partial z^2}. \qquad (4.44)$$

The system of differential equations now consists of the storage equation (4.34) and the equilibrium equations (4.42). These are 4 equations with 4 variables: p, u_x, u_y and u_z. The volume strain ε is not an independent variable, see (4.43).

The initial conditions are that the pore pressure p and the three displacement components are given at a certain time (say $t = 0$). The boundary conditions must be that along the boundary 4 conditions are given. One condition applies to the pore fluid: either the pore pressure or the flow rate normal to the boundary must be specified. The other three conditions refer to the solid material: either the 3 surface tractions or the 3 displacement components must be prescribed (or some combination). Many solutions of the consolidation equations have been published, mainly for bodies of relatively simple geometry (half-spaces, half-planes, cylinders, spheres, etc.) (for references see Schiffman, 1984; Wang, 2000).

4.5 Drained Deformations

In some cases the analysis of consolidation is not really necessary because the duration of the consolidation process is short compared to the time scale of the problem considered. This can be investigated by evaluating the expression $c_v t / h^2$, where h is the average drainage length, and t is a characteristic time. When the value of this parameter is large compared to 1, the consolidation process will be finished after a time t, and consolidation may be disregarded. In such cases the behaviour of the soil is said to be *fully drained*. No excess pore pressures need to be considered for the analysis of the behaviour of the soil. Problems for which consolidation is so fast that it can be neglected are for instance the building of an embankment or a foundation on a sandy subsoil, provided that the smallest dimension of the structure, which determines the drainage length, is not more than say a few meters.

4.6 Undrained Deformations

Quite another class of problems is concerned with the rapid loading of a soil of low permeability (a clay layer). Then it may be that there is hardly any movement of the fluid, and the consolidation process can be simplified in the following way. The basic equation involving the time scale is the storage equation (4.27),

$$\alpha \frac{\partial \varepsilon}{\partial t} + S_p \frac{\partial p}{\partial t} = -\nabla \cdot \mathbf{q}. \tag{4.45}$$

If this equation is integrated over a short time interval Δt one obtains

$$\alpha \varepsilon_0 + S_p p_0 = -\int_0^{\Delta t} \nabla \cdot \mathbf{q} \, dt, \tag{4.46}$$

where ε_0 and p_0 denote the volume strain and the pore pressure immediately after application of the load. The term in the right hand side represents the net outward flow, over a time interval Δt. When the permeability is very small, and the time step Δt is also very small, this term will be very small, and may be neglected. It follows that

$$p_0 = -\frac{\alpha \varepsilon}{S_p} = -\frac{\alpha \varepsilon}{nC_f + (\alpha - n)C_s}. \tag{4.47}$$

This expression enables to eliminate the pore pressure from the other equations, such as the equations of equilibrium (4.42). This gives

$$\left(K_u + \frac{1}{3}G \right) \frac{\partial \varepsilon}{\partial x} + G\nabla^2 u_x + f_x = 0,$$

$$\left(K_u + \frac{1}{3}G \right) \frac{\partial \varepsilon}{\partial y} + G\nabla^2 u_y + f_y = 0, \tag{4.48}$$

$$\left(K_u + \frac{1}{3}G \right) \frac{\partial \varepsilon}{\partial z} + G\nabla^2 u_z + f_z = 0,$$

where

$$K_u = K + \frac{\alpha^2}{S_p} = K + \frac{\alpha^2}{nC_f + (\alpha - n)C_s}, \tag{4.49}$$

the *undrained compression modulus*.

It should be noted that these equations are completely equivalent to the equations of equilibrium for an elastic material, the only difference being that the compression modulus K has been replaced by K_u.

Combination of (4.37) and (4.38) with (4.47) leads to the following relations between the total stresses and the displacements

$$\sigma_{xx} = -\left(K_u - \frac{2}{3}G\right)\varepsilon - 2G\varepsilon_{xx}, \qquad \sigma_{xy} = -2G\varepsilon_{xy}, \qquad \sigma_{xz} = -2G\varepsilon_{xz},$$

$$\sigma_{yy} = -\left(K_u - \frac{2}{3}G\right)\varepsilon - 2G\varepsilon_{yy}, \qquad \sigma_{yz} = -2G\varepsilon_{yz}, \qquad \sigma_{yx} = -2G\varepsilon_{yx}, \quad (4.50)$$

$$\sigma_{zz} = -\left(K_u - \frac{2}{3}G\right)\varepsilon - 2G\varepsilon_{zz}, \qquad \sigma_{zx} = -2G\varepsilon_{zx}, \qquad \sigma_{zy} = -2G\varepsilon_{zy}.$$

These equations also correspond precisely to the standard relations between stresses and displacements from the classical theory of elasticity, again with the exception that K must be replaced by K_u. It may be concluded that the total stresses and the displacements are determined by the equations of the theory of elasticity, except that the compression modulus K must be replaced by K_u. The shear modulus G remains unaffected. This type of approach is called an *undrained analysis*.

If the fluid and the particles are incompressible the storativity of the pore space S_p is zero, see (4.28). In that case the undrained compression modulus is infinitely large, which is in agreement with the physical basis of the original consolidation theory. If the particles and the fluid are incompressible, and the loading process is very fast, no drainage can occur. In that case the soil must indeed be incompressible. In an undrained analysis the material behaves with a shear modulus equal to the drained shear modulus, but with a compression modulus that is practically infinite. In terms of shear modulus and Poisson's ratio, one may say that Poisson's ratio v is (almost) equal to 0.5 when the soil is undrained.

As an example one may consider the case of a rigid circular foundation plate on a semi-infinite elastic porous material, loaded by a total load P. According to the theory of elasticity (Timoshenko and Goodier, 1970) the settlement of the plate is

$$w_\infty = \frac{P(1 - v^2)}{ED}, \tag{4.51}$$

where D is the diameter of the plate. This is the settlement if there were no pore pressures, or when all the pore pressures have been dissipated. In terms of the shear modulus G and Poisson's ratio v this formula may be written as

$$w_\infty = \frac{P(1 - v)}{2GD}. \tag{4.52}$$

This is the settlement after the consolidation process has been completed. At the moment of loading the material reacts as if $v = \frac{1}{2}$, so that the immediate settlement is

$$w_0 = \frac{P}{4GD}. \tag{4.53}$$

This shows that the ratio of the immediate settlement to the final settlement is

$$\frac{w_0}{w_\infty} = \frac{1}{2(1 - v)}. \tag{4.54}$$

Thus the immediate settlement is about 50% of the final settlement, or more, depending upon the value of Poisson's ratio in drained conditions. The consolidation process will account for the remaining part of the settlement, which will be less than 50%.

4.7 Cryer's Problem

A good, and still relatively simple, example of a three-dimensional problem of consolidation is the case of a massive sphere, subjected to an all round pressure at its outer boundary, with drainage to a layer of filter material around the sphere, see Fig. 4.3. The solution of this problem has been given by Cryer (1963) for the case of an incompressible fluid and incompressible solid particles. Here the general case will be considered, which is only slightly more complicated.

Basic Equations

Using a spherical coordinate R, which is appropriate for this problem with spherical symmetry, the storage equation (4.27) can be written as

$$\alpha \frac{\partial \varepsilon}{\partial t} + S_p \frac{\partial p}{\partial t} = \frac{k}{\gamma_w} \left(\frac{\partial^2 p}{\partial R^2} + \frac{2}{R} \frac{\partial p}{\partial R} \right), \tag{4.55}$$

where it has been assumed that the permeability is constant.

The volume strain ε is related to the radial displacement u by

$$\varepsilon = \frac{\partial u}{\partial R} + \frac{2u}{R}. \tag{4.56}$$

The second basic equation is the equation of radial equilibrium, which can be expressed as

$$\frac{\partial \sigma_{RR}}{\partial R} + 2 \frac{\sigma_{RR} - \sigma_{TT}}{R} = 0, \tag{4.57}$$

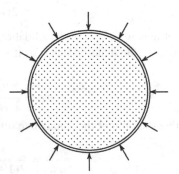

Fig. 4.3 Spherical sample

where σ_{RR} and σ_{TT} are the total stresses in radial and tangential direction. The total stresses can be separated into the effective stresses and the pore pressure by the equations

$$\sigma_{RR} = \sigma'_{RR} + \alpha p, \tag{4.58}$$

$$\sigma_{TT} = \sigma'_{TT} + \alpha p. \tag{4.59}$$

Using these relations the equation of equilibrium can be written as

$$\frac{\partial \sigma'_{RR}}{\partial R} + 2\frac{\sigma'_{RR} - \sigma'_{TT}}{R} + \alpha\frac{\partial p}{\partial R} = 0. \tag{4.60}$$

Using (4.56) and the stress-strain-relations

$$\sigma'_{RR} = -\left(K - \frac{2}{3}G\right)\varepsilon - 2G\frac{\partial u}{\partial R}, \tag{4.61}$$

$$\sigma'_{TT} = -\left(K - \frac{2}{3}G\right)\varepsilon - 2G\frac{u}{R}, \tag{4.62}$$

the equation of equilibrium can be expressed in terms of the volume strain as

$$\left(K + \frac{4}{3}G\right)\frac{\partial \varepsilon}{\partial R} = \alpha\frac{\partial p}{\partial R}. \tag{4.63}$$

In these equations K is the compression modulus of the porous medium in fully drained conditions, and G is its shear modulus.

Boundary Conditions

The problem is further defined by the boundary conditions

$$R = 0 : \frac{\partial p}{\partial R} = 0, \tag{4.64}$$

$$R = 0 : u = 0, \tag{4.65}$$

$$R = a : p = 0, \tag{4.66}$$

$$R = a : \sigma_{RR} = \begin{cases} 0, & \text{if } t < 0, \\ q, & \text{if } t > 0. \end{cases} \tag{4.67}$$

The first boundary condition expresses that there is no flow in the center of the sphere, and the second boundary condition expresses that there can be no radial displacement at the center. The third boundary condition states that the excess pore pressure at the outer boundary of the sphere is zero, assuming perfect drainage. The fourth boundary describes the radial loading at the sphere's circumference. The radius of the sphere is denoted by a.

Initial Response

At the instant of loading there will be a response of the sample, which can be determined by considering the sample to be elastic, with a modified compression modulus K_u, see (4.49),

$$K_u = K + \frac{\alpha^2}{S_p}. \tag{4.68}$$

This leads to a solution in which the state of stress is homogenous, with all normal stresses being equal to the load q. The initial pore pressure then is

$$t = 0 : \quad p = p_0 = \frac{\alpha q}{\alpha^2 + K S_p}. \tag{4.69}$$

The initial volume change is

$$t = 0 : \quad \varepsilon = \varepsilon_0 = \frac{q}{K_u} = \frac{q S_p}{\alpha^2 + K S_p}. \tag{4.70}$$

It may be noted that in the case of an incompressible fluid and incompressible particles $C_f = C_s = 0$, and $\alpha = 1$. In that case $p_0 = q$ and $\varepsilon_0 = 0$, indicating that in this case there is no initial volume change, and the initial pore pressure is equal to the radial load.

Solution of the Problem

The problem is solved using the Laplace transform. For the pore pressure this is defined as

$$\overline{p} = \int_0^\infty p \exp(-st) \, dt, \tag{4.71}$$

where s is the Laplace transform parameter.

The transformed basic equations can easily be solved, involving four integration constants, which can be determined using the boundary conditions (4.64)–(4.67). The final expression for the Laplace transform of the pore pressure is found to be

$$\overline{p} = \frac{q m \beta a^2}{\alpha c (1 + K S_p / \alpha^2)} \frac{\sinh(\lambda a) - (a/R) \sinh(\lambda R)}{[1 + m(\lambda a)^2] \sinh(\lambda a) - (\lambda a) \cosh(\lambda a)}, \tag{4.72}$$

where

$$\lambda^2 = \beta s / c, \tag{4.73}$$

and the following additional parameters have been used

$$c = \frac{k(K + \frac{4}{3}G)}{\gamma_w}, \qquad \beta = \alpha^2 + \left(K + \frac{4}{3}G\right) S_p, \qquad m = \frac{(K + \frac{4}{3}G)(1 + K S_p / \alpha^2}{4G}. \tag{4.74}$$

The coefficient c is the usual coefficient of consolidation. In the case of an incompressible fluid and an incompressible solid particles (the case considered by Cryer, 1963) the Biot coefficient $\alpha = 1$, and the pore space storativity $S_p = 0$. It then follows that $\beta = 1$ and $m = (K + \frac{4}{3}G)/4G$.

Of particular interest is the pore pressure in the center of the sphere, at $R = 0$. If this is denoted by p_c, its Laplace transform is

$$\overline{p}_c = \frac{qm\beta a^2}{\alpha c(1 + K S_p/\alpha^2)} \frac{\sinh(\lambda a) - \lambda a}{[1 + m(\lambda a)^2]\sinh(\lambda a) - (\lambda a)\cosh(\lambda a)}. \tag{4.75}$$

It may be appropriate at this stage to verify the initial condition (4.69), using the fundamental property of the Laplace transform

$$\lim_{t \to 0} p = \lim_{s \to \infty} s\overline{p}. \tag{4.76}$$

With (4.72) this gives

$$p_0 = \frac{q}{\alpha(1 + K S_p/\alpha^2)}. \tag{4.77}$$

This is in agreement with (4.69), thus confirming the derivations.

Inverse Laplace Transformation

Inverse Laplace transformation of (4.75) gives, using Heaviside's inversion theorem,

$$\frac{p_c}{p_0} = 2m \sum_{j=1}^{\infty} \frac{\sin \xi_j - \xi_j}{m\xi_j \cos \xi_j + (2m - 1)\sin \xi_j} \exp(-\xi_j^2 ct/\beta a^2), \tag{4.78}$$

where the coefficients ξ_j are the positive roots of the equation

$$(1 - m\xi_j^2)\sin \xi_j - \xi_j \cos \xi_j = 0. \tag{4.79}$$

This solution agrees with the solution obtained by Cryer (1963) for the case of an incompressible fluid and incompressible particles.

Results

Figure 4.4 shows the results for three values of Poisson's ratio, assuming that the pore fluid and the particles are incompressible. It is interesting to note that for all values of $\nu < \frac{1}{2}$ the pore pressure in the center of the sphere initially increases before it is ultimately reduced to zero. This is caused by the drainage, which starts at the outer shell of the sphere, and which produces a tendency for shrinkage of that

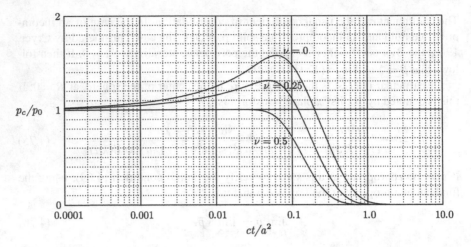

Fig. 4.4 Pore pressure in the center

outer shell. This leads to an additional compressive stress on the practically incompressible core of the sphere, so that an additional pore pressure is generated. The effect can be considered as a consequence of the practically immediate transmission of static stresses, and the gradual progress of the diffusive process of groundwater flow. A similar effect was obtained by Mandel (1953) for the problem of a clay sample compressed between two rigid plates, with lateral drainage. The effect is usually called the *Mandel-Cryer* effect. It has been confirmed experimentally by Gibson et al. (1963) and Verruijt (1965).

4.8 Uncoupled Consolidation

In general the system of equations of three-dimensional consolidation involves solving the storage equation together with the three equations of equilibrium, simultaneously, because these equations are coupled. This may be a formidable task, and it seems worthwhile to try to simplify this procedure. It would be very convenient, for instance, if it could be shown that in the storage equation

$$\alpha \frac{\partial \varepsilon}{\partial t} + S_p \frac{\partial p}{\partial t} = \nabla \cdot \left(\frac{k}{\gamma_w} \nabla p \right) \tag{4.80}$$

the first term can be expressed as

$$\frac{\partial \varepsilon}{\partial t} = C \frac{\partial p}{\partial t}, \tag{4.81}$$

where C is some constant, because then the equation reduces to the form

$$(\alpha C + S_p)\frac{\partial p}{\partial t} = \nabla \cdot \left(\frac{k}{\gamma_w}\nabla p\right), \tag{4.82}$$

which is the classical diffusion equation, for which many analytical solutions are available. The system of equations is then uncoupled, in the sense that first the pore pressure can be determined from (4.82), and then later the deformation problem can be solved using the equations of equilibrium, in which then the gradient of the pore pressure acts as a known body force.

Constant Isotropic Total Stress

There are two possibilities for uncoupling to be realized. The first possibility is obtained by first noting that for an isotropic material the volume strain ε is a function of the isotropic effective stress σ',

$$\sigma' = \frac{\sigma'_{xx} + \sigma'_{yy} + \sigma'_{zz}}{3}. \tag{4.83}$$

For a linear material the relation may be written as

$$\varepsilon = -C_m\sigma', \tag{4.84}$$

where C_m is the compressibility of the porous material, the inverse of its compression modulus, $C_m = 1/K$, and the minus sign is needed because of the different sign conventions used for stresses and strains. The effective stress is the difference between total stress and pore pressure (taking into account Biot's coefficient), and thus one may write

$$\varepsilon = -C_m(\sigma - \alpha p). \tag{4.85}$$

Differentiating this with respect to time gives

$$\frac{\partial \varepsilon}{\partial t} = -C_m\frac{\partial \sigma}{\partial t} + \alpha C_m\frac{\partial p}{\partial t}. \tag{4.86}$$

If it is now assumed, as a first approximation, that isotropic total stress is constant in time, then there indeed appears to be a relation of the type (4.81), with

$$C = \alpha C_m. \tag{4.87}$$

The differential equation now is, with (4.82)

$$(\alpha^2 C_m + S_p)\frac{\partial p}{\partial t} = \nabla \cdot \left(\frac{k}{\gamma_w}\nabla p\right), \tag{4.88}$$

which is indeed a diffusion equation. This simplifying assumption was first suggested by Rendulic (1936). That the isotropic total stress may be constant in certain cases is not unrealistic. In many cases consolidation takes place while the loading of the soil remains constant, and although there may be a certain redistribution of stress, it may well be assumed that the changes in total stress will be small. A proof is impossible to give, however, and it is also difficult to say under what conditions the approximation is acceptable. Various solutions of coupled three-dimensional problems have been obtained, and in many cases a certain difference with the uncoupled solution has been found. Sometimes there is even a very pronounced difference in behaviour for small values of the time, in the sense that sometimes the pore pressures initially show a certain increase, before they dissipate. This is the *Mandel-Cryer* effect, see the previous section, which is a typical consequence of the coupling effect. When the pore pressures at the boundary start to dissipate the local deformation may lead to an immediate effect in other parts of the soil body, and this may lead to an additional pore pressure. In the long run the pore pressures always dissipate, however, and the difference with the uncoupled solution then is often not important. Therefore an uncoupled analysis may be a good first approximation, if it is realized that local errors may occur, especially for short values of time.

Horizontally Confined Deformations

Another important class of problems in which an uncoupled analysis is justified is the case where it can be assumed that the horizontal deformations will be negligible, and the vertical total stress remains constant. In the case of a soil layer of large horizontal extent, loaded by a constant surface load, this may be an acceptable set of assumptions. If the horizontal deformations are set equal to zero, it follows that the volume strain is equal to the vertical strain,

$$\varepsilon = \varepsilon_{zz}. \tag{4.89}$$

For a linear elastic material the vertical strain can be related to the vertical effective stress by the formula

$$\varepsilon_{zz} = -m_v \sigma'_{zz}, \tag{4.90}$$

where m_v is the vertical compressibility of a laterally confined soil sample. Using the effective stress principle this now gives

$$\varepsilon_{zz} = -m_v(\sigma_{zz} - \alpha p), \tag{4.91}$$

and therefore

$$\frac{\partial \varepsilon}{\partial t} = -m_v \frac{\partial \sigma_{zz}}{\partial t} + \alpha m_v \frac{\partial p}{\partial t}. \tag{4.92}$$

Substitution into the storage equation (4.80) gives

$$(\alpha^2 m_v + S_p)\frac{\partial p}{\partial t} = \alpha m_v \frac{\partial \sigma_{zz}}{\partial t} + \nabla \cdot \left(\frac{k}{\gamma_w}\nabla p\right). \tag{4.93}$$

This equation is indeed of the form of a diffusion coefficient if the vertical total stress is constant. It may be concluded that in the case of zero lateral deformation and constant vertical total stress the consolidation equations are uncoupled. If the medium is homogeneous, the coefficient k/γ_w is constant in space, and then the differential equation reduces to the form

$$\frac{\partial p}{\partial t} = c_v \nabla^2 p, \tag{4.94}$$

where ∇^2 is Laplace's operator,

$$\nabla^2 = \frac{\partial}{\partial x^2} + \frac{\partial}{\partial y^2} + \frac{\partial}{\partial z^2}, \tag{4.95}$$

and c_v is the consolidation coefficient,

$$c_v = \frac{k}{(\alpha^2 m_v + S_p)\gamma_w}. \tag{4.96}$$

An equation of the form (4.94) was first derived by Terzaghi (1925), for the one-dimensional case of flow and deformation in the vertical direction only, as occurs in a confined compression test in the laboratory, or in the consolidation of an extensive clay layer in the field, loaded by a uniform surcharge. It was also derived by Jacob (1940), using somewhat different notations, for the case of a compressible aquifer of thickness H, transmissivity T, and storativity S. The consolidation coefficient then can be written as $c_v = T/S$.

In the next section a solution of the differential equation will be presented, for Terzaghi's problem.

4.9 Terzaghi's Problem

The problem first solved by Terzaghi (1925) is that of a layer of thickness $2h$, which is loaded at time $t = 0$ by a load of constant magnitude q. The upper and lower boundaries of the soil layer are fully drained, so that along these boundaries the pore pressure p remains zero.

The differential equation for this case is the fully one-dimensional form of (4.94),

$$\frac{\partial p}{\partial t} = c_v \frac{\partial^2 p}{\partial z^2}. \tag{4.97}$$

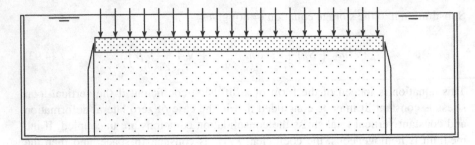

Fig. 4.5 Terzaghi's problem

Because at the moment of loading there can not yet have been any fluid loss from the soil, it follows from (4.93) that the initial pore pressure is

$$t = 0 \; : \; p = p_0 = \frac{\alpha m_v}{\alpha^2 m_v + S_p} q. \tag{4.98}$$

If the fluid and the solid particles are incompressible $\alpha = 1$ and $S_p = 0$, so that then $p_0 = q$, as Terzaghi considered.

The boundary conditions are

$$z = 0 \; : \; p = 0, \tag{4.99}$$

$$z = 2h \; : \; p = 0. \tag{4.100}$$

Solution

The solution of the problem can be obtained by using the mathematical tools supplied by the theory of partial differential equations, for instance the method of separation of variables (see e.g. Wylie, 1960), or, even more conveniently, by the Laplace transform method (see e.g. Churchill, 1972, or Appendix A). The Laplace transform of the pore pressure is defined as

$$\overline{p} = \int_0^\infty p \exp(-st) \, dt, \tag{4.101}$$

where s is a positive parameter. The transformed differential equation is, using the initial condition (4.98)

$$s\overline{p} - p_0 = c_v \frac{d^2 \overline{p}}{dz^2}. \tag{4.102}$$

The partial differential equation has now been reduced to an ordinary differential equation. The solution satisfying the boundary conditions is

$$\frac{\overline{p}}{p_0} = \frac{1}{s} - \frac{\cosh[(h - z)\sqrt{s/c_v}]}{s \cosh[h\sqrt{s/c_v}]}. \tag{4.103}$$

The inverse transform of this expression can be obtained by the complex inversion integral (Churchill, 1972), or in a more simple, although less rigorous way, by application of Heaviside's expansion theorem (Appendix A). This gives, after some elementary mathematical operations,

$$\frac{p}{p_0} = \frac{4}{\pi} \sum_{j=1}^{\infty} \left\{ \frac{(-1)^{j-1}}{2j-1} \cos\left[(2j-1)\frac{\pi}{2}\left(\frac{h-z}{h}\right) \right] \exp\left[-(2j-1)^2 \frac{\pi^2}{4} \frac{c_v t}{h^2} \right] \right\}.$$

(4.104)

This is the analytical solution of the problem. It can be found in many textbooks on theoretical soil mechanics, and also in many textbooks on the theory of heat conduction, as that is governed by the same equations. In the early soil mechanics literature the solution was restricted to the case of incompressible fluid and solids. The only difference with the present solution, in which both the fluid and the solids may be compressible, is in the value of the consolidation coefficient, and in the value of the initial pore pressure p_0.

Because the solution has been derived here by a method that is mathematically perhaps not completely rigorous (Heaviside's expansion theorem, strictly speaking, applies only to a function consisting of the quotient of two polynomials), it is advisable to check whether the solution indeed satisfies all requirements. That it indeed satisfies the differential equation (4.97) can be demonstrated rather easily, because each term satisfies this equation. It can also directly be seen that it satisfies the boundary conditions (4.99) and (4.100) because for $z = 0$ and for $z = 2h$ the function $\cos(\ldots)$ is zero. It is not so easy to verify that the initial condition (4.98) is also satisfied. The simplest method to verify this is to write a computer program that calculates values of the infinite series, and then to show that for any value of z and for very small values of t the value is indeed 1. It will be observed that this requires a very large number of terms. If t is exactly zero, it will even been found that the series does not converge.

The solution (4.104) is shown graphically in Fig. 4.6, for increasing values of $c_v t / h^2$. The number of terms was chosen such that the argument of the exponential function was less than -20. This means that all terms containing a factor $\exp(-20)$, or smaller, are disregarded. The figure also shows that the solution satisfies the boundary conditions, and the initial condition. It does not show, of course, that it is the correct solution. That can only been shown by analytical means, as presented above.

Settlement

The progress of the settlement in time can be obtained from the solution (4.104) by noting that the strain is determined by the effective stress,

$$\varepsilon = -m_v \sigma'_{zz} = -m_v(\sigma_{zz} - \alpha p).$$

(4.105)

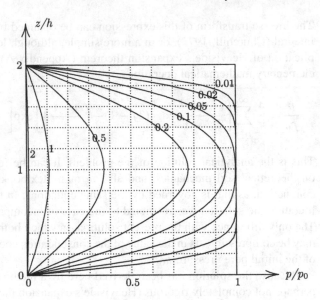

Fig. 4.6 Analytical solution of Terzaghi's problem

The settlement is the integral of this strain over the height of the sample,

$$w = -\int_0^{2h} \varepsilon \, dz = 2m_v h q - \alpha m_v \int_0^{2h} p \, dz. \tag{4.106}$$

The first term in the right hand side is the final settlement, which will be reached when the pore pressures have been completely dissipated. This value will be denoted by w_∞,

$$w_\infty = 2m_v h q. \tag{4.107}$$

Immediately after the application of the load q the pore pressure is equal to p_0, see (4.98). This means that the immediate settlement, at the moment of loading, is

$$w_0 = 2m_v h(q - \alpha p_0). \tag{4.108}$$

In order to describe the settlement as a function of time it is most convenient to introduce the *degree of consolidation U*, defined as

$$U = \frac{w - w_0}{w_\infty - w_0}. \tag{4.109}$$

This quantity will always vary between 0 (at the moment of loading) and 1 (after consolidation has finished). In this case, using the expressions given above, it is found to be related to the pore pressures by

$$U = \frac{1}{2h} \int_0^{2h} \frac{p_0 - p}{p_0} \, dz. \tag{4.110}$$

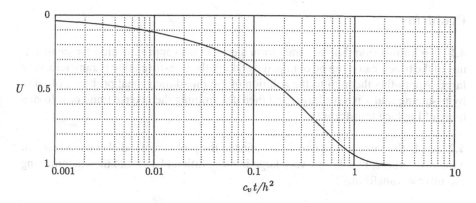

Fig. 4.7 Degree of consolidation

Using the solution (4.104) for the pore pressure distribution the final expression for the degree of consolidation as a function of time is

$$U = 1 - \frac{8}{\pi^2} \sum_{j=1}^{\infty} \frac{1}{(2j-1)^2} \exp\left[-(2j-1)^2 \frac{\pi^2}{4} \frac{c_v t}{h^2}\right]. \qquad (4.111)$$

For $t \rightarrow \infty$ this is indeed 1. For $t = 0$ it is 0, because then the terms in the infinite series add up to $\pi^2/8$. A graphical representation of the degree of consolidation as a function of time is shown in Fig. 4.7.

Theoretically speaking the consolidation phenomenon is finished if $t \rightarrow \infty$. For all practical purposes it can be considered as finished when the argument of the exponential function in the first term of the series is about 4 or 5. This will be the case when

$$\frac{c_v t}{h^2} \approx 2. \qquad (4.112)$$

This is a very useful formula, because it enables to estimate the duration of the consolidation process. It also enables to evaluate the influence of the various parameters on the consolidation process. If the permeability is twice as large, consolidation will take half as long. If the drainage length is reduced by a factor 2, the duration of the consolidation process is reduced by a factor 4. This explains the usefulness of improving the drainage in order to accelerate consolidation. In engineering practice the consolidation process is sometimes accelerated by installing vertical drains. In a thick clay deposit this may be very effective, because it reduces the drainage length from the thickness of the layer to the distance of the drains. As the consolidation is proportional to the square of the drainage length, this may be extremely effective in reducing the consolidation time, and thus accelerating the subsidence due to the construction of an embankment.

Problems

4.1 It is known from Laplace transform theory that an approximation for small values of the time t can often be obtained by taking the transformation parameter s very large. Apply this theorem to the solution of the one-dimensional problem, (4.103), by assuming that s is very large, and then determining the inverse transform from a table of Laplace transforms.

4.2 Apply the same theorem to the solution of the problem of radial consolidation of a massive sphere, by taking s very large in the solution (4.72), and then determining the inverse transform.

Chapter 5
Dynamics of Porous Media

In this chapter the basic equations for the dynamics of a porous medium are presented. They were first derived by De Josselin de Jong (1956) and Biot (1956). These equations can be considered to be the extension of the classical theory of consolidation or poroelasticity to the dynamic case. This chapter will be restricted to the linearized equations, and the applications will be mainly restricted to the one-dimensional case of propagation of plane waves. The basic equations will be derived, and analytical and numerical solutions will be presented.

5.1 Basic Differential Equations

In this first section the basic differential equations of the dynamics of a porous medium will be presented. A porous medium is supposed to be a medium consisting of a solid material with a continuous system of small, interconnected, pores, which are filled with a fluid, for instance water or oil. The fluid may contain some gas bubbles, causing it to be much more compressible than a homogenous liquid. The basic principles are conservation of mass and conservation of momentum.

5.1.1 Conservation of Mass

The first basic equation is the equation of conservation of mass of the pore fluid,

$$\frac{\partial(n\rho_f)}{\partial t} + \frac{\partial(n\rho_f v)}{\partial x} = 0,$$

(5.1)

where n is the porosity, ρ_f is the density of the pore fluid, and v is the velocity of the pore fluid, defined as the average velocity of the fluid particles. The density is supposed to be a function of the fluid pressure, see (4.18),

$$\frac{d\rho_f}{dp} = \rho_f C_f,$$

(5.2)

A. Verruijt, *An Introduction to Soil Dynamics,*
Theory and Applications of Transport in Porous Media 24,
© Springer Science+Business Media B.V. 2010

where C_f is the compressibility of the fluid (perhaps including the compression of gas bubbles in it). With (5.2), (5.1) becomes

$$\frac{\partial n}{\partial t} + nC_f \frac{\partial p}{\partial t} + \frac{\partial (nv)}{\partial x} = 0, \tag{5.3}$$

where a term expressing the product of the fluid velocity and the pressure gradient has been disregarded, assuming that both are small quantities, so that the product is of second order.

The equation of conservation of mass of the solid particles is

$$\frac{\partial [(1-n)\rho_s]}{\partial t} + \frac{\partial [(1-n)\rho_s w]}{\partial x} = 0, \tag{5.4}$$

where ρ_s is the density of the particle material. The density of the particles is supposed to be governed by the isotropic total stress σ and the pore pressure p, as expressed by (4.21),

$$\frac{\partial \rho_s}{\partial t} = \frac{\rho_s C_s}{1-n} \left(\frac{\partial \sigma}{\partial t} - n \frac{\partial p}{\partial t} \right), \tag{5.5}$$

where C_s is the compressibility of the particle material. With (5.5), (5.4) becomes

$$-\frac{\partial n}{\partial t} + C_s \left(\frac{\partial \sigma}{\partial t} - n \frac{\partial p}{\partial t} \right) + \frac{\partial w}{\partial x} - \frac{\partial (nw)}{\partial x} = 0, \tag{5.6}$$

where again a term expressing the product of a velocity and a gradient of stress or pressure has been disregarded.

Elimination of the terms $\partial n/\partial t$ from (5.3) and (5.6) gives

$$n(C_f - C_s) \frac{\partial p}{\partial t} + C_s \frac{\partial \sigma}{\partial t} + \frac{\partial w}{\partial x} + \frac{\partial [n(v-w)]}{\partial x} = 0. \tag{5.7}$$

It may be noted that the term $\partial w/\partial x$ can also be written as $\partial \varepsilon/\partial t$, noting that in this one-dimensional case the volume strain equals the strain in x-direction, $\varepsilon = \varepsilon_{xx}$.

Furthermore, the isotropic total stress can be decomposed into the isotropic effective stress and the pore pressure, $\sigma = \sigma' + \alpha p$, where α is Biot's coefficient, and the isotropic effective stress can be related to the volume strain according to $\partial \varepsilon/\partial t = -C_m \partial \sigma'/\partial t$, where C_m is the compressibility of the porous medium. It follows that (5.7) can also be written as

$$\alpha \frac{\partial w}{\partial x} + S_p \frac{\partial p}{\partial t} = \alpha \frac{\partial \varepsilon}{\partial t} + S_p \frac{\partial p}{\partial t} = -\frac{\partial [n(v-w)]}{\partial x}, \tag{5.8}$$

where, as before, $\alpha = 1 - C_s/C_m$, and S_p is the *storativity* of the pore space,

$$S_p = nC_f + (\alpha - n)C_s. \tag{5.9}$$

Equation (5.8) was also given in the previous chapter, and denoted as the storage equation, see (4.27).

5.1.2 Conservation of Momentum

A second set of basic equations is provided by the equations of conservation of momentum, or the equations of motion. This must be formulated both for the fluid and the particles. The simplest form is to first consider conservation of momentum of the material as a whole, fluid plus particles. The equations are, in the three coordinate directions,

$$-\frac{\partial \sigma_{xx}}{\partial x} - \frac{\partial \sigma_{yx}}{\partial y} - \frac{\partial \sigma_{zx}}{\partial z} = n\rho_f \frac{\partial v_x}{\partial t} + (1-n)\rho_s \frac{\partial w_x}{\partial t},$$

$$-\frac{\partial \sigma_{xy}}{\partial x} - \frac{\partial \sigma_{yy}}{\partial y} - \frac{\partial \sigma_{zy}}{\partial z} = n\rho_f \frac{\partial v_y}{\partial t} + (1-n)\rho_s \frac{\partial w_y}{\partial t}, \qquad (5.10)$$

$$-\frac{\partial \sigma_{xz}}{\partial x} - \frac{\partial \sigma_{yz}}{\partial y} - \frac{\partial \sigma_{zz}}{\partial z} = n\rho_f \frac{\partial v_z}{\partial t} + (1-n)\rho_s \frac{\partial w_z}{\partial t},$$

where the minus signs in the left hand side are a consequence of the convention that compressive stresses are considered as positive.

Because the total stresses can be decomposed as $\sigma_{ij} = \sigma'_{ij} + \alpha p \delta_{ij}$, these equations can also be written as

$$-\frac{\partial \sigma'_{xx}}{\partial x} - \frac{\partial \sigma'_{yx}}{\partial y} - \frac{\partial \sigma'_{zx}}{\partial z} - \alpha \frac{\partial p}{\partial x} = n\rho_f \frac{\partial v_x}{\partial t} + (1-n)\rho_s \frac{\partial w_x}{\partial t},$$

$$-\frac{\partial \sigma'_{xy}}{\partial x} - \frac{\partial \sigma'_{yy}}{\partial y} - \frac{\partial \sigma'_{zy}}{\partial z} - \alpha \frac{\partial p}{\partial y} = n\rho_f \frac{\partial v_y}{\partial t} + (1-n)\rho_s \frac{\partial w_y}{\partial t}, \qquad (5.11)$$

$$-\frac{\partial \sigma'_{xz}}{\partial x} - \frac{\partial \sigma'_{yz}}{\partial y} - \frac{\partial \sigma'_{zz}}{\partial z} - \alpha \frac{\partial p}{\partial z} = n\rho_f \frac{\partial v_z}{\partial t} + (1-n)\rho_s \frac{\partial w_z}{\partial t}.$$

In the equations of conservation of momentum of the components (fluid or solids) the interaction between the two components due to friction must be taken into account. The equations for conservation of momentum of the fluid are assumed to be

$$-n\frac{\partial p}{\partial x} - \frac{n^2 \mu}{\kappa}(v_x - w_x) = n\rho_f \frac{\partial v_x}{\partial t} + \tau n\rho_f \frac{\partial (v_x - w_x)}{\partial t},$$

$$-n\frac{\partial p}{\partial y} - \frac{n^2 \mu}{\kappa}(v_y - w_y) = n\rho_f \frac{\partial v_y}{\partial t} + \tau n\rho_f \frac{\partial (v_y - w_y)}{\partial t}, \qquad (5.12)$$

$$-n\frac{\partial p}{\partial z} - \frac{n^2 \mu}{\kappa}(v_z - w_z) = n\rho_f \frac{\partial v_z}{\partial t} + \tau n\rho_f \frac{\partial (v_z - w_z)}{\partial t},$$

where τ is a tortuosity factor, describing the added mass due to the tortuosity of the flow path, μ is the viscosity of the pore fluid, and κ is the permeability of the porous medium. The tortuosity terms have been included to account for a possible additional force to move the fluid through the tortuous path between the particles.

It may be noted that care has been taken that the equations contain Darcy's law as a special case, when the accelerations are negligible. Actually, in the absence of acceleration terms, (5.12) reduce to the quasi-static case

$$\mathbf{q} = -\frac{\kappa}{\mu}\nabla p, \tag{5.13}$$

where $\mathbf{q}$ is the *specific discharge*, defined as

$$\mathbf{q} = n(\mathbf{v} - \mathbf{w}). \tag{5.14}$$

Equation (5.13) is Darcy's law, in the absence of body forces, such as gravity.

It should also be noted that in (5.12) all interaction terms are expressed in terms of the velocity of the fluid with respect to the solids. When the fluid and the solid have equal velocities these terms vanish.

The equations of conservation of momentum of the solids can be obtained by subtracting (5.12) from (5.11). This gives

$$-\frac{\partial \sigma'_{xx}}{\partial x} - \frac{\partial \sigma'_{yx}}{\partial y} - \frac{\partial \sigma'_{zx}}{\partial z} - (\alpha - n)\frac{\partial p}{\partial x} + \frac{n^2\mu}{\kappa}(v_x - w_x)$$

$$= (1-n)\rho_s\frac{\partial w_x}{\partial t} - \tau n\rho_f\frac{\partial(v_x - w_x)}{\partial t},$$

$$-\frac{\partial \sigma'_{xy}}{\partial x} - \frac{\partial \sigma'_{yy}}{\partial y} - \frac{\partial \sigma'_{zy}}{\partial z} - (\alpha - n)\frac{\partial p}{\partial y} + \frac{n^2\mu}{\kappa}(v_y - w_y)$$

$$\tag{5.15}$$

$$= (1-n)\rho_s\frac{\partial w_y}{\partial t} - \tau n\rho_f\frac{\partial(v_y - w_y)}{\partial t},$$

$$-\frac{\partial \sigma'_{xz}}{\partial x} - \frac{\partial \sigma'_{yz}}{\partial y} - \frac{\partial \sigma'_{zz}}{\partial z} - (\alpha - n)\frac{\partial p}{\partial z} + \frac{n^2\mu}{\kappa}(v_z - w_z)$$

$$= (1-n)\rho_s\frac{\partial w_z}{\partial t} - \tau n\rho_f\frac{\partial(v_z - w_z)}{\partial t}.$$

It can easily be verified that addition of (5.12) and (5.15) yields the equations of total conservation of the mixture (5.11).

5.1.3 Constitutive Equations

The effective stresses determine the deformations of the soil. As a first approximation the effective stresses are now supposed to be related to the strains by the generalized form of Hooke's law. For an isotropic material these relations are, see also (4.38),

$$\sigma'_{xx} = -\left(K - \frac{2}{3}G\right)\varepsilon - 2G\varepsilon_{xx}, \qquad \sigma'_{xy} = -2G\varepsilon_{xy}, \qquad \sigma'_{xz} = -2G\varepsilon_{xz},$$

$$\sigma'_{yy} = -\left(K - \frac{2}{3}G\right)\varepsilon - 2G\varepsilon_{yy}, \qquad \sigma'_{yz} = -2G\varepsilon_{yz}, \qquad \sigma'_{yx} = -2G\varepsilon_{yx}, \quad (5.16)$$

$$\sigma'_{zz} = -\left(K - \frac{2}{3}G\right)\varepsilon - 2G\varepsilon_{zz}, \qquad \sigma'_{zx} = -2G\varepsilon_{zx}, \qquad \sigma'_{zy} = -2G\varepsilon_{zy},$$

where K and G are the elastic coefficients of the material, the compression modulus and the shear modulus, respectively.

Even though this assumption must be considered as a poor representation of the mechanical behaviour of soils, it is essential to note that the governing stress parameter is the effective stress. More complicated stress-strain-relations can be formulated, involving time (to represent creep) and stress history (to represent irreversible plastic deformations). These should all be expressed in terms of the effective stress, however, and should not involve the fluid pressure p, even though the pressure in the fluid will generate an equal stress in the solid particles, which are completely surrounded by the pore fluid.

Equations (5.12), (5.15), (5.16) and (5.8) together form a system of partial differential equations, from which the basic variables, the displacements, the stresses, and the pore pressure must be determined. One of the two sets of equations of balance of momentum can of course be replaced by the total balance equations (5.11).

It is often considered convenient to consider the relative velocity $\mathbf{v} - \mathbf{w}$ as a basic variable, rather than the fluid velocity $\mathbf{v}$, because the relative velocity governs the interaction between the fluid and the solids. In soil mechanics, and in hydrology, it is common practice to express the equations in terms of the particle velocity $\mathbf{w}$ and the *specific discharge*, defined as $\mathbf{q} = n(\mathbf{v} - \mathbf{w})$.

5.2 Propagation of Plane Waves

Because of the complexity of the general system of equations established in the previous section, the much more simple one-dimensional case will be considered in this section. This will enable to study the propagation of plane waves, in a single direction, the x-direction.

From the general system of equations the following equations can be obtained for the one-dimensional case.

$$\alpha\frac{\partial w}{\partial x} + S_p\frac{\partial p}{\partial t} = -\frac{\partial[n(v - w)]}{\partial x}, \tag{5.17}$$

$$m_v\frac{\partial \sigma'}{\partial t} = -\frac{\partial w}{\partial x}, \tag{5.18}$$

$$n\rho_f\frac{\partial v}{\partial t} + (1 - n)\rho_s\frac{\partial w}{\partial t} = -\frac{\partial \sigma'}{\partial x} - \alpha\frac{\partial p}{\partial x}, \tag{5.19}$$

$$n\rho_f \frac{\partial v}{\partial t} + \tau n\rho_f \frac{\partial (v - w)}{\partial t} = -n\frac{\partial p}{\partial x} - \frac{n^2\mu}{\kappa}(v - w). \qquad (5.20)$$

In these equations m_v is the one-dimensional compressibility of the porous medium,

$$m_v = \frac{1}{K + \frac{4}{3}G}. \qquad (5.21)$$

These are the basic equations for the propagation of plane waves in a porous medium, if this is composed of a soft soil, saturated with a compressible fluid. It is useful to realize that (5.17) expresses mass conservation of the fluid and the soil particles, i.e. total mass conservation, (5.18) is the stress-strain relation of the soil skeleton, (5.19) expresses conservation of total momentum, and (5.20) expresses conservation of momentum of the pore fluid, the generalization of Darcy's law to the dynamic case.

5.3 Special Cases

The general equations (5.17)–(5.20) include some interesting special cases, which will be discussed in this section. It will appear later that these special cases are also characteristic for the waves possible in porous media.

5.3.1 Undrained Waves

A special case that can be imagined is when the fluid and the solids move together, $w = v$. This case can be considered to occur when the permeability is very small, see (5.20). The last term in that equation will then dominate, indicating that $v \approx w$. From (5.17) and (5.18) one then obtains, using the relation $\sigma = \sigma' + \alpha p$,

$$p = \frac{\alpha m_v}{\alpha^2 m_v + S_p}\sigma \qquad (5.22)$$

$$\sigma' = \frac{S_p}{\alpha^2 m_v + S_p}\sigma. \qquad (5.23)$$

For a soft saturated soil, when C_f and C_s are small compared to m_v, so that $S_p \ll m_v$ and $\alpha \approx 1$, these equations express that the total stress is carried mainly by the fluid, and that the solid particles carry very little of the total load.

The two remaining equations now are

$$\left(K_u + \frac{4}{3}G\right)\frac{\partial w}{\partial x} = -\frac{\partial \sigma}{\partial t}, \qquad (5.24)$$

$$\rho\frac{\partial w}{\partial t} = -\frac{\partial \sigma}{\partial x}, \qquad (5.25)$$

where K_u is the undrained compression modulus,

$$K_u = K + \frac{\alpha^2}{S_p} = K + \frac{\alpha^2}{nC_f + (\alpha - n)C_s}, \tag{5.26}$$

and ρ is the mass density of the soil as a whole,

$$\rho = n\rho_f + (1 - n)\rho_s. \tag{5.27}$$

In (5.26) K is the compression modulus of the dry soil. This equation is in agreement with (4.49), derived for undrained deformations in the static case.

Equations (5.24) and (5.25) are the familiar standard equations for wave propagation. They admit solutions of the form

$$\sigma - \frac{K_u + \frac{4}{3}G}{c}w = f_1(x + ct), \tag{5.28}$$

$$\sigma + \frac{K_u + \frac{4}{3}G}{c}w = f_2(x - ct), \tag{5.29}$$

where the wave velocity c in this case is

$$c = \sqrt{\frac{K_u + \frac{4}{3}G}{\rho}} = \sqrt{\frac{1}{\rho m_v} + \frac{\alpha^2}{\rho[nC_f + (\alpha - n)C_s]}}. \tag{5.30}$$

This shows that the wave velocity in the undrained case is determined by the classical formula (5.30), with the elastic modulus usually being very large, determined by the undrained condition, and with the density being the total density of the soil.

It should be noted that this simplified solution applies only if the boundary conditions do not violate the assumed relationships. Thus, for instance, a boundary where the fluid and the soil are moving at the same rate satisfies the assumption $v = w$, and at a free boundary, where $p = \sigma' = 0$, the relations (5.22) and (5.23) can be satisfied. When there are other types of boundary conditions the approximation considered here may not be produced.

For a completely saturated soft soil the value of the undrained modulus is approximately $K_u + \frac{4}{3}G \approx 1/nC_f$, see (5.26), because the pore fluid is the stiffest component in this case. The value of C_f is about $C_f = 0.5 \times 10^{-9}$ m^2/N. With $n = 0.40$ and $\rho = 2000$ kg/m^3 one then obtains $c \approx 1600$ m/s. Such wave velocities are indeed often observed in saturated soft soils. In stiffer soils, or saturated rock, the propagation velocity may be considerably larger, up to 2000 m/s or higher.

5.3.2 Rigid Solid Matrix

Another special case that can be imagined is when the solid matrix is very stiff, as in the case of a very stiff porous rock. As a first approximation, the velocity of the solids w can now be assumed to vanish, $w = 0$.

In this case it seems most appropriate to disregard the stress-strain relation (5.18) and the total momentum balance equation (5.19), as this involves the momentum balance of the solid matrix, which are irrelevant when the solid matrix is assumed to be rigid. Thus the two remaining equations are

$$n\frac{\partial v}{\partial x} = -S_p \frac{\partial p}{\partial t}, \tag{5.31}$$

$$(1+\tau)\rho_f \frac{\partial v}{\partial t} = -\frac{\partial p}{\partial x} - \frac{n\mu}{\kappa} v. \tag{5.32}$$

These are two equations in the basic variables for this special case, the pore pressure p and the fluid velocity v.

The behaviour of the material can be investigated by considering the propagation of harmonic waves

$$v = \tilde{v}\exp[i(\lambda x - \omega t)] = \tilde{v}\exp[i\lambda(x - ct)], \tag{5.33}$$

$$p = \tilde{p}\exp[i(\lambda x - \omega t)] = \tilde{p}\exp[i\lambda(x - ct)], \tag{5.34}$$

where λ is the wave number, ω is the frequency of the wave, and c is the wave propagation velocity, $c = \omega/\lambda$. The frequency ω is real, the wave number λ may be complex.

Substitution of (5.33) and (5.34) into (5.31) and (5.32) gives, after combination of the two equations,

$$\frac{(1+\tau)\rho_f S_p}{n}\left[1 + i\frac{n\mu}{(1+\tau)\rho_f\omega\kappa}\right]c^2 = 1. \tag{5.35}$$

The behaviour of the solution is determined by the value of the factor

$$B = \frac{n\mu}{(1+\tau)\rho_f\omega\kappa} = \frac{ng}{(1+\tau)\omega k}, \tag{5.36}$$

where g is the gravity constant ($g \approx 10$ m/s^2), and k is the hydraulic conductivity. For normal soil or rock the permeability is about 10^{-4} m/s, or less, and thus the value of the parameter B is very large, except for extremely rapid fluctuations, say $\omega > 10^5$ s^{-1}. In normal civil engineering practice this may be excluded. Then the (imaginary) second term in the left hand side of (5.36) dominates, and the value of c is determined by

$$c^2 = -i\frac{\omega\kappa}{S_p\mu}. \tag{5.37}$$

Because $c = \omega/\lambda$ it now follows that

$$\lambda^2 = i\frac{S_p\omega\mu}{\kappa}, \tag{5.38}$$

or

$$\lambda = -(1+i)\sqrt{\frac{S_p\omega\mu}{2\kappa}} = -(1+i)\sqrt{\frac{S_p\omega\rho_f g}{2k}}. \tag{5.39}$$

This means that the wave is strongly damped. As an example consider a wave with frequency $\omega = 1$ s^{-1}, in a soil with porosity $n = 0.40$, permeability $k = 10^{-4}$ m/s, completely saturated with water, so that $S_p = nC_f = 0.2 \times 10^{-9}$ m^2/N. In this case one obtains: $\Re(\lambda) = 1$ m^{-1}. This means that the wave will be attenuated very rapidly, in a few meters. If the frequency is higher the attenuation is even stronger. Also, if the permeability is smaller than the (relatively high) value considered here, the wave will be damped in the immediate vicinity of the source. Propagation of this wave over a considerable distance will occur only if the frequency is very low, or the permeability is very high.

In the case of extremely high frequencies the influence of the permeability can be disregarded, and the second term in the left hand side of (5.35) can be disregarded. The wave velocity then is

$$c = \sqrt{n/[(1+\tau)\rho_f S_p]} = \sqrt{1/[(1+\tau)\rho_f C_f]}. \tag{5.40}$$

Apart from the factor $(1+\tau)$ this is simply the propagation velocity of a compression wave in the fluid. As mentioned above, waves of this type will be strongly damped by the friction with the solids.

For a completely saturated soil the value of S_p is the product of the porosity and the compressibility of pure water, which is about $S_p = nC_f = 0.2 \times 10^{-9}$ m^2/N. With $\rho_f = 1000$ kg/m^3 one then obtains $c \approx 1400$ m/s, which is somewhat slower than the undrained wave considered before.

5.4 Analytical Solution

Solutions of the basic differential equations can be obtained using Fourier Analysis. Probably the most simple approach is to start by considering the effect of a general periodic pore pressure at the free end of a very long horizontal column, see Fig. 5.1.

5.4.1 Periodic Solution

In order to derive a basic periodic solution it is assumed that

$$p = P \exp[i(\lambda x + \omega t)], \tag{5.41}$$

$$\sigma' = S \exp[i(\lambda x + \omega t)], \tag{5.42}$$

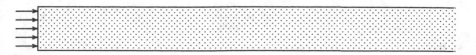

Fig. 5.1 Column with periodic pressure at its end

$$v = V \exp[i(\lambda x + \omega t)], \tag{5.43}$$

$$w = W \exp[i(\lambda x + \omega t)], \tag{5.44}$$

where ω is a given frequency, and λ is an unknown, possibly complex number, characterizing the wave length corresponding to the frequency ω.

Substitution of these expressions into the basic equations (5.17)–(5.20) gives

$$n\lambda V + (\alpha - n)\lambda W = -S_p \omega P, \tag{5.45}$$

$$m_v \omega S = -\lambda W, \tag{5.46}$$

$$n\rho_f \omega V + (1 - n)\rho_s \omega W = -\lambda S - \alpha \lambda P, \tag{5.47}$$

$$(1 + \tau)n\rho_f \omega V - \tau n\rho_f \omega W = -n\lambda P + \frac{in^2\mu}{\kappa} V - \frac{in^2\mu}{\kappa} W. \tag{5.48}$$

These are four equations with four unknowns. It is mathematically more convenient, however, to reduce them to two equations with two unknowns.

Elimination of S and V from (5.45), (5.46) and (5.47) gives

$$(S_p \rho_f \omega^2 - \alpha\lambda^2)m_v \omega P - [(\rho - \alpha\rho_f)m_v \omega^2 - \lambda^2]\lambda W = 0, \tag{5.49}$$

where ρ is the total density,

$$\rho = n\rho_f + (1 - n)\rho_s. \tag{5.50}$$

Elimination of V from (5.45) and (5.48) gives

$$\left[(1 + \tau)S_p \rho_f \omega^2 - iS_p \omega \frac{n\mu}{\kappa} - n\lambda^2\right]P + \left[(\alpha - n + \alpha\tau)\rho_f \omega - i\frac{\alpha n\mu}{\kappa}\right]\lambda W = 0. \tag{5.51}$$

Dimensionless parameters d_f, d_s, a, b, γ are introduced such that

$$d_f = \rho_f/\rho, \tag{5.52}$$

$$d_s = \rho_s/\rho, \tag{5.53}$$

$$a = \frac{n\mu}{\kappa\rho_f\omega} = \frac{ng}{k\omega}, \tag{5.54}$$

$$b = S_p/m_v, \tag{5.55}$$

$$\lambda^2 = \rho m_v \omega^2 \gamma^2. \tag{5.56}$$

The propagation speed c of plane waves in a medium with compressibility m_v and density ρ is defined by

$$c^2 = \frac{1}{\rho m_v}, \tag{5.57}$$

so that

$$\lambda = \omega\gamma/c. \tag{5.58}$$

It should be noted that the definition of the parameter a includes not only the hydraulic conductivity k, but also the frequency ω. This has been done to simplify the further analysis. It may also be noted that the definitions of d_f, d_s and ρ imply that $nd_f + (1-n)d_s = 1$.

Using the dimensionless parameters introduced above, (5.49) and (5.51) can be written as

$$(d_f b - \alpha\gamma^2)cm_v P - (1 - \alpha d_f - \gamma^2)\gamma W = 0, \tag{5.59}$$

$$[(1+\tau)d_f b - iad_f b - n\gamma^2]cm_v P + [(\alpha - n + \alpha\tau)d_f - i\alpha ad_f]\gamma W = 0. \tag{5.60}$$

The homogeneous system of (5.59) and (5.60) has a non-zero solution only if the determinant of the system of equations is zero. This leads to an equation of the form

$$A\gamma^4 + B\gamma^2 + C = 0, \tag{5.61}$$

with

$$A = n, \tag{5.62}$$

$$B = -n(1-n)d_s - [(\alpha - n)^2 + \alpha^2\tau]d_f - (1+\tau)d_f b + iad_f(\alpha^2 + b), \tag{5.63}$$

$$C = [(1-n)d_s + \tau]d_f b - iad_f b. \tag{5.64}$$

The quadratic equation (5.61) has two complex roots, which means that there are four possible values of γ, which are written as

$$\gamma_1 = \pm(q_1 + ir_1), \qquad \gamma_2 = \pm(q_2 + ir_2), \qquad r_1 > 0,\ r_2 > 0, \tag{5.65}$$

where it has been assumed that $r_1 > 0$ and $r_2 > 0$, for definiteness.

Because $\lambda = \omega\gamma/c$ it follows that the four possible values of λ are

$$\lambda_1 = \pm(q_1 + ir_1)\omega/c, \qquad \lambda_2 = \pm(q_2 + ir_2)\omega/c, \qquad r_1 > 0,\ r_2 > 0. \tag{5.66}$$

Restriction will be made to the semi-infinite medium $x > 0$. Then only the roots with a positive imaginary part apply, because only these lead to a finite limit at infinity. This means that the general solutions for the pore pressure and the velocity of the solids can be written as

$$p = A_p \exp[-(\omega/c)(r_1 - iq_1)x]\exp(i\omega t)$$
$$+ B_p \exp[-(\omega/c)(r_2 - iq_2)x]\exp(i\omega t), \tag{5.67}$$

$$w = A_w \exp[-(\omega/c)(r_1 + iq_1)x]\exp(i\omega t)$$
$$+ B_w \exp[-(\omega/c)(r_2 - iq_2)x]\exp(i\omega t), \tag{5.68}$$

where, in order to satisfy (5.59),

$$\frac{A_w}{A_p} = \frac{(d_f b - \alpha\gamma_1^2)cm_v}{(1 - \alpha d_f - \gamma_1^2)\gamma_1}, \tag{5.69}$$

$$\frac{B_w}{B_p} = \frac{(d_f b - \alpha \gamma_2^2) c m_v}{(1 - \alpha d_f - \gamma_2^2) \gamma_2}. \tag{5.70}$$

The general solutions for the effective stress σ' and the velocity of the fluid can be written as

$$\sigma' = A_s \exp[-(\omega/c)(r_1 - iq_1)x] \exp(i\omega t)$$
$$\qquad + B_s \exp[-(\omega/c)(r_2 - iq_2)x] \exp(i\omega t), \tag{5.71}$$
$$v = A_v \exp[-(\omega/c)(r_1 - iq_1)x] \exp(i\omega t)$$
$$\qquad + B_v \exp[-(\omega/c)(r_2 - iq_2)x] \exp(i\omega t). \tag{5.72}$$

In order to satisfy (5.46) and (5.45) the coefficients of these equations must be

$$\frac{A_s}{A_w} = -\frac{\gamma_1}{cm_v}, \tag{5.73}$$

$$\frac{B_s}{B_w} = -\frac{\gamma_2}{cm_v}, \tag{5.74}$$

$$\frac{A_v}{A_w} = -\frac{\alpha - n}{n} - \frac{(1 - \alpha d_f - \gamma_1^2)b}{n(d_f b - \alpha \gamma_1^2)}, \tag{5.75}$$

$$\frac{B_v}{B_w} = -\frac{\alpha - n}{n} - \frac{(1 - \alpha d_f - \gamma_2^2)b}{n(d_f b - \alpha \gamma_2^2)}. \tag{5.76}$$

Equations (5.75) and (5.76) give the ratio of the amplitudes of the displacements of the fluid and the solids, in the two waves.

The boundary conditions for a plane wave applied at the end $x = 0$ of a semi-infinite column of soil, with the wave being applied both to the soil and the fluid, as in the experiments of Van der Grinten (1987) and Smeulders (1992), are, because $\sigma = \sigma' + \alpha p$,

$$x = 0 : \sigma' = (1 - \alpha) p_0 \exp(i\omega t), \tag{5.77}$$

$$x = 0 : p = p_0 \exp(i\omega t). \tag{5.78}$$

From these conditions it follows, with (5.71) and (5.67), that

$$A_s + B_s = (1 - \alpha) p_0, \tag{5.79}$$

$$A_p + B_p = p_0. \tag{5.80}$$

With (5.73) and (5.74), (5.79) gives

$$\gamma_1 A_w + \gamma_2 B_w = -(1 - \alpha) c m_v p_0. \tag{5.81}$$

Using (5.69) and (5.70) this can be transformed into a relation between A_p and B_p,

$$\frac{d_f b - \alpha \gamma_1^2}{1 - \alpha d_f - \gamma_1^2} A_p + \frac{d_f b - \alpha \gamma_2^2}{1 - \alpha d_f - \gamma_2^2} B_p = -(1 - \alpha) p_0. \tag{5.82}$$

The two (complex) constants A_p and B_p may now be solved from (5.80) and (5.82). The result is

$$\frac{A_p}{p_0} = \frac{[d_f b - \gamma_2^2 + (1 - \alpha)(1 - \alpha d_f)](1 - \alpha d_f - \gamma_1^2)}{(\gamma_1^2 - \gamma_2^2)(\alpha - \alpha^2 d_f - b d_f)}, \tag{5.83}$$

$$\frac{B_p}{p_0} = -\frac{[d_f b - \gamma_1^2 + (1 - \alpha)(a - \alpha d_f)](1 - \alpha d_f - \gamma_2^2)}{(\gamma_1^2 - \gamma_2^2)(\alpha - \alpha^2 d_f - b d_f)}. \tag{5.84}$$

Substitution of these expressions for A_p and B_p into (5.67) finalizes the solution for the pore pressure.

5.4.2 Response to a Sinusoidal Load

The solution for a sinusoidal load can be constructed from the general periodic solution considered in the previous section by formulating the boundary condition as

$$x = 0 : p = p_0 \sin(\omega t) = p_0 \Im[\exp(i\omega t)]. \tag{5.85}$$

The solution for this case can immediately be obtained from the solution in the previous section, by taking the imaginary part. Thus, with (5.67),

$$p = \Im\{A_p \exp[-(\omega/c)(r_1 - iq_1)x]\exp(i\omega t) + B_p \exp[-(\omega/c)(r_2 - iq_2)x]\exp(i\omega t)\}, \tag{5.86}$$

or

$$p = \Re\{A_p\} \exp(-\omega r_1 x/c) \sin[\omega(q_1 x/c + t)]$$
$$+ \Im\{A_p\} \exp(-\omega r_1 x/c) \cos[\omega(q_1 x/c + t)]$$
$$+ \Re\{B_p\} \exp(-\omega r_2 x/c) \sin[\omega(q_2 x/c + t)]$$
$$+ \Im\{B_p\} \exp(-\omega r_2 x/c) \cos[\omega(q_2 x/c + t)]. \tag{5.87}$$

It can be observed from this expression that for $x \to \infty$ the pore pressure tends towards zero, because $r_1 > 0$ and $r_2 > 0$. The expected wave character of the solution requires that $q_1 < 0$ and $q_2 < 0$, but this will automatically be satisfied if $r_1 > 0$ and $r_2 > 0$. This property of the solution has been verified by numerical computations of the coefficients for various combinations of the basic parameters.

It can easily be verified numerically that the boundary condition at $x = 0$ is always satisfied, because it appears that $\Re\{A_p\} + \Re\{B_p\} = p_0$ and $\Im\{A_p\} + \Im\{B_p\} = 0$.

In a similar way the response to a load in the form of a cosine function, $\cos(\omega t)$, can be derived, by formulating the boundary condition as

$$x = 0 : \quad p = p_0 \cos(\omega t) = p_0 \Re[\exp(i\omega t)]. \tag{5.88}$$

In this case the solution is

$$
\begin{aligned}
p = \Re\{ & A_p \exp[-(\omega/c)(r_1 - iq_1)x]\exp(i\omega t) \\
& + B_p \exp[-(\omega/c)(r_2 - iq_2)x]\exp(i\omega t)\},
\end{aligned}
\tag{5.89}
$$

or

$$
\begin{aligned}
p = \Re\{A_p\} & \exp(-\omega r_1 x/c)\cos[\omega(q_1 x/c + t)] \\
- \Im\{A_p\} & \exp(-\omega r_1 x/c)\sin[\omega(q_1 x/c + t)] \\
+ \Re\{B_p\} & \exp(-\omega r_2 x/c)\cos[\omega(q_2 x/c + t)] \\
- \Im\{B_p\} & \exp(-\omega r_2 x/c)\sin[\omega(q_2 x/c + t)].
\end{aligned}
\tag{5.90}
$$

This completes the solution for the pore pressures produced by a sinusoidal load.

In the limit $\omega \to 0$ the solution for the steady state problem with the boundary conditions

$$x = 0 : \quad \sigma' = (1 - \alpha)p_0, \qquad x = 0 : \quad p = p_0, \tag{5.91}$$

is obtained as

$$\sigma' = (1 - \alpha)p_0, \quad p = p_0. \tag{5.92}$$

The velocities of the constituents are zero in this case

$$v = w = 0. \tag{5.93}$$

It can easily be verified that this is indeed a solution of the basic equations (5.17)–(5.20), and that it satisfies the boundary conditions (5.91). It should be noted that the vanishing of the velocities does not mean that the deformations and the displacements also vanish. Actually, the effective stress (5.92) implies a uniform deformation.

5.4.3 Approximation of the Solution

For real soils it can be expected that the product of hydraulic conductivity and frequency, $k\omega$, will be rather small, and often very small. This means that the parameter

a, defined in (5.54) as $a = ng/k\omega$, will be very large. In that case the coefficients A, B and C, see (5.62)–(5.64), may be approximated by

$$A = n,\tag{5.94}$$

$$B = iad_f(\alpha^2 + b),\tag{5.95}$$

$$C = -iad_f b.\tag{5.96}$$

The general solution of (5.61),

$$A\gamma^4 + B\gamma^2 + C = 0,\tag{5.97}$$

is

$$\gamma^2 = -\frac{B}{2A}\left\{1 \pm \sqrt{1 - 4AC/B^2}\right\},\tag{5.98}$$

or, because $4AC/B^2 \ll 1$,

$$\gamma^2 = -\frac{B}{2A}\left\{1 \pm [1 - 2AC/B^2]\right\}.\tag{5.99}$$

The two possible solutions for γ^2 now are

$$\gamma_1^2 = -\frac{C}{B} = \frac{b}{\alpha^2 + b}.\tag{5.100}$$

$$\gamma_2^2 = -\frac{B}{A} = -i\frac{ad_f(\alpha^2 + b)}{n} = -i\frac{\rho_f g(\alpha^2 m_v + S_p)}{k\rho m_v \omega}.\tag{5.101}$$

This last expression may also be written as

$$\gamma_2^2 = -i\frac{c^2}{c_v \omega},\tag{5.102}$$

where c is the wave speed in a medium with compressibility m_v and density ρ, see (5.57), and c_v is the one-dimensional consolidation coefficient of the porous medium, defined as

$$c_v = \frac{k}{\rho_f g(\alpha^2 m_v + S_p)},\tag{5.103}$$

see (4.96).

It now follows that the possible solutions for γ are, taking the roots with a negative real part, as only these apply in a semi-infinite beam $x > 0$,

$$\gamma_1 = -\sqrt{\frac{b}{\alpha^2 + b}}.\tag{5.104}$$

$$\gamma_2 = -(1 - i)\sqrt{\frac{c^2}{2c_v \omega}}.\tag{5.105}$$

The propagation speed c of waves in a porous medium is usually of the order of magnitude of 1000 m/s, and the consolidation coefficient c_v is usually of the order of magnitude of 1 m^2/s, or much less, indicating that the parameter $c^2/c_v\omega$ usually is very large. This may also be concluded from the original form of the parameter, which is $g/k\omega$ (and some relative quantities). Because $g = 10$ m/s^2 and the permeability can be assumed to be not larger than $k = 10^{-3}$ m/s, for very coarse sand, it follows that the parameter $g/k\omega$ will be large, except for extremely high frequencies.

It may also be noted that the solution (5.67) contains a factor $\exp(-\omega r x/c)$, where r is the imaginary part of the dimensionless root γ. Now that it has been found that this can be written as $r = \sqrt{c^2/2c_v\omega}$ it follows that the second wave contains a factor $\exp(-x\sqrt{\omega/2c_v})$. This means that this wave is noticeable only for a distance of about

$$L \approx 4\sqrt{2c_v/\omega}. \tag{5.106}$$

For most soils, in which c_v is small compared to 1 m^2/s, this will be a small distance, indicating that the second wave usually will influence only the immediate vicinity of the disturbance. The exception is the case of a very stiff material, of high permeability, for which the consolidation coefficient may be large.

It is also interesting to substitute the results (5.104) or (5.105) into (5.75) and (5.76), which define the relationship between the two velocities. Using the assumption, characterizing the approximation considered here that $a = \frac{ng}{k\omega} \gg 1$, it then follows that

$$\gamma = \gamma_1 : \quad \frac{v}{w} \approx 1, \tag{5.107}$$

$$\gamma = \gamma_2 : \quad \frac{v}{w} \approx -\left[\frac{\alpha - n}{n} + \frac{S_p}{\alpha n m_v}\right]. \tag{5.108}$$

This means that in the first wave the fluid and the solids move together, and that in the second wave the fluid moves out of phase with the solids. This explains why the second wave is so strongly damped. The existence of these two waves, and their characteristic properties, were first noted by De Josselin de Jong (1956) and Biot (1956).

The first wave, in which the soil particles and the pore fluid move at the same velocity, was already considered in Sect. 5.3.1. The order of magnitude of the velocity of this wave was found to be about 1600 m/s for soft soil, and somewhat larger for stiff soils or rock, up to 2000 m/s.

The second wave, characterized by the ratio (5.108) for the two velocities, can be further analyzed by considering the behaviour of the basic differential equations for this ratio. Actually, (5.17) and (5.20) now reduce to

$$[1 - \alpha + S_p/m_v]\frac{\partial w}{\partial x} = S_p \frac{\partial p}{\partial t}, \tag{5.109}$$

$$[\alpha - n + \tau + (1 + \tau)S_p/m_v]\frac{\partial w}{\partial t} = \frac{n}{\rho_f}\frac{\partial p}{\partial x}. \tag{5.110}$$

From these two equations the velocity of this wave can be obtained as

$$c_2 = \left[\frac{1 - \alpha + S_p/m_v}{\alpha - n + \tau + (1+\tau)S_p/m_v} \right] \frac{n}{S_p \rho_f}. \tag{5.111}$$

If $\alpha = 1$ and $\tau = 0$ the storativity is $S_p = nC_f$, where C_f is the compressibility of the fluid. Equation (5.111) then reduces to

$$c_2 = \left[\frac{nC_f/m_v}{1 - n + nC_f/m_v} \right] \frac{1}{C_f \rho_f}. \tag{5.112}$$

The second factor is the velocity of a wave in the pure fluid. The first factor is smaller than 1, so that the velocity of the second wave is somewhat smaller than the velocity of a wave in the fluid.

It may be noted that the solution for a material with incompressible constituents (incompressible particles and incompressible fluid) has been considered by De Boer (2000). In this case the velocity of the first wave is infinitely large, and only the second wave remains.

5.4.4 Numerical Verification

In order to verify the results of the preceding section, the exact results have been obtained by numerically calculating the values of A_v/A_w and B_v/B_w, using (5.75) and (5.76), for a material having the properties listed in Table 5.1.

In this case $(1 - n)/n = 1.5$, $b = S_p/m_v = 1$. The values of the two roots γ_1 and γ_2 are found to be

$$\gamma_1 = -0.707106781 + 0.000000004i \ : \ A_v/A_w = 0.999999981 - 0.000012500i,$$

$$\gamma_2 = -22.439191923 + 22.394414404i \ : \ B_v/B_w = -4.000000019$$

$$- 0.000012500i.$$

Table 5.1 Example properties	Symbol	Property	Value
	ρ_s	Density of solids (kg/m^3)	2650
	ρ_f	Density of fluid (kg/m^3)	1000
	k	Permeability (m/s)	0.001
	n	Porosity (–)	0.400
	τ	Tortuosity (–)	0.000
	α	Biot coefficient (–)	1.000
	m_v	Compressibility of soil (m^2/MN)	0.0002
	C_f	Compressibility of fluid (m^2/MN)	0.0005
	C_s	Compressibility of solids (m^2/MN)	0.0000
	ω	Frequency (1/s)	10

These values compare very well with the approximate results

$$\gamma_1 = -0.7071068 \; : \; A_v/A_w = 1,$$

$$\gamma_2 =: -22.416793 + 22.416793i \; : \; B_v/B_w = -4.$$

The exact results confirm the accuracy of the approximate values. In particular, they confirm that the first wave, in which the solids and the fluid move at practically the same velocity, is only very slightly damped, and that the second wave, in which the solids and the fluid move in opposite directions, is strongly damped.

5.4.5 Response to a Block Wave

On the basis of the response to a single wave of the form $\sin(\omega t)$ or $\cos(\omega t)$, the response to a periodic wave of arbitrary shape can be generated, using Fourier series expansion.

A block wave, of period T, see Fig. 5.2, can be represented by the Fourier series

$$q = \frac{1}{2} + \frac{2}{\pi} \sum_{k=1,3,5,}^{\infty} \frac{\sin(2\pi kt/T)}{k}. \tag{5.113}$$

The response of a column of a porous material, with properties as given in Table 5.1, to a block wave of period T can be generated from the elementary solution given in (5.87), using the Fourier series (5.113), provided that the period of the block wave is large enough for the pore pressures to approach the static value p/p_0 after one half of the period.

The results can be produced by the program SHOCKWAVE. For a point at a distance of 0.2 m from the boundary the pore pressures are shown in Fig. 5.3, taking $T = 0.1$ s. Two waves can be seen to have developed. It can be verified that the first wave, arriving at a distance of 0.2 m after about 0.00009 s, is the undrained wave, in which the velocities of particles and fluid are equal. The propagation velocity of this wave is given by (5.30). It now follows, using the data from Table 5.1, that $c = 2240$ m/s. This value is in good agreement with the value observed in Fig. 5.3 of about 2200 m/s. The second wave observed in this figure is the wave with opposite

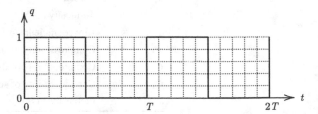

Fig. 5.2 Block wave

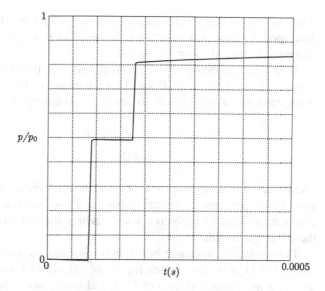

Fig. 5.3 Shock wave in a porous medium—analytical solution

velocities, for which the propagation velocity is found to be, with (5.112) and the data from Table 5.1, $c_2 = 1120$ m/s, which is a factor 2 slower than the first wave. This is in good agreement with the results shown in Fig. 5.3.

It can be verified that in the special (and unrealistic) case of an incompressible fluid only one wave, the wave in which the velocities of the fluid and the solids move in opposite directions, can be observed. This is a highly damped wave. This case can be investigated by taking a very small value for the compressibility C_f of the fluid, for instance a factor 1000 smaller than the value given in Table 5.1.

5.5 Numerical Solution

For a numerical solution by finite differences it is convenient to rewrite (5.17)–(5.20) in the following form,

$$\left[1 + \tau \left\{1 + \frac{n\rho_f}{(1-n)\rho_s}\right\}\right] \frac{\partial v}{\partial t}$$

$$= -\frac{1}{\rho_f} \frac{\partial p}{\partial x} - \frac{n\mu}{\kappa \rho_f}(v - w) - \frac{\tau}{(1-n)\rho_s}\left\{\frac{\partial \sigma'}{\partial x} + \alpha \frac{\partial p}{\partial x}\right\}, \quad (5.114)$$

$$\frac{\partial w}{\partial t} = -\frac{n\rho_f}{(1-n)\rho_s} \frac{\partial v}{\partial t} - \frac{1}{(1-n)\rho_s}\left\{\frac{\partial \sigma'}{\partial x} + \alpha \frac{\partial p}{\partial x}\right\}, \quad (5.115)$$

$$\frac{\partial p}{\partial t} = -\frac{n}{S_p} \frac{\partial v}{\partial x} - \frac{\alpha - n}{S_p} \frac{\partial w}{\partial x}, \quad (5.116)$$

$$\frac{\partial \sigma'}{\partial t} = -\frac{1}{m_v} \frac{\partial w}{\partial x}. \quad (5.117)$$

When written in this form a numerical solution by finite differences can easily be developed, because new values for the variables v, w, p and σ' can be calculated successively from the four equations.

As an example the problem of propagation of an under water shock wave will be considered. At time $t = 0$ all variables are assumed to be zero, and at that time a shock wave hits the end $x = 0$, so that the boundary conditions are

$$x = 0 : p = q, \tag{5.118}$$

$$x = 0 : \sigma' = (1 - \alpha)q. \tag{5.119}$$

The shock wave is supposed to act both in the total stress and in the pore water pressure. This means that the effective stress at the surface remains zero. This situation can be considered to apply to a wave reaching the soil through a layer of water to the left of the boundary $x = 0$.

The numerical procedure now can be that new values for v are first calculated from (5.114), then new values for w are calculated from (5.115), next new values for p are calculated from (5.116), and finally new values for σ' are calculated from (5.117). This completes the calculations in a time step. The same calculations can then be repeated for a new time step, and so the process can be solved in successive time steps.

The main part of the program SHOCKWAVENUM (in C) that may be used to perform the calculations, is reproduced below.

```
NN=2000;N=4000;RS=2650.0;RF=1000.0;PORO=0.4;PERM=0.001;GRAVITY=10.0;
SP=0.5*0.001*0.001*0.001*PORO;MV=0.2*0.001*0.001*0.001;TAU=0;ALPHA=1;
XX=1.0;DX=XX/NN;XP=0.2;RR=PORO*RF+(1.0-PORO)*RS;EE=1.0/(MV+ALPHA*ALPHA*SP);
CC=sqrt(EE/RR);C2=sqrt((1-ALPHA+SP/MV)*EF/(RF*(ALPHA-PORO+TAU+(1+TAU)*SP/MV)));
if (C2>CC) CC=C2;TC=DX/CC;DT=0.4*TC;IP=XP/DX;
for (i=0;i<=N;i++) {V[i]=0.0;W[i]=0.0;P[i]=0.0;S[i]=0.0;F[i]=0.0;}
a1=1.0+TAU*(1.0+PORO*RF/((1.0-PORO)*RS));a2=1.0/(RF*DX);
a3=PORO*GRAVITY/PERM;a4=TAU/((1.0-PORO)*RS*DX);
b1=PORO*RF/((1.0-PORO)*RS);b2=1.0/((1.0-PORO)*RS*DX);
c1=PORO/(SP*DX);c2=(ALPHA-PORO)/(SP*DX);d1=1.0/(MV*DX);
for (j=1;j<=N;j++)
  {
  if (j<20) P[0]=j*0.05;else P[0]=1.0;
  for (i=1;i<=j;i++)
    {
    aa=-(a2*(P[i]-P[i-1])+a3*(V[i]-W[i])+a4*(S[i]-S[i-1]+ALPHA*P[i]-ALPHA*P[i-1]))/a1;
    V[i]=V[i]+aa*DT;W[i]=W[i]-(b1*aa+b2*(S[i]-S[i-1]+ALPHA*P[i]-ALPHA*P[i-1]))*DT;
    }
    for (i=1;i<j;i++)
    {
    P[i]=P[i]-(c1*(V[i+1]-V[i])+c2*(W[i+1]-W[i]))*DT;
    S[i]=S[i]-d1*(W[i+1]-W[i])*DT;
    }
    F[j]=P[IP];
  }
```

The program applies to a soil column of 1000 mm length. The column is subdivided into 2000 elements of 0.5 mm length. The data describing the problem are defined in the first 6 lines. They are in agreement with the data given in Table 5.1. The time step is determined such that the two waves will not lead to instabilities. The arrays $P[i]$, $S[i]$, $V[i]$, $W[i]$, $F[i]$ must be defined so that they can store values from $i=0$ to $i=4000$. The values of the array $F[j]$ denote the relative pore pressure at a depth XP.

Fig. 5.4 Shock wave in a porous medium–numerical solution

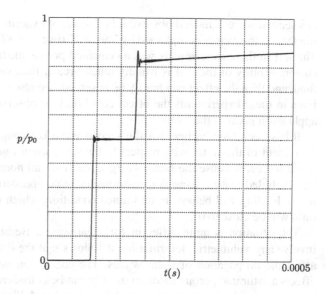

The results of the computations of the pore pressure as a function of time are shown in Fig. 5.4.

The results of the numerical calculation agree very well with the analytical results shown in Fig. 5.3. As in the analytical solution, it appears that two waves are generated in the column. The arrival time of the first wave corresponds with that of the undrained wave, in which the soil particles move with the pore fluid. The second wave is typical of porous media in which the compressibility of the fluid and of the solids have the same order of magnitude (Van der Grinten, 1987). In this wave the fluid particles move with respect to the soil particles. This wave is strongly damped, because of the friction between the fluid and the solid particles. The effect of this wave can only be observed near the surface (in the example this is at a depth of 200 mm). At large depths it has been dissipated.

It may be noted that the high frequency oscillations observed in Fig. 5.4 are a result of the numerical process. In the analytical solution they do not appear. Actually, these fluctuations have been largely suppressed by approximating the step load at the boundary by a load that gradually increases in steps of 5% of the total load, in 20 very short time steps. If the load is applied in a single step the fluctuations are much larger.

5.6 Conclusion

It has been seen in this chapter that in a saturated porous medium two compressive waves can be generated, one in which the particles and the fluid move together, and one in which they move in opposite directions. As could be expected, this second wave is strongly damped, because of the friction between the soil particles and the fluid in the small pores. Actually, it is not so easy to choose the data such that this second wave is indeed observed. In a series of experiments at the University of Eindhoven this was accomplished by Van der Grinten (1987) and Smeulders (1992). The soil in these experiments was a cemented porous medium. In normal soils the compressibility of the soil is usually much greater than the compressibility of the fluid, and then the effect can hardly be observed, see also (5.106) and its derivation. Even in these experiments the effect could only be observed in the vicinity of the application point of the load.

It has been seen earlier that the second wave is strongly damped, because the movement of the water with respect to the soil particles generates such large frictional forces. Because the factor $k\omega/g$ is small for all normal saturated soils it can be concluded that in plane deformations, or in compression, these soils under dynamic loading will behave in undrained condition, which means that its Poisson's ratio will be close to 0.5.

The situation is quite different for shear waves. Because pure shear does not involve any volumetric deformation, it follows that the fluid in the pores does not affect the propagation of shear waves. The conclusion must be that for dynamic effects a saturated porous medium usually can be considered as a soil in undrained conditions, with its compression modulus given by (4.49),

$$K_u = K + \frac{\alpha^2}{S_p}, \tag{5.120}$$

where K is the compression modulus, and S_p is the storativity, see (4.28). Because for a saturated soft soil the fluid compressibility and the compressibility of the soil particles are very small, this means that the soil is almost incompressible.

Chapter 6
Cylindrical Waves

In this chapter a number of cylindrically symmetric problems from the theory of elasticity are considered, both for the static case and the dynamic case. These problems can be considered as first approximations for the analysis of the influence of a local disturbance in a very large homogeneous elastic plate, or for the case of deformation of bodies bounded by very long cylindrical surfaces. Certain problems of this kind are known as the expansion of cylindrical cavities.

6.1 Static Problems

6.1.1 Basic Equations

Figure 6.1 shows an element of material in a cylindrical coordinate system. If the radial coordinate is denoted by r, and the tangential coordinate by θ, then the area of the element is $r\,dr\,d\theta$. If it is assumed that the displacement field is cylindrically symmetrical it may be assumed that there are no shear stresses acting upon the element, and that the normal stresses σ_{rr} and σ_{tt} are independent of the tangential coordinate θ. The stresses acting upon the element are indicated in the figure.

The only non-trivial equation of equilibrium now is the one in radial direction,

$$\frac{d\sigma_{rr}}{dr} + \frac{\sigma_{rr} - \sigma_{tt}}{r} = 0. \tag{6.1}$$

The deformations are related to the stresses by Hooke's law. If the body considered is a thick plate, it may be assumed that the plate deforms in a state of plane strain. In that case Hooke's law states, in its inverse form,

$$\sigma_{rr} = \lambda e + 2\mu\varepsilon_{rr}, \tag{6.2}$$

$$\sigma_{tt} = \lambda e + 2\mu\varepsilon_{tt}, \tag{6.3}$$

where e is the volume strain,

$$e = \varepsilon_{rr} + \varepsilon_{tt}, \tag{6.4}$$

A. Verruijt, *An Introduction to Soil Dynamics*,
Theory and Applications of Transport in Porous Media 24,
© Springer Science+Business Media B.V. 2010

Fig. 6.1 Element in circular coordinates

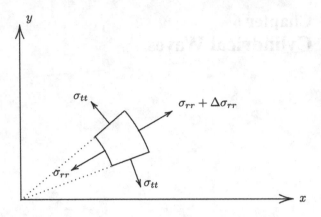

and λ and μ are the elastic coefficients (Lamé constants),

$$\lambda = \frac{\nu E}{(1+\nu)(1-2\nu)}.$$

(6.5)

$$\mu = \frac{E}{2(1+\nu)}.$$

(6.6)

The strains ε_{rr} and ε_{tt} can be related to the radial displacement u by the relations

$$\varepsilon_{rr} = \frac{du}{dr},$$

(6.7)

$$\varepsilon_{tt} = \frac{u}{r}.$$

(6.8)

Substitution of (6.2)–(6.8) into (6.1) gives

$$(\lambda + 2\mu)\left\{\frac{d^2u}{dr^2} + \frac{1}{r}\frac{du}{dr} - \frac{u}{r^2}\right\} = 0,$$

(6.9)

or

$$\frac{d^2u}{dr^2} + \frac{1}{r}\frac{du}{dr} - \frac{u}{r^2} = 0.$$

(6.10)

This is the basic differential equation for radially symmetric elastic deformations. It is remarkable that all terms appear to have a coefficient $(\lambda + 2\mu)$, which means that the equation is independent of the elastic properties of the material. Hence, if the boundary conditions can all be expressed in terms of the displacement u, then the solution will be independent of the elastic properties.

The stresses can be expressed into the radial displacement by substitution of (6.7) and (6.8) into (6.2) and (6.3). This gives

$$\sigma_{rr} = (\lambda + 2\mu)\frac{du}{dr} + \lambda\frac{u}{r},$$

(6.11)

$$\sigma_{tt} = (\lambda + 2\mu)\frac{u}{r} + \lambda\frac{du}{dr}. \tag{6.12}$$

6.1.2 General Solution

The general solution of the differential equation (6.10) is

$$u = Ar + \frac{B}{r}, \tag{6.13}$$

where A and B are integration constants, to be determined from the boundary conditions.

The general expression for the volume strain, corresponding to the solution (6.13) is

$$e = 2A. \tag{6.14}$$

It appears that the deformation field corresponding to the basic solution B/r is *isochoric*, i.e. of constant volume.

The stresses σ_{rr} and σ_{tt} can be expressed as

$$\sigma_{rr} = 2(\lambda + \mu)A - 2\mu\frac{B}{r^2}, \tag{6.15}$$

$$\sigma_{tt} = 2(\lambda + \mu)A + 2\mu\frac{B}{r^2}. \tag{6.16}$$

Some examples will be given in the next section.

6.1.3 Examples

Example 1: Cylinder Under External Pressure

Of the two basic solutions the solution with the coefficient B is singular in the origin. Thus for a massive cylinder, which includes the axis $r = 0$, this solution must vanish, to prevent the stresses and displacements from becoming singular. If the boundary condition at the outer boundary of the cylinder is

$$r = a \; : \; \sigma_{rr} = -p, \tag{6.17}$$

see Fig. 6.2, then the two constants are

$$A = -\frac{p}{2(\lambda + \mu)}, \tag{6.18}$$

$$B = 0. \tag{6.19}$$

Fig. 6.2 Cylinder under
external pressure

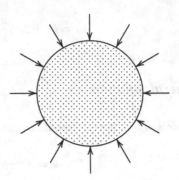

Fig. 6.3 Hollow cylinder
under external pressure

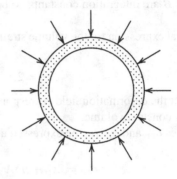

The stresses now are, in the entire cylinder,

$$\sigma_{rr} = \sigma_{tt} = -p. \tag{6.20}$$

The displacement field is

$$u = -\frac{pr}{2(\lambda + \mu)}. \tag{6.21}$$

Thus the stresses (and the strains) in the cylinder are homogeneous, and the displacement field is such that the radial displacement increases linearly with the distance from the origin.

Example 2: Hollow Cylinder Under External Pressure

For a hollow cylinder under external pressure, see Fig. 6.3, the boundary conditions are

$$r = a \; : \; \sigma_{rr} = 0, \tag{6.22}$$

$$r = b \; : \; \sigma_{rr} = -p, \tag{6.23}$$

where it has been assumed that $a < b$.

In this case the constants A and B are found to be

$$A = -\frac{pb^2}{2(\lambda + \mu)(b^2 - a^2)}, \tag{6.24}$$

$$B = -\frac{pa^2b^2}{2\mu(b^2 - a^2)}. \tag{6.25}$$

The stresses now are

$$\sigma_{rr} = -p\frac{1 - a^2/r^2}{1 - a^2/b^2}, \tag{6.26}$$

$$\sigma_{tt} = -p\frac{1 + a^2/r^2}{1 - a^2/b^2}. \tag{6.27}$$

It can easily be seen that the expression (6.26) satisfies the boundary conditions (6.22) and (6.23).

A special case occurs when the outer boundary is located at infinity. This is the case of a very large plate with a small circular hole. In this case $b \to \infty$. At infinity ($r \to \infty$) all stresses then approach the limiting value $-p$. At the boundary of the hole the radial stresses σ_{rr} are zero, but the tangential stress σ_{tt} at the boundary of the hole then is $-2p$. The multiplication factor 2 is called a *stress concentration factor*. This solution also applies to a plate under tension, of course. The only difference then is that p is negative. The stresses along the hole then are twice as large as the stresses at infinity. If the material has a limited range in which it can withstand stresses, as most materials do, the material will start to crack or yield at the boundary of the hole. More refined studies for other cases, such as a plate with an elliptical hole, have shown that the stress concentration factor can be much larger than 2, for instance near the corner points of a square hole.

Example 3: Cylinder with Rigid Inclusion

For a cylinder under external pressure, with a rigid circular inclusion, see Fig. 6.4, the boundary conditions are

$$r = a : u_r = 0, \tag{6.28}$$

$$r = b : \sigma_{rr} = -p. \tag{6.29}$$

In this case the constants A and B are found to be

$$A = -\frac{p}{2(\lambda + \mu) + 2\mu a^2/b^2}, \tag{6.30}$$

$$B = +\frac{pa^2}{2(\lambda + \mu) + 2\mu a^2/b^2}. \tag{6.31}$$

Fig. 6.4 Cylinder with rigid inclusion

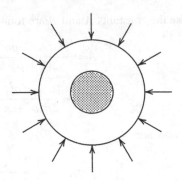

The stresses now are

$$\sigma_{rr} = -p \frac{2(\lambda + \mu) + 2\mu a^2/r^2}{2(\lambda + \mu) + 2\mu a^2/b^2}, \tag{6.32}$$

$$\sigma_{tt} = -p \frac{2(\lambda + \mu) - 2\mu a^2/r^2}{2(\lambda + \mu) + 2\mu a^2/b^2}. \tag{6.33}$$

Again the case of an infinite plate is of some special interest. In this case the radial stress is not zero at the boundary of the inclusion, of course. Actually, the stresses at the boundary of the inclusion are

$$a \to \infty, \ r = a \ : \ \sigma_{rr} = -p \frac{\lambda + 2\mu}{\lambda + \mu}, \tag{6.34}$$

$$a \to \infty, \ r = a \ : \ \sigma_{tt} = -p \frac{\lambda}{\lambda + \mu}. \tag{6.35}$$

One might suppose that the solutions for an infinite plate with a circular hole and for an infinite plate with a rigid circular inclusion would become identical when the radius of the hole and the inclusion tends to zero. This is not the case, however. Apparently the special behaviour near the boundary of the hole or the boundary of the inclusion can still be felt, even when the radius is infinitely small. Especially for the case of a rigid inclusion this seems strange, because one would expect that the solution for this case approaches the solution for a homogeneous plate if the radius of the inclusion tends towards zero. The explanation for this paradox is that the limiting cases should be approached more carefully. Actually, the limiting case for $a \to 0$ should be considered first for the integration constants A and B, see (6.30) and (6.31). These then correctly reduce to the values given in (6.18) and (6.19). The procedure of first setting $r = a$ and then taking $a = 0$ in (6.32) and (6.33) leads to a result differing from the one obtained by first setting $a = 0$ and then letting $r \to 0$.

Fig. 6.5 Cavity expansion

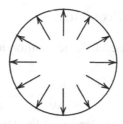

Example 4: Cavity Expansion

A case of special interest is the expansion of a cylindrical cavity in an infinite body, see Fig. 6.5. In this case the boundary conditions are

$$r \to \infty \ : \ \sigma_{rr} = 0, \tag{6.36}$$

$$r = a \ : \ \sigma_{rr} = -p. \tag{6.37}$$

The constants A and B are found to be

$$A = 0, \tag{6.38}$$

$$B = \frac{pa^2}{2\mu}. \tag{6.39}$$

The stresses now are

$$\sigma_{rr} = -p\frac{a^2}{r^2}, \tag{6.40}$$

$$\sigma_{tt} = p\frac{a^2}{r^2}. \tag{6.41}$$

The displacement field is

$$u = \frac{pa^2}{2\mu r}. \tag{6.42}$$

The displacement at the boundary of the cavity is of particular interest,

$$u_a = \frac{pa}{2\mu}. \tag{6.43}$$

If this displacement can be measured, the value of the shear modulus μ can be obtained. This is the basis of the pressuremeter test, developed by Ménard.

6.2 Dynamic Problems

In the dynamic case the basic equilibrium equation (6.1) must be extended with an inertia term,

$$\frac{\partial \sigma_{rr}}{\partial r} + \frac{\sigma_{rr} - \sigma_{tt}}{r} = \rho \frac{\partial^2 u}{\partial t^2}, \tag{6.44}$$

where ρ is the mass density of the material. Substitution of (6.2)–(6.8) into this equation gives

$$\frac{\partial^2 u}{\partial r^2} + \frac{1}{r}\frac{\partial u}{\partial r} - \frac{u}{r^2} = \frac{1}{c^2}\frac{\partial^2 u}{\partial t^2}, \tag{6.45}$$

where c is the propagation velocity of compression waves,

$$c = \sqrt{(\lambda + 2\mu)/\rho}. \tag{6.46}$$

Equation (6.45) is the basic differential equation for cylindrically symmetric dynamic elastic deformations. For certain cases solutions of this differential equation may be obtained by separation of variables or by the Laplace transform method.

6.2.1 Sinusoidal Vibrations at the Cavity Boundary

As a first example the case of a sinusoidal variation of the displacements at the boundary of a cylindrical cavity in an infinite medium will be considered. The boundary condition is supposed to be

$$r = a \ : \ u = u_0 \sin(\omega t), \tag{6.47}$$

where ω is the frequency of the vibration.

The solution may be obtained by separation of variables. In this method it is assumed that the solution of the differential equation (6.45) can be written as

$$u = \mathrm{Re}\{F(r)\exp(i\omega t)\}. \tag{6.48}$$

Substitution into (6.45) shows that this is the case if the function $F(r)$ satisfies the equation

$$\frac{d^2 F}{dr^2} + \frac{1}{r}\frac{dF}{dr} + \left(\frac{\omega^2}{c^2} - \frac{1}{r^2}\right)F = 0. \tag{6.49}$$

The solution of this differential equation can be expressed in terms of Bessel functions (Abramowitz and Stegun, 1964). The general solution is

$$F = A J_1(\omega r/c) + B Y_1(\omega r/c), \tag{6.50}$$

where $J_1(x)$ and $Y_1(x)$ are the Bessel functions of the first and second kind, of order one, see Abramowitz and Stegun (1964).

In many problems of mathematical physics in an infinite region one of the two fundamental solutions can be excluded because of its behaviour near infinity. In the radial case this is not so, because both fundamental solutions $J_1(x)$ and $Y_1(x)$ behave in about the same way as $x \to \infty$. Therefore the condition at infinity has to be formulated in a more refined way, namely by specifying that at infinity the behaviour of the solution must be such that it corresponds to an outgoing wave. This is known as the *radiation condition*, as formulated by Sommerfeld (1949), and used implicitly by Lord Rayleigh (1894) and Lamb (1904).

At very large distances the Bessel functions $J_1(x)$ and $Y_1(x)$ may be approximated by the asymptotic expansions

$$x \to \infty : J_1(x) \approx -\sqrt{2/\pi x} \cos\left(x + \frac{1}{4}\pi\right), \tag{6.51}$$

$$x \to \infty : Y_1(x) \approx -\sqrt{2/\pi x} \sin\left(x + \frac{1}{4}\pi\right). \tag{6.52}$$

This means that for very large values of r the radial displacement will be

$$r \to \infty : u \approx \operatorname{Re}\left[\sqrt{c/2\pi \omega r}\left\{(A - iB)\exp\left[i\omega(t + r/c) + \frac{1}{4}\pi i\right]\right.\right.$$
$$\left.\left. + (A + iB)\exp\left[i\omega(t - r/c) - \frac{1}{4}\pi i\right]\right\}\right]. \tag{6.53}$$

The first term in the right hand side represents a wave traveling from infinity towards the origin, whereas the second term represents an outgoing wave, traveling towards infinity. This is the only acceptable term, and thus the radiation condition in this case requires that

$$A = iB. \tag{6.54}$$

The solution for the function $F(r)$ now is

$$F = iBJ_1(\omega r/c) + BY_1(\omega r/c). \tag{6.55}$$

The coefficient B must be determined from the condition at the inner boundary $r = a$, see (6.47). The result is

$$B = -u_0 \frac{J_1(\omega a/c) + iY_1(\omega a/c)}{J_1^2(\omega a/c) + Y_1^2(\omega a/c)}. \tag{6.56}$$

Using this expression for the coefficient B the final solution for the radial displacement is

$$\frac{u}{u_0} = \frac{J_1(\omega a/c)J_1(\omega r/c) + Y_1(\omega a/c)Y_1(\omega r/c)}{J_1^2(\omega a/c) + Y_1^2(\omega a/c)} \sin(\omega t)$$
$$- \frac{J_1(\omega a/c)Y_1(\omega r/c) - Y_1(\omega a/c)J_1(\omega r/c)}{J_1^2(\omega a/c) + Y_1^2(\omega a/c)} \cos(\omega t). \tag{6.57}$$

It can easily be verified that this solution satisfies the differential equation (6.44), because it consists of products of Bessel functions and circular functions, and that it satisfies the boundary condition (6.47), because for $r = a$ only the first term remains, and its coefficient reduces to 1. For large values of the radial coordinate r the solution can be approximated by

$$r \to \infty : \quad \frac{u}{u_0} \approx -\frac{J_1(\omega a/c)\sqrt{2c/\pi \omega r}}{J_1^2(\omega a/c) + Y_1^2(\omega a/c)} \sin\left[\omega(t - r/c) - \frac{1}{4}\pi\right]$$

$$-\frac{Y_1(\omega a/c)\sqrt{2c/\pi \omega r}}{J_1^2(\omega a/c) + Y_1^2(\omega a/c)} \cos\left[\omega(t - r/c) - \frac{1}{4}\pi\right]. \quad (6.58)$$

Because the time t appears in this expression only in the form of the factor $(t - r/c)$ it can be seen that the solution indeed satisfies the radiation condition that at infinity only an outgoing wave remains.

One of the most important aspects of the solution (6.57) and its approximation (6.58) is that the amplitude of the vibrations tends towards zero for $r \to \infty$ as the factor $\sqrt{1/r}$. This is in contrast with the static case, in which the solution tends towards zero at infinity as $1/r$, which is much faster. This means that dynamic effects are attenuated in space much slower than static effects.

The amplitude of the solution is shown graphically in Fig. 6.6, as a function of the distance r. The value of the parameter $\omega a/c$ has been taken as 0.2. The figure indeed shows that the amplitudes at great distances from the inner boundary approach zero rather slowly. Even at a distance of 50 times the radius of the cavity the amplitude of the wave is still about 10% of amplitude at the cavity boundary. In the static case this would only be about 2%.

The displacements in the wave, as a function of the radial coordinate r, are shown in Fig. 6.7, for a value of time such that $\sin(\omega t) = 1$, for instance $\omega t = \pi/2$. Again it can be observed that the damping of the maximum displacements in space is rather slow.

From the solution (6.57) the radial stress σ_{rr} and the tangential stress σ_{tt} can be determined, using (6.11) and (6.12). This gives

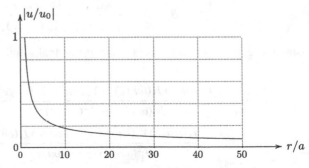

Fig. 6.6 Amplitude of wave, $\omega a/c = 0.2$

Fig. 6.7 Radial displacement, $\omega a/c = 0.2$, $\omega t/\pi = 0.5$

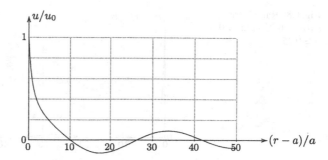

$$\frac{\sigma_{rr}a}{2\mu u_0} = \frac{a}{r}\frac{J_1(\omega a/c)J_1^r(\omega r/c) + Y_1(\omega a/c)Y_1^r(\omega r/c)}{J_1^2(\omega a/c) + Y_1^2(\omega a/c)}\sin(\omega t)$$

$$-\frac{a}{r}\frac{J_1(\omega a/c)Y_1^r(\omega r/c) - Y_1(\omega a/c)J_1^r(\omega r/c)}{J_1^2(\omega a/c) + Y_1^2(\omega a/c)}\cos(\omega t), \quad (6.59)$$

$$\frac{\sigma_{tt}a}{2\mu u_0} = \frac{a}{r}\frac{J_1(\omega a/c)J_1^t(\omega r/c) + Y_1(\omega a/c)Y_1^t(\omega r/c)}{J_1^2(\omega a/c) + Y_1^2(\omega a/c)}\sin(\omega t)$$

$$-\frac{a}{r}\frac{J_1(\omega a/c)Y_1^t(\omega r/c) - Y_1(\omega a/c)J_1^t(\omega r/c)}{J_1^2(\omega a/c) + Y_1^2(\omega a/c)}\cos(\omega t), \quad (6.60)$$

where

$$J_1^r(x) = J_1(x) + \frac{\lambda + 2\mu}{2\mu}x J_0(x), \quad (6.61)$$

$$Y_1^r(x) = Y_1(x) + \frac{\lambda + 2\mu}{2\mu}x Y_0(x), \quad (6.62)$$

$$J_1^t(x) = J_1(x) + \frac{\lambda}{2\mu}x J_0(x), \quad (6.63)$$

$$Y_1^t(x) = Y_1(x) + \frac{\lambda}{2\mu}x Y_0(x). \quad (6.64)$$

Here $J_0(x)$ and $Y_0(x)$ are the Bessel functions of the first and second kind, of order zero, see Abramowitz and Stegun (1964).

The radial stress σ_{rr} is shown as a function of $(r - a)/a$ in Fig. 6.8, at a moment in time for which $\omega t = \pi/2$, and assuming that $\omega a/c = 0.2$ and $\nu = 0.25$. It appears that the stresses tend to zero much faster than the displacements, as a result of the factor a/r in the expression (6.59). In the elastostatic case the convergence is even faster, of order a^2/r^2, see (6.40).

Of particular interest is the radial stress at the inner boundary. If this is denoted by $-p$,

$$r = a : \sigma_{rr} = -p, \quad (6.65)$$

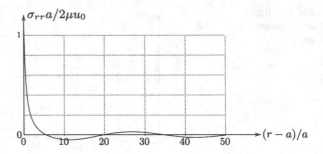

the relation of this boundary pressure with the displacement u_0 is found to be

$$p = \frac{2\mu u_0}{a}[F_1(\omega a/c)\sin(\omega t) + F_2(\omega a/c)\cos(\omega t)], \tag{6.66}$$

where

$$F_1(x) = 1 - x\frac{\lambda + 2\mu}{2\mu}\frac{J_1(x)J_0(x) + Y_1(x)Y_0(x)}{J_1^2(x) + Y_1^2(x)}, \tag{6.67}$$

$$F_2(x) = x\frac{\lambda + 2\mu}{2\mu}\frac{J_1(x)Y_0(x) - Y_1(x)J_0(x)}{J_1^2(x) + Y_1^2(x)}. \tag{6.68}$$

The expression for $F_2(x)$ can be simplified using the relation for the Wronskian determinant (Abramowitz and Stegun, 1964, formula 9.1.16)

$$J_1(x)Y_0(x) - Y_1(x)J_0(x) = \frac{2}{\pi x}. \tag{6.69}$$

This gives

$$F_2(x) = \frac{\lambda + 2\mu}{\pi\mu}\frac{1}{J_1^2(x) + Y_1^2(x)}. \tag{6.70}$$

If the frequency ω is very small, the static solution is recovered,

$$\omega \to 0 : p = \frac{2\mu u_0}{a}. \tag{6.71}$$

This is in agreement with the results obtained for cavity expansion in the static case, see (6.43).

The absolute value of the expression between square brackets in (6.66) gives the multiplication factor for the amplitude of the radial pressure in case of a dynamic load of given displacement. Its inverse represents the multiplication factor for the dynamic displacements for a given pressure at the inner boundary,

$$\left|\frac{u_d}{u_s}\right| = \frac{1}{\sqrt{F_1^2(\omega a/c) + F_2^2(\omega a/c)}}. \tag{6.72}$$

Fig. 6.9 Dynamic
amplification factor

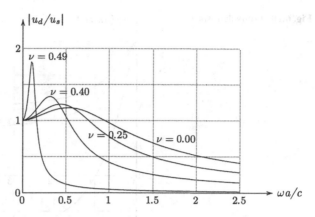

This dynamic amplification factor is shown in graphical form in Fig. 6.9, for various
values of the Poisson ratio ν. It appears from the figure that for large frequencies
the dynamic amplitude is very small, especially for values of Poisson's ratio approaching the incompressible limit $\nu = 0.5$. This means that then the response is
very stiff. This phenomenon is often observed in dynamics. The reason for it is that
it is very difficult to move the mass of the material in a very short time interval
(this is called *inertia* of the material). In this case of cylindrically symmetric deformations the static response is governed by the shear modulus μ only, see (6.43).
In the dynamic case, however, the other elastic parameter, Poisson's ratio ν or the
compression modulus K, also influences the response, which indicates that in the
dynamic case the wave not only involves shear, but also compression. The solution
of the problem also indicates that at large distances from the inner boundary, where
the disturbance is generated, the compression wave dominates, because the wave
velocity at infinity is that of a compression wave.

6.2.2 Equivalent Spring and Damping

It is convenient to write the relation (6.66) between the pressure p at the cavity
boundary $r = a$ and the displacement u_0 of that boundary in the form

$$2\pi a L p = \{K \sin(\omega t) + \omega C \cos(\omega t)\} u_0, \qquad (6.73)$$

where L is the thickness of the plate, K is an equivalent spring stiffness and C is
an equivalent damping, see also Sect. 1.5. The expressions for these quantities are,
with (6.66), (6.67) and (6.70),

$$\frac{K}{4\pi \mu L} = F_1(\omega a/c), \qquad (6.74)$$

Fig. 6.10 Equivalent spring

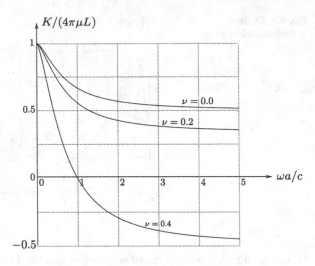

and

$$\frac{Cc}{2\pi(\lambda+2\mu)La} = \frac{2}{\pi}\frac{c/\omega a}{J_1^2(\omega a/c)+Y_1^2(\omega a/c)}. \tag{6.75}$$

The equivalent dynamic stiffness is shown, as a function of the dimensionless frequency $(\omega a/c)$, in Fig. 6.10. For small frequencies the spring constant is practically equal to the static value, but for large frequencies the spring is much more flexible. Actually, for very large frequencies the function $F_1(x)$ may be approximated by the following relation, which may be derived by using asymptotic expansions for the Bessel functions (Abramowitz and Stegun, 1964). The result is

$$\frac{\omega a}{c} \gg 1 : \quad \frac{K}{4\pi\mu L} \approx \frac{2\pi(\lambda+2\mu)La}{c}. \tag{6.76}$$

If $\nu > \frac{1}{3}$ this is negative, indicating that the force and the displacement are out of phase. This is confirmed by the curve for $\nu = 0.4$ in Fig. 6.10.

The value of the equivalent damping C is shown in Fig. 6.11. For large values of the dimensionless frequency $\omega a/c$ the damping is practically constant,

$$\frac{\omega a}{c} \gg 1 : \quad C \approx \frac{2\pi(\lambda+2\mu)La}{c}. \tag{6.77}$$

This approximation can be derived by using the asymptotic expansions of the Bessel functions appearing in (6.75).

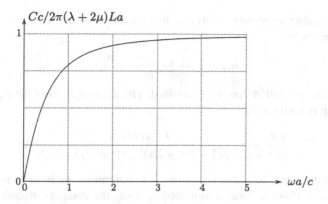

Fig. 6.11 Equivalent damping

6.3 Propagation of a Shock Wave

6.3.1 Solution by Laplace Transform

Another interesting problem concerns a shock wave propagating from a cavity in an infinite medium. In this case the Laplace transform method (Churchill, 1972) seems well suited to solve the problem. The Laplace transform of the displacement is defined by

$$\bar{u} = \int_0^\infty u \exp(-st)\, dt, \tag{6.78}$$

where s is the Laplace transform parameter, which is supposed to be sufficiently large so that all transforms exist.

Application of the Laplace transform to the differential equation (6.45) gives

$$\frac{d^2\bar{u}}{dr^2} + \frac{1}{r}\frac{d\bar{u}}{dr} - \left(\frac{s^2}{c^2} + \frac{1}{r^2}\right)\bar{u} = 0, \tag{6.79}$$

where it has been assumed that the initial values of the displacement and the velocity are zero. The solution of the differential equation (6.79), vanishing at infinity, is

$$\bar{u} = A K_1(sr/c), \tag{6.80}$$

where $K_1(x)$ is the modified Bessel function of the second kind, and of order one. The boundary condition is supposed to be that from time $t = 0$ on a compressive stress p is acting inside the cavity,

$$r = a,\ t > 0 : \sigma_{rr} = -p. \tag{6.81}$$

The condition for the transformed problem is

$$r = a : \bar{\sigma}_{rr} = -\frac{p}{s}. \tag{6.82}$$

Using the boundary condition (6.79) and the expression for the radial stress in terms of the displacement,

$$\sigma_{rr} = (\lambda + 2\mu)\frac{\partial u}{\partial r} + \lambda\frac{u}{r}, \tag{6.83}$$

the integration constant A can be determined. The final solution of the transformed problem then is found to be

$$\bar{u} = \frac{pa}{2\mu s}\frac{K_1(sr/c)}{K_1(sa/c) + [(\lambda + 2\mu)/2\mu](sa/c)K_0(sa/c)}. \tag{6.84}$$

The mathematical problem now remaining is to determine the inverse transform of this expression. This can be accomplished by using the complex inversion integral (Churchill, 1972), but will not be elaborated further here, because of the mathematical complexities, see Miklowitz (1978).

An approximate solution valid for small values of the time may be obtained by using the theorem (Churchill, 1972) that by assuming that the Laplace transform parameter s is very large, the Laplace transform

$$\bar{u}(s) = \int_0^\infty u(t)\exp(-st)\,dt, \tag{6.85}$$

practically contains contributions of $u(t)$ only for very small values of t. If in (6.84) the parameter s is assumed to be very large, the Bessel functions may be approximated by their asymptotic expansions,

$$\frac{sa}{c} \gg 1 \ : \ K_0\left(\frac{sa}{c}\right) \approx \sqrt{\frac{\pi c}{2sa}}\exp\left(-\frac{sa}{c}\right), \tag{6.86}$$

$$\frac{sa}{c} \gg 1 \ : \ K_1\left(\frac{sa}{c}\right) \approx \sqrt{\frac{\pi c}{2sa}}\exp\left(-\frac{sa}{c}\right). \tag{6.87}$$

The expression (6.84) then reduces to

$$\bar{u} = \frac{pc}{(\lambda + 2\mu)s^2}\sqrt{\frac{a}{r}}\exp[-s(r-a)/c]. \tag{6.88}$$

The inverse Laplace transformation now is simple, with the result

$$u = \frac{pc[t - (r-a)/c]}{\lambda + 2\mu}\sqrt{\frac{a}{r}}H[t - (r-a)/c], \tag{6.89}$$

where $H(t - t_0)$ is Heaviside's unit step function,

$$H(t - t_0) = \begin{cases} 0, & \text{if } t < t_0, \\ 1, & \text{if } t > t_0. \end{cases} \tag{6.90}$$

The approximate solution (6.89) indicates that for small values of time, i.e. shortly after the application of the shock, the response of the system is comparable to that of a one dimensional system (a pile) in which a compression wave is generated. The factor $\sqrt{a/r}$ indicates that the displacements are decreasing in radial direction, as could be expected. The approximate solution also has the convenient, and expected, property that the displacements are zero if $t < (r - a)/c$, i.e. when the wave has not yet arrived.

Another approximation can be obtained by using the property of the Laplace transform that

$$\lim_{t\to\infty} u(t) = \lim_{s\to 0} s\bar{u}(s), \tag{6.91}$$

see e.g. Churchill (1972). Application of this theorem to (6.84) now gives

$$\lim_{t\to\infty} u(t) = \frac{p a^2}{2\mu r}. \tag{6.92}$$

This is the static solution, see (6.42). It appears that the dynamic solution indeed approaches the static solution for very large values of time.

6.3.2 Solution by Fourier Series

The problem of the propagation of a shock wave from a cylindrical cavity can also be solved, and perhaps in a simpler way, by determining the response to a block wave of sufficiently long duration, considered as the summation of many sinusoidal variations.

A block wave, of period T, see Fig. 6.12, can be represented by the Fourier series

$$\frac{p}{p_0} = \frac{1}{2} + \frac{2}{\pi} \sum_{k=1,3,5,}^{\infty} \frac{\sin(2\pi kt/T)}{k}. \tag{6.93}$$

The basic element of the solution by Fourier analysis is the response of the medium to a sinusoidal variation of the radial stress at the cavity boundary,

$$r = a : \sigma_{rr} = -p_0 \sin(\omega t). \tag{6.94}$$

The displacement of the boundary due to this load can be written as

$$r = a : u = \frac{p_0 a}{2\mu}[G_1(\omega a/c)\sin(\omega t) + G_2(\omega a/c)\cos(\omega t)], \tag{6.95}$$

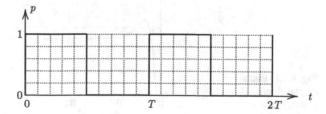

Fig. 6.12 Block wave

where the functions $G_1(x)$ and $G_2(x)$ are the inverse forms of the functions $F_1(x)$ and $F_2(x)$, defined in (6.67) and (6.70),

$$G_1(x) = \frac{F_1(x)}{F_1^2(x) + F_2^2(x)}, \tag{6.96}$$

$$G_2(x) = -\frac{F_2(x)}{F_1^2(x) + F_2^2(x)}. \tag{6.97}$$

The expression for the radial displacement as a function of r and t is, generalizing equation (6.57), and replacing the reference parameter u_0 by $p_0 a/2\mu$,

$$\frac{2\mu u}{p_0 a} = \frac{J_1(\omega a/c)J_1(\omega r/c) + Y_1(\omega a/c)Y_1(\omega r/c)}{J_1^2(\omega a/c) + Y_1^2(\omega a/c)} A(\omega t)$$

$$- \frac{J_1(\omega a/c)Y_1(\omega r/c) - Y_1(\omega a/c)J_1(\omega r/c)}{J_1^2(\omega a/c) + Y_1^2(\omega a/c)} B(\omega t), \tag{6.98}$$

where

$$A(\omega t) = G_1(\omega a/c)\sin(\omega t) + G_2(\omega a/c)\cos(\omega t), \tag{6.99}$$

$$B(\omega t) = G_1(\omega a/c)\cos(\omega t) - G_2(\omega a/c)\sin(\omega t). \tag{6.100}$$

The expression (6.93) consists of a constant load $p_0/2$, the average pressure, and a number of sinusoidal loads, with amplitude $2p_0/k\pi$. The first term gives rise to a solution of the form $u = p_0 a^2/(4\mu r)$, in agreement with the solution (6.42) for elastostatic cavity expansion. The other terms lead to partial solutions of the form (6.98), with $\omega t = 2\pi kt/T$, or $\omega = 2\pi k/T$. The results of a Fourier series solution, taking 500 terms, are shown in Fig. 6.13. The value of T has been chosen very large, so that $cT/a = 100$, to ensure that in half a period of the block wave the steady state solution will be reached. The other parameters in the solution are Poisson's ratio, and the distance from the cavity for which the radial displacement is plotted. The figure also shows, by dots, the solution for small values of time, as given by (6.89), and the steady state solution. The Fourier series solution appears to agree very well with these approximations for small and large values of time.

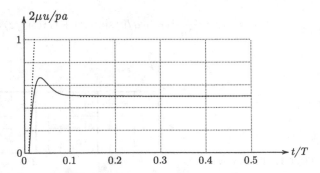

Fig. 6.13 Radial displacement, $cT/a = 100$, $\nu = 0$, $r/a = 2$

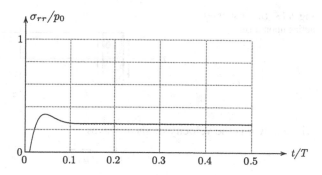

Fig. 6.14 Radial stress, $cT/a = 100$, $v = 0.25$, $r/a = 2$

The shape of the curve is also in agreement with a solution by Miklowitz (1978), obtained using the Laplace transform and numerical integration.

The expression for the radial stress σ_{rr} as a function of r and t caused by a single component of the Fourier series is, generalizing equation (6.59), and replacing the reference parameter u_0 by $p_0a/2\mu$,

$$\frac{\sigma_{rr}}{p_0} = \frac{a}{r} \frac{J_1(\omega a/c)J_1^r(\omega r/c) + Y_1(\omega a/c)Y_1^r(\omega r/c)}{J_1^2(\omega a/c) + Y_1^2(\omega a/c)} A(\omega t)$$

$$- \frac{a}{r} \frac{J_1(\omega a/c)Y_1^r(\omega r/c) - Y_1(\omega a/c)J_1^r(\omega r/c)}{J_1^2(\omega a/c) + Y_1^2(\omega a/c)} B(\omega t), \quad (6.101)$$

where $A(\omega t)$ and $B(\omega t)$ are defined by (6.99) and (6.100).

In this case the constant load term $p_0/2$ in the Fourier series (6.93) gives rise to a solution of the form $\sigma_{rr} = -p_0a^2/(2r^2)$, as given by the solution (6.40) for elastostatic cavity expansion. The other terms lead to solutions of the form (6.101), taking into account the amplitude $2p_0/k\pi$, and writing $\omega t = 2\pi kt/T$. The results of a Fourier series solution for the radial stress, taking 500 terms, are shown in Fig. 6.14. The dots in the figure indicate the elastostatic solution, that should be obtained for $t/T \to \infty$.

6.4 Radial Propagation of Shear Waves

For the analysis of the transmission of vertical forces from a foundation pile to the surrounding soil, it may be interesting to consider the propagation of shear waves through the soil, in radial direction. If it is assumed that there are no vertical deformations in the layer, and that its only mode of displacement is a vertical displacement w, which is a function of the radial distance r and the time t only, the basic differential equation is

$$\frac{\partial^2 w}{\partial r^2} + \frac{1}{r}\frac{\partial w}{\partial r} = \frac{1}{c^2}\frac{\partial^2 w}{\partial t^2}, \quad (6.102)$$

Fig. 6.15 Shear stresses
acting upon a ring

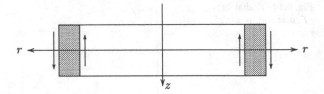

where now c is the velocity of shear waves,

$$c = \sqrt{\mu/\rho}.$$ (6.103)

Equation (6.102) can be derived from the equation of conservation of momentum, in vertical direction, of a ring of radius r, see Fig. 6.15, which requires that

$$\frac{\partial(2\pi r \tau)}{\partial r} = 2\pi r \rho \frac{\partial^2 w}{\partial t^2}.$$ (6.104)

Using an appropriate form of Hooke's law for this schematization,

$$\tau = \mu \frac{\partial w}{\partial r},$$ (6.105)

to relate the shear stress τ to the vertical displacement w, (6.102) is obtained. It should be noted that these equations are valid only if the vibrating soil layer is not supported at its base. The only means of stress transfer is in radial direction, through shear stresses.

6.4.1 Sinusoidal Vibrations at the Cavity Boundary

As an example the case of a sinusoidal variation of the displacements at the boundary of a cylindrical cavity in an infinite medium may be considered. The boundary condition is supposed to be

$$r = a \; : \; w = w_0 \sin(\omega t),$$ (6.106)

where ω is the frequency of the vibration.

As in the case of radial compression waves, considered in the previous section, the solution may be obtained by separation of variables. The solution proceeds in very much the same way, except that the Bessel functions now are of order zero. The determination of the integration constants again requires the radiation condition at infinity in order to eliminate the incoming wave. The final solution is

$$\frac{w}{w_0} = \frac{J_0(\omega a/c)J_0(\omega r/c) + Y_0(\omega a/c)Y_0(\omega r/c)}{J_0^2(\omega a/c) + Y_0^2(\omega a/c)} \sin(\omega t)$$

$$- \frac{J_0(\omega a/c)Y_0(\omega r/c) - Y_0(\omega a/c)J_0(\omega r/c)}{J_0^2(\omega a/c) + Y_0^2(\omega a/c)} \cos(\omega t).$$ (6.107)

For large values of the radial coordinate r the solution can be approximated by

$$r \to \infty : \quad \frac{w}{w_0} \approx -\frac{J_0(\omega a/c)\sqrt{2c/\pi \omega r}}{J_0^2(\omega a/c) + Y_0^2(\omega a/c)} \sin\left[\omega(t - r/c) + \frac{1}{4}\pi\right]$$
$$-\frac{Y_0(\omega a/c)\sqrt{2c/\pi \omega r}}{J_0^2(\omega a/c) + Y_0^2(\omega a/c)} \cos\left[\omega(t - r/c) + \frac{1}{4}\pi\right]. \quad (6.108)$$

Because the time t appears in this expression only in the form of the factor $(t - r/c)$ it can be seen that the solution indeed satisfies the radiation condition that it represents an outgoing wave at infinity.

The shear stress τ can be found from the relation (6.105). For the total force T, acting on a pile of length L, this gives

$$T = \{K \sin(\omega t) + \omega C \cos(\omega t)\} w_0, \quad (6.109)$$

where

$$K = 2\pi \mu L \left(\frac{\omega a}{c}\right) \frac{J_0(\omega a/c)J_1(\omega a/c) + Y_0(\omega a/c)Y_1(\omega a/c)}{J_0^2(\omega a/c) + Y_0^2(\omega a/c)}, \quad (6.110)$$

and

$$C = \left(\frac{4\mu L}{\omega}\right) \frac{1}{J_0^2(\omega a/c) + Y_0^2(\omega a/c)}. \quad (6.111)$$

These quantities can be considered as the equivalent dynamic spring and damping of the soil system, as acting upon the mass of the pile. They are shown in graphical form, as a function of the frequency ω, in Figs. 6.16 and 6.17. It appears from Fig. 6.16 that for high frequencies the equivalent stiffness is

$$\frac{\omega a}{c} \gg 1 : \quad K \approx \pi \mu L. \quad (6.112)$$

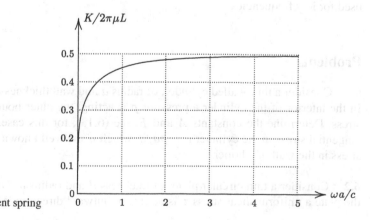

Fig. 6.16 Equivalent spring

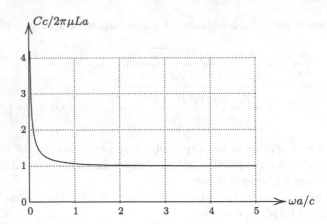

Fig. 6.17 Equivalent damping

This relation can also be obtained by an asymptotic expansion of the general formula (6.110) for large values of the parameter $\omega a/c$. For small frequencies, which correspond to the static case, the equivalent spring stiffness is zero. This is due to the circumstance that the circular plate is not supported.

The equivalent damping can be approximated for large values of the frequency by

$$\frac{\omega a}{c} \gg 1 : C \approx 2\pi \mu La/c, \tag{6.113}$$

which is a constant. It appears from Fig. 6.17 that this approximation can be used for practically all values of the frequency with reasonable accuracy. Only for very small frequencies the damping is larger.

It may be noted that the approximations obtained above indicate that in the analysis of the behaviour of foundation piles the effect of the generation of shear waves due to the transmission of friction to the ground can be approximated, at least for high frequencies, by a spring and a damper, with constant properties, as indicated by (6.112) and (6.113). It should also be noted that these approximations cannot be used for low frequencies.

Problems

6.1 Consider a thin-walled cylinder, of radius a and wall thickness d, with $d \ll a$. In the interior of the cylinder a pressure p is acting, the outer boundary is free of stress. Determine the constants A and B, see (6.13), for this case. Show that the tangential stress in the cylinder is $-pa/d$, which is the well known formula for the stress in the wall of a boiler.

6.2 Consider a thin circular plate, of thickness d and radius a. On the surface of the plate a uniform shear stress τ is acting, in outward direction. Modify the basic equation (6.9) to take this shear stress into account, and solve the modified differential equation. Calculate the stress in the center of the plate.

6.3   Figure 6.6 was prepared using the value 0.2 for the parameter $\omega a/c$. Construct a similar figure using a different value for this parameter, for instance $\omega a/c = 0.1$, or $\omega a/c = 1$.

Chapter 7
Spherical Waves

In this chapter a number of spherically symmetric problems from the theory of elasticity are considered, especially the problem of the expansion of a spherical cavity. This can be considered as a first approximation for the analysis of the influence of a local disturbance in an infinite homogeneous elastic body. Both static and dynamic loading will be considered.

7.1 Static Problems

7.1.1 Basic Equations

Figure 7.1 shows an element of material in a spherical coordinate system. If the radial coordinate is denoted by r, the angle in the x, y-plane by θ, and the angle with the vertical axis by ψ, then the volume of the element is $r^2 \, dr \, d\theta \, d\psi \, \cos(\psi)$. It is assumed that the displacement field is spherically symmetrical, so that there are no shear stresses acting upon the element, and the tangential stress σ_{tt} is independent of the orientation of the plane. The stresses acting upon the element are indicated in the figure. The only non-trivial equation of equilibrium now is the one in radial direction,

$$\frac{\partial \sigma_{rr}}{\partial r} + \frac{2(\sigma_{rr} - \sigma_{tt})}{r} = 0. \tag{7.1}$$

The stresses can be related to the strains by Hooke's law,

$$\sigma_{rr} = \lambda e + 2\mu\varepsilon_{rr}, \tag{7.2}$$

$$\sigma_{tt} = \lambda e + 2\mu\varepsilon_{tt}, \tag{7.3}$$

where e is the volume strain,

$$e = \varepsilon_{rr} + 2\varepsilon_{tt}, \tag{7.4}$$

A. Verruijt, *An Introduction to Soil Dynamics*,
Theory and Applications of Transport in Porous Media 24,
© Springer Science+Business Media B.V. 2010

Fig. 7.1 Element in spherical coordinates

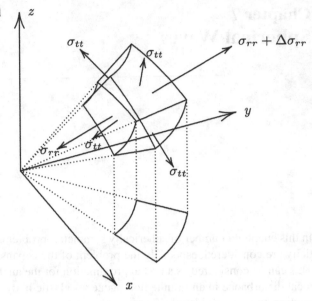

and λ and μ are the elastic coefficients (Lamé constants),

$$\lambda = \frac{\nu E}{(1+\nu)(1-2\nu)},$$ (7.5)

$$\mu = \frac{E}{2(1+\nu)}.$$ (7.6)

The strains ε_{rr} and ε_{tt} can be related to the radial displacement u_r by the relations

$$\varepsilon_{rr} = \frac{\partial u_r}{\partial r},$$ (7.7)

$$\varepsilon_{tt} = \frac{u_r}{r}.$$ (7.8)

The volume strain can now also be written as

$$e = \frac{\partial u_r}{\partial r} + \frac{2u_r}{r}.$$ (7.9)

Substitution of (7.2)–(7.9) into (7.1) gives

$$\frac{\partial^2 u}{\partial r^2} + \frac{2}{r}\frac{\partial u}{\partial r} - \frac{2u}{r^2} = 0.$$ (7.10)

This is the basic differential equation for spherically symmetric elastic deformations. As in the case of cylindrical deformations the elastic properties of the material do not appear in this equation. This means that if the boundary conditions can all be expressed in terms of the displacement, the solution will be independent of the elastic properties.

7.1.2 General Solution

The general solution of the differential equation (7.10) is

$$u = Ar + \frac{B}{r^2}, \tag{7.11}$$

where A and B are integration constants, to be determined from the boundary conditions.

The general expression for the volume strain, corresponding to the solution (7.11) is

$$e = 3A. \tag{7.12}$$

Because the volume strain appears to be independent of the constant B it can be concluded that the deformation field corresponding to the basic solution B/r^2 is *isochoric*, i.e. of constant volume. The deformation field corresponding to the other basic solution Ar appears to lead to a homogeneous volume strain, independent of the radial coordinate r. Thus the volume change is always homogeneous throughout the spherical body.

The stresses σ_{rr} and σ_{tt} can be expressed as

$$\sigma_{rr} = (3\lambda + 2\mu)A - 4\mu\frac{B}{r^3}, \tag{7.13}$$

$$\sigma_{tt} = (3\lambda + 2\mu)A + 2\mu\frac{B}{r^3}, \tag{7.14}$$

Some examples will be given in the next section.

7.1.3 Examples

Example 1: Sphere Under External Pressure

Of the two basic solutions the solution with the coefficient B is singular in the origin. Thus for a massive sphere, which includes the origin $r = 0$, this solution must vanish, to prevent the stresses and displacements from becoming singular. If the boundary condition at the outer boundary of the sphere is

$$r = a \ : \ \sigma_{rr} = -p, \tag{7.15}$$

see Fig. 7.2, then the two constants are

$$A = -\frac{p}{(3\lambda + 2\mu)}, \tag{7.16}$$

$$B = 0. \tag{7.17}$$

Fig. 7.2 Sphere under external pressure

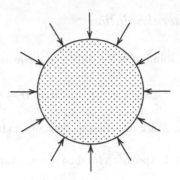

The stresses now are, in the entire sphere,

$$\sigma_{rr} = \sigma_{tt} = -p. \tag{7.18}$$

The displacement field is

$$u = -\frac{pr}{(3\lambda + 2\mu)}. \tag{7.19}$$

Thus the stresses (and the strains) in the sphere are homogeneous, and the displacement field is such that the radial displacement increases linearly with the distance from the origin.

Example 2: Hollow Sphere Under External Pressure

For a hollow sphere under external pressure, the boundary conditions are

$$r = a \; : \; \sigma_{rr} = 0, \tag{7.20}$$

$$r = b \; : \; \sigma_{rr} = -p, \tag{7.21}$$

where it has been assumed that $a < b$.

In this case the constants A and B are found to be

$$A = -\frac{pb^3}{(3\lambda + 2\mu)(b^3 - a^3)}, \tag{7.22}$$

$$B = -\frac{pa^3b^3}{4\mu(b^3 - a^3)}. \tag{7.23}$$

The stresses now are

$$\sigma_{rr} = -p\frac{1 - a^3/r^3}{1 - a^3/b^3}, \tag{7.24}$$

$$\sigma_{tt} = -p\frac{1 + a^3/(2r^3)}{1 - a^3/b^3}. \tag{7.25}$$

Fig. 7.3 Hollow sphere
under external pressure

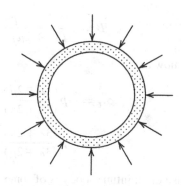

Fig. 7.4 Sphere with rigid
inclusion

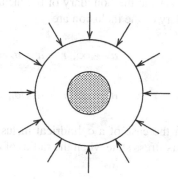

It can easily be seen that the expression (7.24) satisfies the boundary conditions
(7.20) and (7.21).

A special case occurs when the outer boundary is located at infinity. This is the
case of a very large body with a small spherical hole. In this case $b \to \infty$. At infinity
($r \to \infty$) all stresses then approach the limiting value $-p$. At the boundary of the
hole the radial stresses σ_{rr} are zero, but the tangential stress σ_{tt} at the boundary of
the hole then is $-1.5p$. Thus the stress concentration factor in this case is 1.5.

Example 3: Sphere with Rigid Inclusion

For a sphere under external pressure, with a rigid spherical inclusion, see Fig. 7.4,
the boundary conditions are

$$r = a \; : \; u_r = 0, \tag{7.26}$$

$$r = b \; : \; \sigma_{rr} = -p. \tag{7.27}$$

In this case the constants A and B are found to be

$$A = -\frac{p}{(3\lambda + 2\mu) + 4\mu a^3/b^3}, \tag{7.28}$$

$$B = +\frac{pa^3}{(3\lambda + 2\mu) + 4\mu a^3/b^3}. \tag{7.29}$$

The stresses now are

$$\sigma_{rr} = -p\frac{(3\lambda + 2\mu) + 4\mu a^3/r^3}{(3\lambda + 2\mu) + 4\mu a^3/b^3}, \tag{7.30}$$

$$\sigma_{tt} = -p\frac{(3\lambda + 2\mu) - 2\mu a^3/r^3}{(3\lambda + 2\mu) + 4\mu a^3/b^3}. \tag{7.31}$$

Again the case of an infinite body is of some special interest. In this case the radial stress is not zero at the boundary of the inclusion, of course. Actually, the stresses at the boundary of the inclusion are

$$a \to \infty, \ r = a \ : \ \sigma_{rr} = -p\frac{3\lambda + 6\mu}{3\lambda + 2\mu}, \tag{7.32}$$

$$a \to \infty, \ r = a \ : \ \sigma_{tt} = -p\frac{3\lambda}{3\lambda + 2\mu}. \tag{7.33}$$

Again, as in the case of a cylindrical inclusion, the solution does not tend to the homogeneous stress state when the radius of the inclusion becomes infinitely small.

Example 4: Cavity Expansion

An interesting problem is the case of expansion of a spherical cavity in an infinite body, see Fig. 7.5. In this case the boundary conditions are

$$r \to \infty \ : \ \sigma_{rr} = 0, \tag{7.34}$$

$$r = a \ : \ \sigma_{rr} = -p. \tag{7.35}$$

The constants A and B are now found to be

$$A = 0, \tag{7.36}$$

$$B = \frac{pa^3}{4\mu}. \tag{7.37}$$

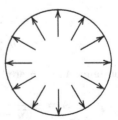

Fig. 7.5 Cavity expansion

The stresses now are

$$\sigma_{rr} = -p\frac{a^3}{r^3}, \tag{7.38}$$

$$\sigma_{tt} = p\frac{a^3}{2r^3}. \tag{7.39}$$

The displacement field is

$$u = \frac{pa^3}{4\mu r^2}. \tag{7.40}$$

Of special interest is the displacement at the boundary of the cavity,

$$u_a = \frac{pa}{4\mu}. \tag{7.41}$$

If this displacement can be measured, the value of the shear modulus μ can be obtained.

It may be noted that in this case the volume change is zero, because the constant A is zero, see (7.12). For soil mechanics practice this implies that in a water-saturated linear elastic medium no pore water pressures will be generated.

An interesting aspect of the solution for this problem of cavity expansion is that for $r \to \infty$ the displacements tend to zero as $1/r^2$, and the stresses tend to zero as $1/r^3$. This means that the displacements, and especially the stresses, decrease very fast with the radial distance. At a distance of 4 times the radius of the cavity, for instance, the stresses are a factor 64 times smaller than the stress at the boundary of the cavity, or, in other words, the stresses have been reduced to a level of about 1.5%. It will appear later that in the dynamic case this is quite different.

7.2 Dynamic Problems

In the dynamic case the basic equilibrium equation (7.1) must be extended with an inertia term,

$$\frac{\partial \sigma_{rr}}{\partial r} + \frac{2(\sigma_{rr} - \sigma_{tt})}{r} = \rho\frac{\partial^2 u}{\partial t^2}. \tag{7.42}$$

Substitution of (7.2)–(7.8) into this equation gives

$$\frac{\partial^2 u}{\partial r^2} + \frac{2}{r}\frac{\partial u}{\partial r} - \frac{2u}{r^2} = \frac{1}{c_p^2}\frac{\partial^2 u}{\partial t^2}, \tag{7.43}$$

where

$$c_p^2 = \frac{\lambda + 2\mu}{\rho}. \tag{7.44}$$

Equation (7.43) is the basic differential equation for spherically symmetric dynamic elastic deformations. The differential equation can be solved by various methods, such as separation of variables or the Laplace transform method. In this way general types of boundary value problems can be solved. It should be noted that the quantity c_p, as defined by (7.44), is the propagation velocity of compression waves in an elastic medium. It should not come as a surprise that for spherically symmetric waves the characteristic velocity is the velocity of compression waves.

7.2.1 Propagation of Waves

A simple method of solution for wave propagation problems has been given by Hopkins (1960). This solution can be obtained by observing that the displacement field is irrotational, so that one may introduce a displacement potential ϕ such that

$$u = \frac{\partial \phi}{\partial r}. \tag{7.45}$$

Substitution of (7.45) into (7.43) shows that ϕ must satisfy the equation

$$\frac{\partial^3 \phi}{\partial r^3} + \frac{2}{r}\frac{\partial^2 \phi}{\partial r^2} - \frac{2}{r^2}\frac{\partial \phi}{\partial r} = \frac{1}{c_p^2}\frac{\partial^3 \phi}{\partial r \partial t^2}. \tag{7.46}$$

This equation can be integrated once with respect to r, which gives

$$\frac{\partial^2 \phi}{\partial r^2} + \frac{2}{r}\frac{\partial \phi}{\partial r} = \frac{1}{c_p^2}\frac{\partial^2 \phi}{\partial t^2}. \tag{7.47}$$

Equation (7.47) can also be written as

$$\frac{\partial^2 (r\phi)}{\partial r^2} = \frac{1}{c_p^2}\frac{\partial^2 (r\phi)}{\partial t^2}. \tag{7.48}$$

This is the standard form of the one-dimensional wave equation. The solution can immediately be written down as

$$r\phi = f(r - c_p t) + g(r + c_p t), \tag{7.49}$$

where f and g are arbitrary functions, to be determined from the boundary conditions. The solution (7.49) represents two waves, a diverging and a converging one. Of special interest is the solution that travels in outward direction from a certain local disturbance,

$$r\phi = f(r - c_p t). \tag{7.50}$$

The corresponding displacement field is

$$u = \frac{1}{r}\frac{df}{dr} - \frac{f}{r^2}. \tag{7.51}$$

Some examples of solutions of particular problems using this general form of the solution, will be presented below.

7.2.2 Sinusoidal Vibrations at the Cavity Boundary

As an example the case of sinusoidal variations of the displacements at the boundary of a spherical cavity will be elaborated. Therefore the boundary condition is assumed to be

$$r = a : u = u_0 \sin(\omega t), \tag{7.52}$$

where ω is the frequency of the variation. The other boundary is supposed to be at infinity.

The solution of the problem can be described by a single function f, as in (7.50). It can be expected that this function can be written as

$$f = A \sin(\omega t - kr) + B \cos(\omega t - kr), \tag{7.53}$$

where A and B are constants, and $k = \omega/c_p$. The factor kr can also be written as $2\pi r/\lambda$, where now λ is the wave length. It appears that

$$k = 2\pi/\lambda. \tag{7.54}$$

The radial displacement corresponding to the solution (7.53) is, with (7.51),

$$u = -\frac{1}{r^2}\{(A - Bkr)\sin(\omega t - kr) + (B + Akr)\cos(\omega t - kr)\}. \tag{7.55}$$

The constants A and B can now be determined from the boundary condition (7.52). The result is

$$A = -u_0 a^2 \frac{\cos(ka) + (ka)\sin(ka)}{1 + (ka)^2}, \tag{7.56}$$

$$B = -u_0 a^2 \frac{\sin(ka) - (ka)\cos(ka)}{1 + (ka)^2}. \tag{7.57}$$

Substitution of these values into (7.55) gives, after some elementary mathematical operations,

$$\frac{u}{u_0} = \frac{a^2}{r^2[1 + (ka)^2]}\{[1 + (ka)(kr)]\sin[\omega t - k(r - a)]$$
$$+ k(r - a)\cos[\omega t - k(r - a)]\}. \tag{7.58}$$

This solution can also be written in the standard form

$$\frac{u}{u_0} = f\left(\frac{r}{a}\right)\sin(\omega t - \psi), \tag{7.59}$$

where $f(r/a)$ is a dimensionless damping factor and ψ is a phase angle. It can be shown that

$$f\left(\frac{r}{a}\right) = \left(\frac{a}{r}\right)^2 \sqrt{\frac{1 + (ka)^2(r/a)^2}{1 + (ka)^2}}, \tag{7.60}$$

and

$$\psi = ka\left(\frac{r-a}{a}\right) - \arctan\left[\frac{ka(r-a)/a}{1 + (ka)^2(r/a)}\right]. \tag{7.61}$$

The volume strain e can be obtained from the relation (7.9),

$$e = \frac{u_0(ka)^2}{r[1 + (ka)^2]}\{\sin[\omega t - k(r-a)] - (ka)\cos[\omega t - k(r-a)]\}. \tag{7.62}$$

The radial stress σ_{rr} can be obtained from (7.2),

$$\sigma_{rr} = \frac{(\lambda + 2\mu)u_0(ka)^2}{r[1 + (ka)^2]}\{\sin[\omega t - k(r-a)] - (ka)\cos[\omega t - k(r-a)]\}$$

$$- \frac{4\mu u_0 a^2}{r^3[1 + (ka)^2]}\{[1 + (ka)(kr)]\sin[\omega t - k(r-a)]$$

$$+ k(r-a)\cos[\omega t - k(r-a)]\}. \tag{7.63}$$

One of the most striking features of this solution is that at large distances from the cavity (i.e. for large values of r/a) the displacements and the stresses are much larger than in the static case. Both the radial displacement and the radial stress are of the order $O(1/r)$. This is in sharp contrast with the static case, in which the displacement and the stress tend to zero much faster, with a factor $1/r^3$.

The static solution can be obtained from the dynamic solution by assuming that the frequency ω is so small that ka and kr tend to zero. The solutions then reduce to

$$\omega \to 0 : \frac{u}{u_0} = \frac{a^2}{r^2}\sin(\omega t), \tag{7.64}$$

$$\omega \to 0 : \sigma_{rr} = -\frac{4\mu u_0 a^2}{r^3}\sin(\omega t). \tag{7.65}$$

This is in agreement with the static solution, as expressed by (7.38) and (7.40).

The relation between the pressure at the cavity boundary ($p = -\sigma_{rr}$) and the displacement of that boundary is, in the static case,

$$p = \frac{4\mu u_0}{a}. \tag{7.66}$$

In engineering practice it is often convenient to describe the response of a linear elastic system by a spring constant, writing

$$u = \frac{p}{c_e},$$ (7.67)

where c_e is the spring constant. It appears that in this case the equivalent spring constant is

$$c_e = \frac{4\mu}{a}.$$ (7.68)

Thus the equivalent spring constant is proportional to the shear modulus of the material, and inversely proportional to the radius of the cavity. The larger the cavity, the smaller the spring constant. A uniform pressure inside a large cavity results in a large displacement.

It is also interesting to investigate the behaviour of the pressure at the cavity boundary, in relation to the amplitude of the displacement u_0 for the general dynamic case. If the pressure inside the cavity is again denoted by p, which is the opposite of the stress σ_{rr} at the cavity boundary (i.e. for $r = a$), one obtains, from (7.63),

$$p = \frac{4\mu u_0}{a} \left\{ \left[1 - \left(\frac{\lambda + 2\mu}{4\mu} \right) \left(\frac{(ka)^2}{1 + (ka)^2} \right) \right] \sin(\omega t) \right.$$

$$\left. + \left(\frac{\lambda + 2\mu}{4\mu} \right) \left(\frac{(ka)^3}{1 + (ka)^2} \right) \cos(\omega t) \right\}.$$ (7.69)

The coefficients of the trigonometric functions in (7.69) are now written as a_1 and a_2, respectively,

$$a_1 = 1 - \left(\frac{\lambda + 2\mu}{4\mu} \right) \left(\frac{(ka)^2}{1 + (ka)^2} \right),$$ (7.70)

$$a_2 = \left(\frac{\lambda + 2\mu}{4\mu} \right) \left(\frac{(ka)^3}{1 + (ka)^2} \right).$$ (7.71)

The expression (7.69) can then also be written in the standard form

$$p = \frac{4\mu u_0}{a} \{ a_1 \sin(\omega t) + a_2 \cos(\omega t) \} = \frac{u_0}{c_d} \sin(\omega t + \psi),$$ (7.72)

where now c_d is the dynamic spring stiffness,

$$c_d = \frac{4\mu}{a} \sqrt{a_1^2 + a_2^2},$$ (7.73)

and ψ is the phase angle, defined by

$$\tan \psi = \frac{a_2}{a_1}.$$ (7.74)

Fig. 7.6 Dynamic response, amplitude

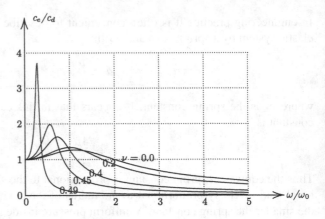

The dynamic spring constant c_d and the phase angle ψ depend upon the frequency ω through the parameter ka, which can also be written as

$$ka = \omega a / c_p = \omega / \omega_0, \tag{7.75}$$

where now

$$\omega_0 = \frac{c_p}{a} = \sqrt{\frac{\lambda + 2\mu}{\rho a^2}}. \tag{7.76}$$

This is a characteristic frequency of the system, depending upon the ratio of the elastic stiffness and the mass density, and upon the radius of the cavity. It has the form of the square root of the ratio of an elastic stiffness and a mass. A characteristic frequency of this type often exists in dynamic systems.

The dynamic spring constant c_d is shown, in the form of the ratio c_e/c_d, in Fig. 7.6, as a function of the dimensionless frequency ω/ω_0, or ka. This can also be interpreted as the amplitude of the dynamic response of the displacement of the cavity boundary to a periodic pressure of constant amplitude. The response curves have been plotted for various values of the Poisson ratio ν, which determines the ratio $(\lambda + 2\mu)/4\mu$. It appears that the response tends to zero when the frequency is very high, but for a certain low frequency there is a form of resonance, especially if Poisson's ratio is large. The order of magnitude of this resonance frequency is ω_0.

The phase angle ψ is shown graphically in Fig. 7.7, as a function of the frequency, and for various values of the Poisson ratio ν. It follows from the figure, but also from the formula (7.74), that for large values of the frequency (that is for very rapid vibrations), the phase angle tends towards $\pi/2$, indicating a very large amount of damping. In this case the damping cannot be the result of viscous or hysteretic damping of the material, as these effects have not been included. The cause of the damping must be the spreading of the energy over ever larger areas when the waves travel from the cavity. This form of damping is called *radiation damping*. In engineering practice this is often one of the most important causes of damping.

Fig. 7.7 Dynamic response, phase angle

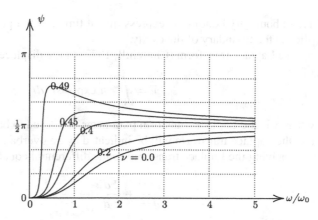

7.2.3 Propagation of a Shock Wave

Another approach to problems of elastodynamics is by using the Laplace transform method. This seems especially suited for the analysis of the propagation of a shock wave. This method will be used in this section to solve the problem of a shock wave propagated from a spherical cavity.

The problem is described by the equation of motion (7.43),

$$\frac{\partial^2 u}{\partial r^2} + \frac{2}{r}\frac{\partial u}{\partial r} - \frac{2u}{r^2} = \frac{1}{c_p^2}\frac{\partial^2 u}{\partial t^2}, \tag{7.77}$$

where, as before, c_p is the propagation velocity of compression waves,

$$c_p^2 = \frac{\lambda + 2\mu}{\rho}. \tag{7.78}$$

The stresses can be related to the radial displacement by the equations

$$\sigma_{rr} = \lambda e + 2\mu\frac{\partial u_r}{\partial r}, \tag{7.79}$$

$$\sigma_{tt} = \lambda e + 2\mu\frac{u_r}{r}, \tag{7.80}$$

where e is the volume strain,

$$e = \frac{\partial u_r}{\partial r} + \frac{2u_r}{r}. \tag{7.81}$$

The boundary conditions now are supposed to be:

$$r = a, \ t > 0 : \sigma_{rr} = -p, \tag{7.82}$$

$$r \to \infty, \ t > 0 : \sigma_{rr} = 0. \tag{7.83}$$

These boundary conditions express that at time $t = 0$ a pressure p is suddenly applied at the boundary of the cavity.

The Laplace transform (Churchill, 1972) of the displacement is defined by

$$\bar{u} = \int_0^\infty u \exp(-st)\,dt, \tag{7.84}$$

where s is the Laplace transform parameter, supposed to be positive. For all quantities the Laplace transform will be indicated by an overbar.

Applying the Laplace transform to the differential equation (7.77) gives

$$\frac{d^2\bar{u}}{dr^2} + \frac{2}{r}\frac{d\bar{u}}{dr} - \frac{2\bar{u}}{r^2} = k^2\bar{u}, \tag{7.85}$$

where

$$k = s/c_p. \tag{7.86}$$

The general solution of this differential equation is

$$\bar{u} = A\frac{1+kr}{(kr)^2}\exp(-kr) + B\frac{1-kr}{(kr)^2}\exp(+kr). \tag{7.87}$$

Because of the boundary condition at infinity the coefficient B can be assumed to be zero, so that the solution reduces to

$$\bar{u} = A\frac{1+kr}{(kr)^2}\exp(-kr). \tag{7.88}$$

The volume strain corresponding to this solution is

$$\bar{e} = -\frac{A}{r}\exp(-kr). \tag{7.89}$$

And the stress components are found to be

$$\bar{\sigma}_{rr} = -4\mu kA\frac{1+(kr)+m(kr)^2}{(kr)^3}\exp(-kr), \tag{7.90}$$

$$\bar{\sigma}_{tt} = 2\mu kA\frac{1+(kr)-(2d-1)(kr)^2}{(kr)^3}\exp(-kr), \tag{7.91}$$

where d is an additional elastic coefficient defined by

$$d = \frac{\lambda + 2\mu}{4\mu} = \frac{1-\nu}{2(1-2\nu)}. \tag{7.92}$$

The coefficient A must be determined from the boundary condition (7.82). The transformed boundary condition is

$$r = a \; : \; \bar{\sigma}_{rr} = -\frac{p}{s}. \tag{7.93}$$

With (7.90) the value of A is found to be

$$A = \frac{p}{4\mu ks} \frac{(ka)^3}{1 + (ka) + d(ka)^2} \exp(ka). \tag{7.94}$$

The Laplace transform of the radial displacement now is, with (7.88) and (7.94),

$$\bar{u} = \frac{pa^3}{4\mu sr^2} \frac{1 + sr/c_p}{1 + sa/c_p + d(sa/c_p)^2} \exp[-s(r-a)/c_p], \tag{7.95}$$

or

$$\bar{u} = \frac{pa^3}{4\mu sr^2} \frac{1 + xs}{1 + bs + db^2 s^2} \exp[-s(x-b)], \tag{7.96}$$

where $b = a/c_p$ and $x = r/c_p$.

Using the value (7.94) for the constant A the expressions for the transformed stresses are

$$\bar{\sigma}_{rr} = -\frac{pa^3}{sr^3} \frac{1 + xs + dx^2 s^2}{1 + bs + db^2 s^2} \exp[-s(x-b)], \tag{7.97}$$

$$\bar{\sigma}_{tt} = \frac{pa^3}{2sr^3} \frac{1 + xs - (2d-1)x^2 s^2}{1 + bs + db^2 s^2} \exp[-s(x-b)]. \tag{7.98}$$

The mathematical problem now remaining is to find the inverse transform of the expressions (7.96), (7.97) and (7.98). The displacement will first be elaborated.

Displacement

In order to perform the inverse Laplace transformation of the expression (7.96) it is first required to decompose the denominator into the form of a product of single factors, by writing

$$1 + bs + db^2 s^2 = db^2(s + s_1)(s + s_2), \tag{7.99}$$

where

$$s_1 = (1 + i\alpha)/2db, \tag{7.100}$$

$$s_2 = (1 - i\alpha/2db, \tag{7.101}$$

with

$$\alpha = \sqrt{4d - 1} = 1/\sqrt{1 - 2v}. \tag{7.102}$$

Using this decomposition equation (7.96) can be written as

$$\bar{u} = -\frac{pa^3}{4\mu db^2 r^2} \left\{ \frac{C_1}{s} + \frac{C_2}{s + s_1} + \frac{C_3}{s + s_2} \right\} \exp[-s(x-b)], \tag{7.103}$$

where

$$C_1 = \frac{1}{s_1 s_2} = db^2, \tag{7.104}$$

$$C_2 = \frac{1 - x s_1}{s_1(s_1 - s_2)}, \tag{7.105}$$

$$C_3 = \frac{1 - x s_2}{s_2(s_2 - s_1)}. \tag{7.106}$$

Inverse Laplace transformation is now a relatively simple operation, because the expression (7.103) consists of a summation of elementary fractions. The process is somewhat laborious, however, because the coefficients C_2 and C_3 are complex. After separation into real and imaginary parts the final result is

$$u = \frac{pa^3}{4\mu r^2} \left\{ 1 - \left[\cos\left(\frac{\alpha c_p \tau}{2da}\right) - \frac{2r - a}{\alpha a} \sin\left(\frac{\alpha c_p \tau}{2da}\right) \right] \exp\left(-\frac{c_p \tau}{2da}\right) \right\} H(\tau), \tag{7.107}$$

where

$$\tau = t - (r - a)/c_p. \tag{7.108}$$

The appearance of the Heaviside step function $H(\tau)$ in the solution (7.107) indicates that, as expected, a shock wave travels through the medium, with a velocity c_p, the velocity of compression waves.

The displacement u_0 of the inner boundary of the medium, at the radius of the circular cavity ($r = a$), is of particular interest. This is found to be

$$u_0 = \frac{pa}{4\mu} \left\{ 1 - \left[\cos\left(\frac{\alpha c_p \tau}{2da}\right) - \frac{1}{\alpha} \sin\left(\frac{\alpha c_p \tau}{2da}\right) \right] \exp\left(-\frac{c_p \tau}{2da}\right) \right\} H(t). \tag{7.109}$$

This function is shown, for three values of Poisson's ratio, in Fig. 7.8.

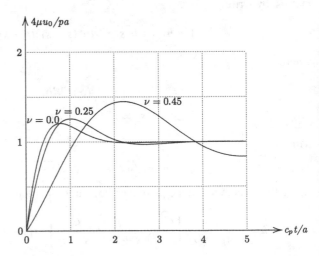

Fig. 7.8 Displacement of boundary

It may be interesting to further investigate the behaviour of the solution (7.107). It can be seen, for instance, that for large values of time the solution approaches the static solution $u = pa^3/4\mu r^2$. It also appears from the solution that at the arrival of the shock wave the displacement is continuous, but shortly after this arrival there is a relatively large effect, as indicated by the factor r/a in the term between brackets. This shows that during the passage of the shock wave the displacements are considerably larger than in the static case. It is left to the reader to plot the behaviour of the solution (7.107) for various values of the distance from the cavity.

Stresses

Using the decomposition (7.99), (7.97) and (7.98) can be written as

$$\bar{\sigma}_{rr} = -\frac{pa^3}{db^2r^3}\left\{\frac{D_1}{s} + \frac{D_2}{s+s_1} + \frac{D_3}{s+s_2}\right\}\exp[-s(x-b)], \qquad (7.110)$$

$$\bar{\sigma}_{tt} = \frac{pa^3}{2db^2r^3}\left\{\frac{E_1}{s} + \frac{E_2}{s+s_1} + \frac{E_3}{s+s_2}\right\}\exp[-s(x-b)], \qquad (7.111)$$

where

$$D_1 = \frac{1}{s_1 s_2} = db^2, \qquad (7.112)$$

$$D_2 = \frac{1 - xs_1 + dx^2 s_1^2}{s_1(s_1 - s_2)}, \qquad (7.113)$$

$$D_3 = \frac{1 - xs_2 + dx^2 s_2^2}{s_2(s_2 - s_1)}, \qquad (7.114)$$

$$E_1 = \frac{1}{s_1 s_2} = db^2, \qquad (7.115)$$

$$E_2 = \frac{1 - xs_1 - (2d-1)x^2 s_1^2}{s_1(s_1 - s_2)}, \qquad (7.116)$$

$$E_3 = \frac{1 - xs_2 + (2d-1)x^2 s_2^2}{s_2(s_2 - s_1)}. \qquad (7.117)$$

The inverse transformation of the expressions (7.110) and (7.111) is now relatively simple. After some mathematical manipulations the final results are

$$\sigma_{rr} = -\frac{pa^3}{r^3}\left\{1 + \left[\left(\frac{r^2 - a^2}{a^2}\right)\cos\left(\frac{\alpha c_p \tau}{2da}\right)\right.\right.$$
$$\left.\left. - \left(\frac{r-a}{a}\right)^2 \frac{1}{\alpha}\sin\left(\frac{\alpha c_p \tau}{2da}\right)\right]\exp\left(-\frac{c_p \tau}{2da}\right)\right\}H(\tau), \qquad (7.118)$$

$$\sigma_{tt} = \frac{pa^3}{2r^3}\left\{1 + \left[\left\{\left(\frac{r^2-a^2}{a^2}\right) - \left(\frac{3d-1}{d}\right)\left(\frac{r^2}{a^2}\right)\right\}\cos\left(\frac{\alpha c_p \tau}{2da}\right)\right.\right.$$
$$\left.\left. - \left\{\left(\frac{r-a}{a}\right)^2 - \left(\frac{3d-1}{d}\right)\left(\frac{r^2}{a^2}\right)\right\}\frac{1}{\alpha}\sin\left(\frac{\alpha c_p \tau}{2da}\right)\right]\exp\left(-\frac{c_p \tau}{2da}\right)\right\}H(\tau),$$

$$(7.119)$$

where, as before,

$$\alpha = \sqrt{4d - 1} = 1/\sqrt{1 - 2v}. \tag{7.120}$$

$$\tau = t - (r - a)/c_p. \tag{7.121}$$

The solutions indicate again that a shock wave is propagated through the medium at velocity c_p. Before the arrival of the shock wave the stresses are zero. The values of the stresses at the front of the wave can be obtained by letting $\tau \downarrow 0$. This gives

$$\tau \downarrow 0 : \sigma_{rr} = -\frac{pa}{r}, \tag{7.122}$$

$$\tau \downarrow 0 : \sigma_{tt} = -\frac{(2d-1)pa}{dr}. \tag{7.123}$$

Again, as in the case of a sinusoidal vibration, it is found that the dynamic stresses tend to zero as $1/r$ when $r \to \infty$.

After a very long time the influence of the shock wave has been attenuated. The stresses then are

$$\tau \to \infty : \sigma_{rr} = -\frac{pa^3}{r^3}, \tag{7.124}$$

$$\tau \to \infty : \sigma_{tt} = \frac{pa^3}{2r^3}. \tag{7.125}$$

These expressions are in agreement with the static solution. The static stresses tend to zero as $1/r^3$ at infinity. Again it may be noted that the dynamic stresses far from the cavity are much larger than the static stresses.

Problems

7.1   Consider a thin-walled sphere, of radius a and wall thickness d, with $d \ll a$. In the interior of the sphere a pressure p is acting, the outer boundary is free of stress. Determine the constants A and B, see (7.11), for this case, and determine the tangential stress in the wall of the sphere. Note that this is the problem of a pressurized balloon.

7.2   From (7.74) derive an asymptotic expression valid for large values of the frequency ω. Express the material constant in this expression into the Poisson ratio v.

7.3 In Fig. 7.8 the displacement of the cavity boundary is plotted as a function of time, using the dimensionless parameter $c_p t/a$. Replot this figure, now using a time scale based on Young's modulus E, i.e. using a dimensionless parameter $c_1 t/a$, where c_1 is defined by $c_1 = \sqrt{E/\rho}$. It may appear that the waves in the plots now have approximately equal periods.

7.4 The solution (7.118) contains a damping factor $\exp(-c_p \tau/2ma)$. Normal values for the propagation speed of compression waves in soils are of the order 1000 m/s. Now estimate the duration of the shock generated from a cavity of radius 1 m.

7.5 In the solution (7.118) assume that $r \gg a$. Sketch the stress at a certain point as a function of time.

7.6 Redraw Fig. 7.6, using the parameter ω/ω_s as the independent variable, with $\omega_s = \sqrt{\mu/\rho a^2}$. If the resonance frequency now appears to be independent of μ it has been found that resonance is determined by the velocity of shear waves.

Chapter 8
Elastostatics of a Half Space

In soil mechanics it is often required to determine the stresses and deformations of a soil deposit under the influence of loads applied on the upper surface. As a first approximation it may be useful to consider an elastic half space, or, in the case of plane strain deformations, an elastic half plane, loaded on its upper surface, see Fig. 8.1. In this chapter some solutions are derived, for vertical loads. For the sake of completeness the basic equations of the theory of linear elasticity are included. The examples to be presented are the classical solutions for a point load and a line load on a half space (the problems of Boussinesq and Flamant), the solution for a uniform load on a circular area, and some mixed boundary value problems. The problems can be solved effectively by using Fourier transforms or Hankel transforms. These methods will be described briefly.

It can be expected that for the class of problems considered here, an elastic half space loaded by vertical loads on its surface, the vertical displacements are more important than the horizontal displacements. On the basis of this expectation, which is also confirmed by the analytical solutions that can be obtained for certain problems, an approximate method of solution can be developed by neglecting all horizontal displacements. This approximate method, which was first proposed by Westergaard (1938), is also presented in this chapter, and its results are compared with the complete analytical solution.

It should be noted that throughout this chapter the material is supposed to be homogeneous and isotropic, and linear elastic, so that its mechanical properties can

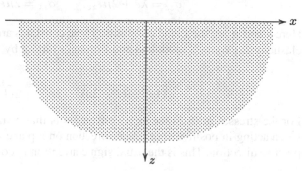

Fig. 8.1 Half space

A. Verruijt, *An Introduction to Soil Dynamics*,
Theory and Applications of Transport in Porous Media 24,
© Springer Science+Business Media B.V. 2010

be fully characterized by an elastic modulus E and Poisson's ratio ν, or some other combination. The strains are assumed to be small compared to 1.

8.1 Basic Equations of Elastostatics

The basic equations of the theory of elasticity are the conditions on the stresses, the strains, and the displacements in a linear elastic continuum. These are the conditions of equilibrium, the constitutive relations, and the compatibility conditions.

Let the stresses and displacements be described in a Cartesian coordinate system x, y, z. The components of the displacement vector in the three coordinate directions are denoted by u_x, u_y and u_z. If it is assumed that the displacement gradients are small compared to 1, then the expressions for the strains are

$$
\begin{aligned}
\varepsilon_{xx} &= \frac{\partial u_x}{\partial x}, & \varepsilon_{xy} &= \frac{1}{2}\left(\frac{\partial u_x}{\partial y} + \frac{\partial u_y}{\partial x}\right), \\
\varepsilon_{yy} &= \frac{\partial u_y}{\partial y}, & \varepsilon_{yz} &= \frac{1}{2}\left(\frac{\partial u_y}{\partial z} + \frac{\partial u_z}{\partial y}\right), \\
\varepsilon_{zz} &= \frac{\partial u_z}{\partial z}, & \varepsilon_{zx} &= \frac{1}{2}\left(\frac{\partial u_z}{\partial x} + \frac{\partial u_x}{\partial z}\right).
\end{aligned} \tag{8.1}
$$

The three normal strains ε_{xx}, ε_{yy} and ε_{zz} express the relative elongation of line elements in the three coordinate directions ($\Delta l/l$), and the three shear strains ε_{xy}, ε_{yz} and ε_{zx} express the deformation of right angles. The volume strain $e = \Delta V/V$ is the sum of the normal strains in the three coordinate directions,

$$
e = \varepsilon_{xx} + \varepsilon_{yy} + \varepsilon_{zz}. \tag{8.2}
$$

The stresses can be expressed into the strains by the generalized form of Hooke's law. For an isotropic material this is

$$
\begin{aligned}
\sigma_{xx} &= \lambda e + 2\mu\varepsilon_{xx}, & \sigma_{xy} &= 2\mu\varepsilon_{xy}, \\
\sigma_{yy} &= \lambda e + 2\mu\varepsilon_{yy}, & \sigma_{yz} &= 2\mu\varepsilon_{yz}, \\
\sigma_{zz} &= \lambda e + 2\mu\varepsilon_{zz}, & \sigma_{zx} &= 2\mu\varepsilon_{zx}.
\end{aligned} \tag{8.3}
$$

Here λ and μ are the Lamé constants. These constants are related to the modulus of elasticity E (Young's modulus) and Poisson's ratio ν by

$$
\lambda = \frac{\nu E}{(1+\nu)(1-2\nu)}, \qquad \mu = \frac{E}{2(1+\nu)}. \tag{8.4}
$$

For the stresses in (8.3) the sign convention is that a stress component is positive when acting in positive coordinate direction on a plane with an outward normal in positive direction. This is the usual sign convention in continuum mechanics, which

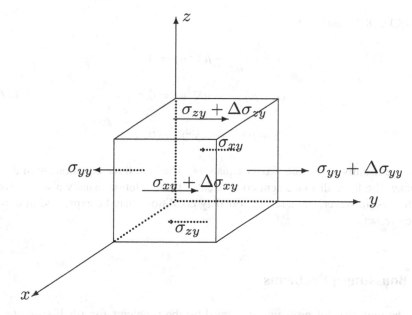

Fig. 8.2 Equilibrium of element

implies that tensile stresses are positive. It may be noted that in soil mechanics the sign convention is often just opposite, with compressive stresses being considered positive.

The stresses must satisfy the equations of equilibrium. In the absence of body forces these are

$$\frac{\partial \sigma_{xx}}{\partial x} + \frac{\partial \sigma_{yx}}{\partial y} + \frac{\partial \sigma_{zx}}{\partial z} = 0, \qquad \sigma_{xy} = \sigma_{yx},$$

$$\frac{\partial \sigma_{xy}}{\partial x} + \frac{\partial \sigma_{yy}}{\partial y} + \frac{\partial \sigma_{zy}}{\partial z} = 0, \qquad \sigma_{yz} = \sigma_{zy}, \qquad (8.5)$$

$$\frac{\partial \sigma_{xz}}{\partial x} + \frac{\partial \sigma_{yz}}{\partial y} + \frac{\partial \sigma_{zz}}{\partial z} = 0, \qquad \sigma_{zx} = \sigma_{xz}.$$

The second of these equations is illustrated in Fig. 8.2.

The stresses, strains, and displacements in an isotropic linear elastic body must satisfy all the equations given above, and in addition must satisfy the conditions on the boundary, which may specify the surface stress or the surface displacement, or a combination. For general methods of analysis the reader is referred to textbooks on the theory of elasticity, e.g. by Timoshenko and Goodier (1970), or Sokolnikoff (1956). In the next sections some special solutions will be presented.

For the purpose of future reference it is convenient to express the equations of equilibrium in terms of the displacements. If it is assumed that the parameters λ and μ are constants (which means that the material is homogeneous), one obtains from (8.1), (8.3) and (8.5),

$$(\lambda + \mu)\frac{\partial e}{\partial x} + \mu \nabla^2 u_x = 0,$$

$$(\lambda + \mu)\frac{\partial e}{\partial y} + \mu \nabla^2 u_y = 0, \tag{8.6}$$

$$(\lambda + \mu)\frac{\partial e}{\partial z} + \mu \nabla^2 u_z = 0.$$

These are usually called the Navier equations. They are three equations with three
variables, the three displacement components. Their solution usually also involves
the stresses, however, because the boundary conditions may be expressed in terms
of the stresses.

8.2 Boussinesq Problems

An important class of problems is formed by the problems for a half space ($z >$
0), bounded by the plane $z = 0$, loaded by vertical normal stresses on the surface
only, see Fig. 8.3. This is called the class of Boussinesq problems, after the French
scientist who published several solutions of such problems in 1885.

This type of problem can be solved conveniently by introducing a specially chosen displacement function ϕ (Green and Zerna, 1954), from which the displacements can be derived by the formulas

$$u_x = (1 - 2v)\frac{\partial \phi}{\partial x} + z\frac{\partial^2 \phi}{\partial x \partial z}, \tag{8.7}$$

$$u_y = (1 - 2v)\frac{\partial \phi}{\partial y} + z\frac{\partial^2 \phi}{\partial y \partial z}, \tag{8.8}$$

$$u_z = -2(1 - v)\frac{\partial \phi}{\partial z} + z\frac{\partial^2 \phi}{\partial z^2}. \tag{8.9}$$

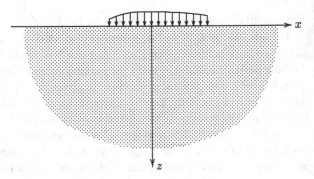

Fig. 8.3 Boussinesq problem

Substitution of these expressions into (8.6) shows that all three equations of equilibrium are identically satisfied, provided that the function ϕ satisfies the Laplace equation

$$\nabla^2 \phi = 0. \tag{8.10}$$

The advantage of the introduction of the function ϕ is that there now is only a single unknown function, which must satisfy a relatively simple differential equation, (8.10), for which many particular solutions and several general solution methods are available. That the solutions are useful appears when the stresses are expressed in the function ϕ. With (8.1), (8.3) and (8.10) one obtains for the normal stresses

$$\frac{\sigma_{xx}}{2\mu} = (1 - 2v)\frac{\partial^2 \phi}{\partial x^2} + z\frac{\partial^3 \phi}{\partial x^2 \partial z} - 2v\frac{\partial^2 \phi}{\partial z^2}, \tag{8.11}$$

$$\frac{\sigma_{yy}}{2\mu} = (1 - 2v)\frac{\partial^2 \phi}{\partial y^2} + z\frac{\partial^3 \phi}{\partial y^2 \partial z} - 2v\frac{\partial^2 \phi}{\partial z^2}, \tag{8.12}$$

$$\frac{\sigma_{zz}}{2\mu} = -\frac{\partial^2 \phi}{\partial z^2} + z\frac{\partial^3 \phi}{\partial z^3}. \tag{8.13}$$

For the shear stresses the following expressions are obtained

$$\frac{\sigma_{xy}}{2\mu} = (1 - 2v)\frac{\partial^2 \phi}{\partial x \partial y} + z\frac{\partial^3 \phi}{\partial x \partial y \partial z}, \tag{8.14}$$

$$\frac{\sigma_{yz}}{2\mu} = z\frac{\partial^3 \phi}{\partial y \partial z^2}, \tag{8.15}$$

$$\frac{\sigma_{zx}}{2\mu} = z\frac{\partial^3 \phi}{\partial x \partial z^2}. \tag{8.16}$$

From the last two equations it can be seen that on the surface $z = 0$ the shear stresses are always zero,

$$z = 0 : \sigma_{zx} = \sigma_{zy} = 0. \tag{8.17}$$

This means that the function ϕ can only be used for problems for which the plane $z = 0$ is free from shear stresses. This is an essential restriction. On the other hand, this restriction appears to lead to a relatively simple mathematical problem, namely the solution of the Laplace equation (8.10). On the boundary $z = 0$ the stress σ_{zz} may be prescribed, or the displacement u_z. On the surface $z = 0$ the expression for the vertical displacement reduces to

$$z = 0 : u_z = -2(1 - v)\frac{\partial \phi}{\partial z}, \tag{8.18}$$

and the expression for the vertical normal stress reduces to

$$z = 0 : \sigma_{zz} = -2\mu\frac{\partial^2 \phi}{\partial z^2}. \tag{8.19}$$

Thus, if the displacement or the stress on the surface is given, this means that either the first or the second derivative of the displacement function ϕ is known. In the next sections a number of solutions will be presented.

8.2.1 Concentrated Force

A classical problem, the solution of which was first given by Boussinesq, is the problem of a concentrated point force on the half space $z > 0$, see Fig. 8.4.
The solution is assumed to be given by the function

$$\phi = -\frac{P}{4\pi\mu}\ln(z + R), \tag{8.20}$$

where

$$R = \sqrt{x^2 + y^2 + z^2}. \tag{8.21}$$

It can easily be verified that this function indeed satisfies the differential equation (8.10). That it satisfies the correct boundary conditions is not immediately obvious, but may be verified by considering the stress field.

Differentiation of ϕ with respect to z gives

$$\frac{\partial\phi}{\partial z} = -\frac{P}{4\pi\mu}\frac{1}{R}, \tag{8.22}$$

$$\frac{\partial^2\phi}{\partial z^2} = \frac{P}{4\pi\mu}\frac{z}{R^3}, \tag{8.23}$$

$$\frac{\partial^3\phi}{\partial z^3} = \frac{P}{4\pi\mu}\left(\frac{1}{R^3} - 3\frac{z^2}{R^5}\right). \tag{8.24}$$

The vertical normal stress σ_{zz} is now found to be, with (8.13),

$$\sigma_{zz} = -\frac{3P}{2\pi}\frac{z^3}{R^5}. \tag{8.25}$$

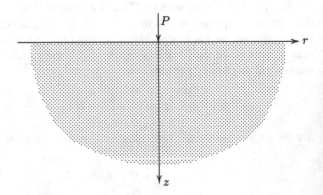

Fig. 8.4 Concentrated force
on half space

On the surface $z = 0$ this stress is zero everywhere, except in the origin, where the stress is infinitely large. That the solution is indeed the correct one can be verified by integration of the stress over the surface are. This gives

$$\int_{-\infty}^{\infty} \int_{-\infty}^{\infty} \sigma_{zz} \, dx \, dy = -P. \tag{8.26}$$

Every horizontal plane transfers a vertical force P, as required.

The vertical displacement is, with (8.8),

$$u_z = \frac{P}{4\pi \mu R} \left[2(1 - v) + \frac{z^2}{R^2} \right]. \tag{8.27}$$

On the surface $z = 0$ the displacement is, expressed in terms of E and v,

$$z = 0 : u_z = \frac{P(1 - v)}{2\pi \mu r} = \frac{P(1 - v^2)}{\pi Er}, \tag{8.28}$$

where $r = \sqrt{x^2 + y^2}$. In the origin the displacement is singular, as might be expected in this case of a concentrated force.

All the other stresses and displacements can of course also be derived from the solution (8.20). This is left as an exercise.

8.2.2 Uniform Load on a Circular Area

Starting from the elementary solution (8.20) many other interesting solutions can be obtained, see the literature (Timoshenko and Goodier, 1970; Sokolnikoff, 1956). As an example the displacement of the center of a circular area carrying a uniform load will be derived, see Fig. 8.5. The starting point of the considerations is the observation that a load of magnitude $p \, dA$ at a distance r from the origin results in a vertical displacement at the origin of

$$\frac{p \, dA(1 - v^2)}{\pi Er},$$

as follows immediately from the formula (8.28).

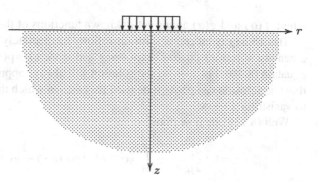

Fig. 8.5 Uniform load on circular area

The displacement due to a uniform load over a circular area with radius a can be obtained by integration over that area. Because $dA = r\,dr\,d\theta$ one obtains, after integration over θ from $\theta = 0$ to $\theta = 2\pi$, and integration over r from $r = 0$ to $r = a$,

$$r = 0, \ z = 0 \ : \ u_z = \frac{2pa(1 - v^2)}{E}. \tag{8.29}$$

This is a well known and useful formula. If the formula is expressed in the total load $P = \pi a^2 p$ it reads

$$r = 0, \ z = 0 \ : \ u_z = \frac{2P(1 - v^2)}{\pi E a}. \tag{8.30}$$

This shows that the displacement of a foundation plate can be reduced by making it larger, as one would expect intuitively. The relationship appears to be that the displacement is inversely proportional to the radius a of the plate, and not to the area of the plate, as one might perhaps have expected.

Actually, this result can also be obtained by considering the physical dimensions of the parameters of the problem. It can be expected that the displacement will be proportional to the load P, because the theory is linear, and it can also be expected that the displacement then will be inversely proportional to the modulus of elasticity. The only possibility to obtain a quantity having the dimension of a length then is that the displacement is proportional to P/Ea.

8.3 Fourier Transforms

A class of solutions can be found by the use of Fourier transforms (Sneddon, 1951). This method will be presented here, for the case of plane strain deformations $(u_y = 0)$.

The solution is sought in the form

$$\phi = \int_0^\infty \{f(\alpha)\cos(\alpha x) + g(\alpha)\sin(\alpha x)\} \exp(-\alpha z)\,d\alpha, \tag{8.31}$$

where $f(\alpha)$ and $g(\alpha)$ are as yet unknown functions of the variable α.

That (8.31) is indeed a solution follows immediately by substitution of the elementary solutions $\cos(\alpha x)\exp(-\alpha z)$ and $\sin(\alpha x)\exp(-\alpha z)$ into the differential equation (8.10). For $z \to \infty$ the solution will always approach zero, which suggests that this solution can perhaps be used for cases in which the stresses can be expected to vanish for $z \to \infty$.

With (8.13) one now obtains

$$z = 0 : \frac{\sigma_{zz}}{2\mu} = -\int_0^\infty \alpha^2 \{f(\alpha)\cos(\alpha x) + g(\alpha)\sin(\alpha x)\}\,d\alpha. \tag{8.32}$$

Suppose that the boundary condition is

$$z = 0, \quad -\infty < x < \infty \; : \; \sigma_{zz} = q(x), \tag{8.33}$$

in which $q(x)$ is a given function. Then the condition is that

$$\int_0^\infty \{A(\alpha)\cos(\alpha x) + B(\alpha)\sin(\alpha x)\}\, d\alpha = q(x), \tag{8.34}$$

where

$$A(\alpha) = -2\mu\alpha^2 f(\alpha), \tag{8.35}$$

$$B(\alpha) = -2\mu\alpha^2 g(\alpha). \tag{8.36}$$

The problem of determining the functions $A(\alpha)$ en $B(\alpha)$ from (8.34) is exactly the standard problem from the theory of Fourier transforms. The solution is given by the inversion theorem, which will not be derived here, see the literature on Fourier analysis (e.g. Sneddon, 1951). The final result is

$$A(\alpha) = \frac{1}{\pi} \int_{-\infty}^\infty q(t)\cos(\alpha t)\, dt, \tag{8.37}$$

$$B(\alpha) = \frac{1}{\pi} \int_{-\infty}^\infty q(t)\sin(\alpha t)\, dt. \tag{8.38}$$

The problem has now been solved, at least in principle, for an arbitrary surface load $q(x)$. In a specific case, with a given surface load $q(x)$ the integrals (8.37) and (8.38) must be evaluated, and then the results must be substituted into the general solution (8.31). Depending on the load function this may be a difficult mathematical problem. In the next section a simple example is given, in which all integrals can be evaluated analytically.

8.3.1 Line Load

As a first example the case of a line load on a half space will be considered (Flamant's problem), see Fig. 8.6. In this case the load function is

$$q(x) = \begin{cases} -F/(2\epsilon), & |x| < \epsilon, \\ 0, & |x| > \epsilon, \end{cases} \tag{8.39}$$

where it will later be assumed that $\epsilon \to 0$. From (8.37) and (8.38) it follows that

$$A(\alpha) = -\frac{F}{\pi\epsilon}\frac{\sin(\alpha\epsilon)}{\alpha},$$

$$B(\alpha) = 0.$$

Fig. 8.6 Line load on half space

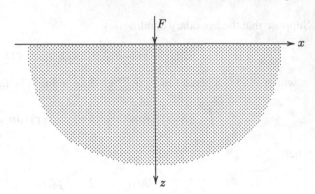

If $\epsilon \to 0$ this reduces to

$$A(\alpha) = -F/\pi, \tag{8.40}$$

$$B(\alpha) = 0. \tag{8.41}$$

With (8.35) and (8.36) one obtains

$$f(\alpha) = \frac{F}{2\pi \mu \alpha^2}, \tag{8.42}$$

$$g(\alpha) = 0. \tag{8.43}$$

The solution of the problem therefore is

$$\phi = \frac{F}{2\pi \mu} \int_0^\infty \frac{\cos(\alpha x) \exp(-\alpha z)}{\alpha^2} d\alpha. \tag{8.44}$$

Although this integral does not converge, due to the behavior of the factor α^2 in the denominator near $\alpha \to 0$, the result may well be useful, because the relevant quantities are derived expressions, such as the stresses, which require differentiation. It is found, for instance, that

$$\frac{\partial^2 \phi}{\partial x^2} = -\frac{F}{2\pi \mu} \int_0^\infty \cos(\alpha x) \exp(-\alpha z) d\alpha,$$

and this integral converges. The result is

$$\frac{\partial^2 \phi}{\partial x^2} = -\frac{F}{2\pi \mu} \frac{z}{x^2 + z^2}. \tag{8.45}$$

In a similar way one obtains

$$\frac{\partial^2 \phi}{\partial z^2} = \frac{F}{2\pi \mu} \frac{z}{x^2 + z^2}. \tag{8.46}$$

After another differentiation one obtains

$$\frac{\partial^3 \phi}{\partial z^3} = \frac{F}{2\pi\mu} \frac{x^2 - z^2}{(x^2 + z^2)^2} \tag{8.47}$$

and

$$\frac{\partial^3 \phi}{\partial x^2 \partial z} = -\frac{F}{2\pi\mu} \frac{x^2 - z^2}{(x^2 + z^2)^2}. \tag{8.48}$$

The expressions for the stresses are finally, using (8.11), (8.13) and (8.16),

$$\sigma_{xx} = -\frac{2F}{\pi} \frac{x^2 z}{(x^2 + z^2)^2}, \tag{8.49}$$

$$\sigma_{zz} = -\frac{2F}{\pi} \frac{z^3}{(x^2 + z^2)^2}, \tag{8.50}$$

$$\sigma_{xz} = -\frac{2F}{\pi} \frac{xz^2}{(x^2 + z^2)^2}. \tag{8.51}$$

These are usually called the Flamant formulas. Their form is somewhat simpler when using polar coordinates $x = r\cos\theta$ and $z = r\sin\theta$,

$$\sigma_{xx} = -\frac{2F}{\pi r} \sin\theta \cos^2\theta, \tag{8.52}$$

$$\sigma_{zz} = -\frac{2F}{\pi r} \sin^3\theta, \tag{8.53}$$

$$\sigma_{xz} = -\frac{2F}{\pi r} \sin^2\theta \cos\theta. \tag{8.54}$$

When the stress components are also transformed into polar coordinates the formulas are even simpler,

$$\sigma_{rr} = -\frac{2F}{\pi r} \sin\theta = -\frac{2Fz}{\pi r^2}, \tag{8.55}$$

$$\sigma_{\theta\theta} = 0, \tag{8.56}$$

$$\sigma_{r\theta} = 0. \tag{8.57}$$

It appears that the only non-vanishing stress is the radial stress, and that it decreases inversely proportional to the distance from the origin, and with the sine of the angle with the horizontal axis.

8.3.2 Strip Load

Another classical example is the case of a strip load on a half space, see Fig. 8.7. The solution of this problem can be found in many textbooks on theoretical soil mechanics. Here the solution will be determined using the Fourier cosine transform. In this case the boundary condition is

$$q(x) = \begin{cases} -p, & |x| < a, \\ 0, & |x| > a. \end{cases} \tag{8.58}$$

The displacement function ϕ now is found to be

$$\phi = \frac{p}{\pi\mu} \int_0^\infty \frac{\sin(\alpha a)\cos(\alpha x)\exp(-\alpha z)}{\alpha^3} \, d\alpha, \tag{8.59}$$

or

$$\phi = \frac{p}{\pi\mu} \int_0^\infty \frac{\{\sin[\alpha(x+a)] - \sin[\alpha(x-a)]\}\exp(-\alpha z)}{\alpha^3} \, d\alpha. \tag{8.60}$$

Again, this integral does not converge, but its second and third derivatives, which are needed to determine the stresses, do converge. Expressions for the stresses can be obtained using (8.11), (8.13) and (8.16), and a table of integral transforms. The result is

$$\sigma_{xx} = -\frac{p}{\pi}\left\{ \arctan\left(\frac{x+a}{z}\right) - \arctan\left(\frac{x-a}{z}\right) - \frac{(x+a)z}{(x+a)^2 + z^2} \right.$$
$$\left. + \frac{(x-a)z}{(x-a)^2 + z^2} \right\}, \tag{8.61}$$

$$\sigma_{zz} = -\frac{p}{\pi}\left\{ \arctan\left(\frac{x+a}{z}\right) - \arctan\left(\frac{x-a}{z}\right) + \frac{(x+a)z}{(x+a)^2 + z^2} \right.$$
$$\left. - \frac{(x-a)z}{(x-a)^2 + z^2} \right\}, \tag{8.62}$$

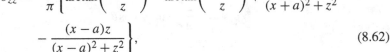

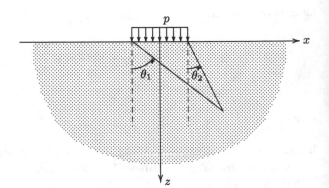

Fig. 8.7 Strip load on half space

$$\sigma_{xz} = \frac{p}{\pi} \left\{ \frac{z^2}{(x+a)^2 + z^2} - \frac{z^2}{(x-a)^2 + z^2} \right\}. \tag{8.63}$$

These are well known formulas, see for instance Sneddon (1951). They can also be written in the form

$$\sigma_{xx} = -\frac{p}{\pi} \{\theta_1 - \theta_2 - \sin\theta_1 \cos\theta_1 + \sin\theta_2 \cos\theta_2\}, \tag{8.64}$$

$$\sigma_{zz} = -\frac{p}{\pi} \{\theta_1 - \theta_2 + \sin\theta_1 \cos\theta_1 - \sin\theta_2 \cos\theta_2\}, \tag{8.65}$$

$$\sigma_{xz} = \frac{p}{\pi} \{\cos^2\theta_1 - \cos^2\theta_2\}, \tag{8.66}$$

where the angles θ_1 and θ_2 are indicated in Fig. 8.7.

8.4 Axially Symmetric Problems

Problems for an elastic half space loaded by a radially symmetric normal stress on the surface $z = 0$ can conveniently be solved by the Hankel transform method (Sneddon, 1951). The problem can be formulated in terms of the displacement function ϕ introduced in (8.8). This function must satisfy the Laplace equation (8.10). In axially symmetric coordinates this equation is

$$\frac{\partial^2 \phi}{\partial r^2} + \frac{1}{r} \frac{\partial \phi}{\partial r} + \frac{\partial^2 \phi}{\partial z^2} = 0. \tag{8.67}$$

The Hankel transform of the function ϕ is defined as

$$\Phi(\xi, z) = \int_0^\infty r \, \phi(r, z) \, J_0(r\xi) \, dr, \tag{8.68}$$

where $J_0(x)$ is the Bessel function of the first kind and order zero. The inverse transformation is (Sneddon, 1951)

$$\phi(r, z) = \int_0^\infty \xi \, \Phi(\xi, z) \, J_0(\xi r) \, d\xi. \tag{8.69}$$

The advantage of the Hankel transformation is that the operator

$$\frac{\partial^2}{\partial r^2} + \frac{1}{r} \frac{\partial}{\partial r}$$

is transformed into multiplication by $-\xi^2$. This means that the differential equation (8.67) becomes, after application of the Hankel transform,

$$\frac{d^2 \Phi}{dz^2} - \xi^2 \Phi = 0, \tag{8.70}$$

which is an ordinary differential equation. The general solution of this equation is

$$\Phi = A \exp(\xi z) + B \exp(-\xi z), \tag{8.71}$$

where the integration constants A and B may depend upon the transformation parameter ξ. In the half space $z > 0$ the constant A can be assumed to vanish.

If the boundary condition is

$$z = 0 : \sigma_{zz} = q(r), \tag{8.72}$$

then we obtain, with (8.19) and (8.71),

$$-2\mu B \xi^2 = \int_0^\infty r q(r) J_0(\xi r) \, dr, \tag{8.73}$$

from which the value of B can be determined. In the next two sections some examples will be given.

8.4.1 Uniform Load on a Circular Area

A well known classical problem is the problem of a uniform load over a circular area. This problem was already considered above, where the displacement of the origin was derived from a particular solution, see (8.29). Here the complete solution will be derived by a straightforward analysis.

In this case the load function $q(r)$ is

$$q(r) = \begin{cases} -p, & r < a, \\ 0, & r > a. \end{cases} \tag{8.74}$$

Substitution of this function into the general expression (8.73) gives

$$B = \frac{p}{2\mu\xi^2} \int_0^a r J_0(\xi r) \, dr. \tag{8.75}$$

This is a well known integral (Abramowitz and Stegun, 1964, 11.3.20). The result is

$$B = \frac{pa}{2\mu\xi^3} J_1(\xi a), \tag{8.76}$$

where $J_1(x)$ is the Bessel function of the first kind and order one.

The displacement function ϕ now is

$$\phi = \frac{pa}{2\mu} \int_0^\infty \frac{J_1(\xi a) \exp(-\xi z) J_0(\xi r)}{\xi^2} \, d\xi. \tag{8.77}$$

Although this integral itself cannot be evaluated, because of the logarithmic singularity in the origin, certain useful results can still be derived from it, because the physical quantities such as the displacements and the stresses must be derived from it by differentiation, and after differentiation the integrals may well converge, as indeed they do.

The vertical displacement of the surface can be obtained from the formula (8.18). With (8.77) this gives

$$z = 0 : u_z = \frac{pa(1-v)}{\mu} \int_0^\infty \frac{J_1(\xi a) J_0(\xi r)}{\xi} \, d\xi. \tag{8.78}$$

This integral is given in Appendix A, see (A.69). The result is

$$z = 0 : u_z = \frac{2pa(1-v)}{\pi \mu} \begin{cases} E(r^2/a^2), & r < a, \\ F(r^2/a^2), & r > a, \end{cases} \tag{8.79}$$

where

$$F(x) = \sqrt{x}[E(1/x) - (1 - 1/x)K(1/x)], \tag{8.80}$$

and where $K(x)$ and $E(x)$ are complete elliptic integrals of the first and second kind, respectively. A short list of values, adapted from Abramowitz and Stegun (1964), is given in Table A.2, in Appendix A. For $x = 0$ both $K(x)$ and $E(x)$ are equal to $\pi/2$. The result (8.79) is also given by Timoshenko and Goodier (1970).

Figure 8.8 shows the displacements of the surface in this case in graphical form. The displacement of the origin is of special interest. This is found to be

$$r = 0, \ z = 0 : u_z = u_0 = \frac{pa(1-v)}{\mu}, \tag{8.81}$$

which agrees with the expression (8.29) found before.

It should be noted that in this section the symbol E is used for the complete elliptic integral, whereas it has also been used earlier for Young's modulus of elasticity. The reader should carefully distinguish between the symbol E for Young's modulus, and the function $E(x)$, which denotes the complete elliptic integral of the second kind.

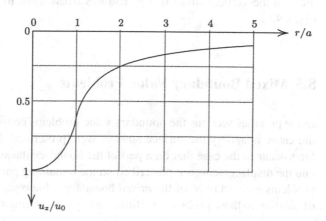

Fig. 8.8 Displacements of the surface

Fig. 8.9 Vertical stress σ_{zz} for $r = 0$

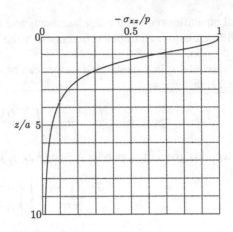

The vertical normal stress σ_{zz} is, with (8.13) and (8.77),

$$\frac{\sigma_{zz}}{p} = -\int_0^\infty a(1 + \xi z) J_1(\xi a) \exp(-\xi z) J_0(\xi r) \, d\xi. \tag{8.82}$$

Along the vertical axis, for $r = 0$, this integral reduces to

$$r = 0 : \quad \frac{\sigma_{zz}}{p} = -\int_0^\infty a(1 + \xi z) J_1(\xi a) \exp(-\xi z) \, d\xi, \tag{8.83}$$

which can be evaluated using a table of Laplace transforms (Churchill, 1972). The result is

$$r = 0 : \quad \frac{\sigma_{zz}}{p} = -1 + \frac{z^3}{(a^2 + z^2)^3}. \tag{8.84}$$

This is a well known formula, see e.g. Timoshenko and Goodier (1970). Just under the load the vertical stress is $-p$, and this stress tends to zero when $z \to \infty$, see Fig. 8.9.

8.5 Mixed Boundary Value Problems

In the previous sections the boundary value problems considered were such that on the entire boundary the surface stresses were prescribed. More complicated problems occur in the case that on a part of the boundary the surface stresses are given, and the displacements are prescribed on the remaining part of the boundary. These problems are said to be of the *mixed* boundary value type. In this section a method of solution to these problems is illustrated by considering some examples.

8.5.1 Rigid Circular Plate

In the first example a vertical load P is applied to a half space by a rigid circular plate of radius a, see Fig. 8.10. In this case the boundary conditions on the upper surface are that the shear stress $\sigma_{zx} = 0$ along the entire surface, and that

$$z = 0 : u_z = w_0, \quad 0 \leq r < a, \tag{8.85}$$

$$z = 0 : \sigma_{zz} = 0, \quad r > a, \tag{8.86}$$

where w_0 is the given vertical displacement of the plate.

If the elasticity equations are formulated using a potential function ϕ, as in the previous sections, the general solution for the half plane $z > 0$ is, with (8.69) and (8.71),

$$\phi(r, z) = \int_0^\infty \xi B(\xi) \exp(-\xi z) J_0(\xi r) \, d\xi, \tag{8.87}$$

where $B(\xi)$ is an unknown function, that should be determined from the boundary conditions. Using (8.18) and (8.19), the boundary conditions (8.85) and (8.86) can be expressed as

$$z = 0 : u_z = -2(1 - v) \frac{\partial \phi}{\partial z} = w_0, \quad 0 \leq r < a, \tag{8.88}$$

$$z = 0 : \sigma_{zz} = -2\mu \frac{\partial^2 \phi}{\partial z^2} = 0, \quad r > a. \tag{8.89}$$

With (8.87) these conditions can also be written as

$$\int_0^\infty \xi^2 B(\xi) J_0(\xi r) \, d\xi = f(r) = \frac{w_0}{2(1 - v)}, \quad 0 \leq r < a, \tag{8.90}$$

$$\int_0^\infty \xi^3 B(\xi) J_0(\xi r) \, d\xi = 0, \quad r > a. \tag{8.91}$$

A system of this form is denoted as a pair of *dual integral equations*. For the solution a method described by Sneddon (1966) will be used here, see also Selvadurai (1979).

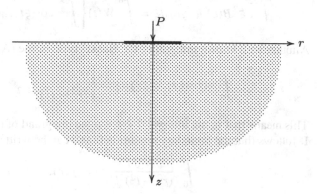

Fig. 8.10 Rigid circular plate on half space

In the example the function $f(r)$ is a constant, but the method applies equally well to the more general case that $f(r)$ is an arbitrary function. Some general aspects of the solution of dual integral equations are given in Appendix B.

The solution method consists of two steps, each addressing one of the dual integral equations. The first step is that it is assumed that the function $\xi^2 B(\xi)$ can be represented by the finite Fourier transform

$$\xi^2 B(\xi) = \int_0^a W(t) \cos(\xi t)\, dt, \tag{8.92}$$

where $W(t)$ is a new unknown function, defined in the interval $0 < t < a$. Substitution of (8.92) into the integral appearing in (8.91) gives

$$\int_0^\infty \xi^3 B(\xi) J_0(\xi r)\, d\xi = \int_0^\infty \xi \left\{ \int_0^a W(t) \cos(\xi t)\, dt \right\} J_0(\xi r)\, d\xi, \tag{8.93}$$

or, using partial integration,

$$\int_0^\infty \xi^3 B(\xi) J_0(\xi r)\, d\xi = W(a) \int_0^\infty \sin(\xi a) J_0(\xi r)\, d\xi$$
$$- \int_0^a W'(t) \left\{ \int_0^\infty \sin(\xi t) J_0(\xi r)\, d\xi \right\} dt. \tag{8.94}$$

A well known integral of the Hankel type is, see (A.76),

$$\int_0^\infty \sin(\xi t) J_0(\xi r)\, d\xi = \begin{cases} 0, & r > t, \\ (t^2 - r^2)^{-1/2}, & 0 \le r < t. \end{cases} \tag{8.95}$$

Because in the last integral in (8.94) the value of t is restricted to the interval $0 < t < a$, it follows that both integrals in the right hand side are zero if $r > a$, so that it can be concluded that the second boundary condition (8.91) is automatically satisfied, whatever the function $W(t)$ is.

The second step in the solution method is to determine the function $W(t)$ from the first boundary condition, (8.90). Substitution of (8.92) into the integral in this condition gives, again assuming that the order of integration may be interchanged,

$$\int_0^\infty \xi^2 B(\xi) J_0(\xi r)\, d\xi = \int_0^a W(t) \left\{ \int_0^\infty \cos(\xi t) J_0(\xi r)\, d\xi \right\} dt. \tag{8.96}$$

Another well known integral of the Hankel type is, see (A.77),

$$\int_0^\infty \cos(\xi t) J_0(\xi r)\, d\xi = \begin{cases} 0, & 0 \le r < t, \\ (r^2 - t^2)^{-1/2}, & r > t. \end{cases} \tag{8.97}$$

This means that in the interval $0 < t < a$ the integrand of (8.96) is zero if $r < t < a$. It follows that the boundary condition (8.90) can be written as

$$\int_0^r \frac{W(t)}{(r^2 - t^2)^{1/2}}\, dt = f(r), \quad 0 \le r < a. \tag{8.98}$$

This is an Abel integral equation. Its solution is (Sneddon, 1966, p. 42)

$$W(t) = \frac{2}{\pi} \frac{d}{dt} \int_0^t \frac{rf(r)}{(t^2 - r^2)^{1/2}} \, dr, \quad 0 < t < a. \tag{8.99}$$

In the example of a uniform displacement of a rigid plate the function $f(r)$ is, see (8.90),

$$f(r) = \frac{w_0}{2(1 - \nu)} \quad 0 \le r < a. \tag{8.100}$$

In this case the function $W(t)$ is found to be the constant

$$W(t) = \frac{w_0}{(1 - \nu)\pi}. \tag{8.101}$$

The function $\xi^2 B(\xi)$ now is, with (8.92),

$$\xi^2 B(\xi) = \frac{w_0}{(1 - \nu)\pi} \frac{\sin(\xi a)}{\xi}. \tag{8.102}$$

The solution for the potential function ϕ is, with (8.87)

$$\phi = \frac{w_0}{(1 - \nu)\pi} \int_0^\infty \frac{\sin(\xi a)}{\xi^2} \exp(-\xi z) J_0(\xi r) \, d\xi. \tag{8.103}$$

Of particular interest is the vertical normal stress at the surface. With (8.19) this is found to be

$$z = 0 : \sigma_{zz} = -2\mu \frac{\partial^2 \phi}{\partial z^2} = \frac{E w_0}{\pi (1 - \nu^2)} \int_0^\infty \sin(\xi a) J_0(\xi r) \, d\xi. \tag{8.104}$$

Using the integral (8.95) the boundary stress is

$$z = 0 : \sigma_{zz} = \begin{cases} 0, & r > a, \\ \frac{P}{2\pi a(a^2 - r^2)^{1/2}}, & 0 \le r < a, \end{cases} \tag{8.105}$$

where

$$P = \frac{2Eaw_0}{1 - \nu^2}, \tag{8.106}$$

the total force on the plate. The first part of (8.105) confirms the second boundary condition (8.86). The second part is a well known result of the theory of elasticity, see e.g. Timoshenko and Goodier (1970).

Another quantity of special interest is the vertical displacement of the surface. With (8.18) and (8.103) this is

$$z = 0 : u_z = \frac{2w_0}{\pi} \int_0^\infty \frac{\sin(\xi a)}{\xi} \exp(-\xi z) J_0(\xi r) \, d\xi. \tag{8.107}$$

Fig. 8.11 Surface
displacements, rigid circular
plate

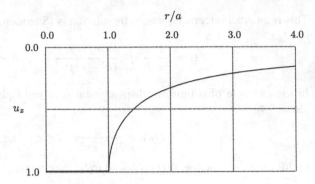

Using the integral (A.78) the displacement of the boundary is found to be

$$z = 0 : u_z = \begin{cases} w_0, & r < a, \\ \frac{2}{\pi} w_0 \arcsin(a/r), & r > a. \end{cases} \tag{8.108}$$

The first part of (8.108) confirms the first boundary condition (8.85). The second
part is a well known result of the theory of elasticity, see e.g. Sneddon (1951). The
surface displacements are shown, as a function of r/a, in Fig. 8.11.

Alternative Derivation

An alternative for the second step of the derivation, avoiding the Abel integral integration, is as follows.

The first boundary condition is, see (8.90),

$$\int_0^\infty \xi^2 B(\xi) J_0(\xi r) \, d\xi = f(r) = \frac{w_0}{2(1 - \nu)}, \quad 0 \le r < a. \tag{8.109}$$

The Bessel function $J_0(\xi r)$ can now be eliminated form this equation by using the
integral

$$\int_0^s \frac{r}{(s^2 - r^2)^{1/2}} J_0(\xi r) \, dr = \frac{\sin(\xi s)}{\xi}, \tag{8.110}$$

which can be considered to be the inverse form of the integral (8.95) when this is
considered as the Hankel transform of the function $\sin(\xi t)/\xi$.

It follows that (8.109) can also be written as

$$\int_0^\infty \xi B(\xi) \sin(\xi t) \, d\xi = \int_0^t \frac{r f(r) \, dr}{(t^2 - r^2)^{1/2}}, \quad 0 < t < a, \tag{8.111}$$

or, using the representation (8.92), and interchanging the order of integration,

$$\int_0^a W(s) \left\{ \int_0^\infty \frac{\sin(\xi t) \cos(\xi s)}{\xi} \, d\xi \right\} ds = \int_0^t \frac{r f(r) \, dr}{(t^2 - r^2)^{1/2}}, \quad 0 < t < a. \tag{8.112}$$

The integral between brackets is a well known Fourier integral,

$$\int_0^\infty \frac{\sin(\xi t)\cos(\xi s)}{\xi}\,d\xi = \begin{cases} \frac{\pi}{2}, & s < t, \\ 0, & s > t. \end{cases} \tag{8.113}$$

This means that (8.112) reduces to

$$\int_0^t W(s)\,ds = \frac{2}{\pi}\int_0^t \frac{rf(r)\,dr}{(t^2 - r^2)^{1/2}}, \quad 0 < t < a. \tag{8.114}$$

Differentiation with respect to t gives

$$W(t) = \frac{2}{\pi}\frac{d}{dt}\int_0^t \frac{rf(r)\,dr}{(t^2 - r^2)^{1/2}}, \quad 0 < t < a, \tag{8.115}$$

which is the same as the solution (8.99) derived above.

8.5.2 Penny Shaped Crack

Another class of problems involving mixed boundary conditions is concerned with the stress distribution in an elastic medium with a circular (*penny shaped*) crack, see e.g. Kassir and Sih (1975). For the problem of a crack in an infinite elastic plate, loaded by a uniform internal pressure p in the crack, the problem can be schematized as a problem for a half plane (see Fig. 8.12), with the boundary conditions

$$z = 0 : u_z = -2(1-v)\frac{\partial\phi}{\partial z} = 0, \quad r > a, \tag{8.116}$$

$$z = 0 : \sigma_{zz} = -2\mu\frac{\partial^2\phi}{\partial z^2} = -p, \quad 0 \le r < a. \tag{8.117}$$

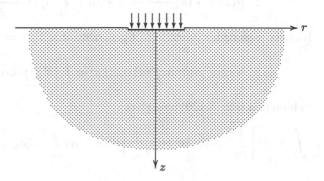

Fig. 8.12 Penny shaped crack

If the elasticity equations are again formulated using a potential function ϕ, the general solution for the half plane $z > 0$ is, with (8.69) and (8.71),

$$\phi(r, z) = \int_0^\infty \xi B(\xi) \exp(-\xi z) J_0(\xi r) \, d\xi, \tag{8.118}$$

where $B(\xi)$ is an unknown function, that should be determined from the boundary conditions. With (8.118) these conditions can also be written as

$$\int_0^\infty \xi^2 B(\xi) J_0(\xi r) \, d\xi = 0, \quad r > a, \tag{8.119}$$

$$\int_0^\infty \xi^3 B(\xi) J_0(\xi r) \, d\xi = g(r), \quad 0 \leq r < a, \tag{8.120}$$

where in the example considered $g(r) = p/2\mu$.

In order to solve the system of dual integral equations (see also Appendix B), in two steps, it is first assumed that the function $\xi^2 B(\xi)$ can be represented by the finite Fourier transform

$$\xi^2 B(\xi) = \int_0^a V(t) \sin(\xi t) \, dt, \tag{8.121}$$

where $V(t)$ is a new unknown function, defined in the interval $0 < t < a$. Substitution of (8.121) into the integral appearing in (8.119) gives, if the order if integration is interchanged,

$$\int_0^\infty \xi^2 B(\xi) J_0(\xi r) \, d\xi = \int_0^a V(t) \left\{ \int_0^\infty \sin(\xi t) J_0(\xi r) \, d\xi \right\} dt. \tag{8.122}$$

In the integral the variable t is always smaller than a, so that for $r > a$ it is certain that $r > t$, and then the integral is zero, see (A.76). This means that the boundary condition (8.119) is automatically satisfied by the representation (8.121).

In the second step of the solution the unknown function $V(t)$ is determined from the remaining boundary condition (8.120). For this purpose the definition (8.121) is first rewritten, by using integration by parts, as

$$\xi^2 B(\xi) = V(a) \frac{\cos(\xi a)}{\xi} - V(0) + \int_0^a V'(t) \frac{\cos(\xi t)}{\xi} \, dt. \tag{8.123}$$

It can be assumed, without loss of generality, that $V(0) = 0$, so that

$$\xi^3 B(\xi) = V(a) \cos(\xi a) + \int_0^a V'(t) \cos(\xi t) \, dt. \tag{8.124}$$

Substitution into (8.120) now gives

$$\int_0^a V'(t) \left\{ \int_0^\infty J_0(\xi r) \cos(\xi t) \, d\xi \right\} dt - V(a) \int_0^\infty J_0(\xi r) \cos(\xi a) \, d\xi = \frac{p}{2\mu}, \tag{8.125}$$

where it should be noted that $r < a$. The integrals are of the form of (A.76), hence

$$\int_0^\infty \cos(\xi t) J_0(\xi r) \, d\xi = \begin{cases} 0, & t > r, \\ (r^2 - t^2)^{-1/2}, & 0 < t < r. \end{cases} \tag{8.126}$$

This means that (8.125) reduces to

$$\int_0^r \frac{V'(t)}{(r^2 - t^2)^{1/2}} dt = g(r). \tag{8.127}$$

This is again an Abel integral equation. Its solution is, as before, see (8.99),

$$V'(t) = \frac{2}{\pi} \frac{d}{dt} \int_0^t \frac{rg(r)}{(t^2 - r^2)^{1/2}} dr, \quad 0 < t < a. \tag{8.128}$$

Integrating this equation gives, taking into account that it has already been assumed that $V(0) = 0$,

$$V(t) = \frac{2}{\pi} \int_0^t \frac{rg(r)}{(t^2 - r^2)^{1/2}} dt, \quad 0 < t < a. \tag{8.129}$$

In the example considered here $g(r) = p/2\mu$. In that case the result is

$$V(t) = \frac{pt}{\pi \mu}, \quad 0 < t < a. \tag{8.130}$$

In this case the function $B(\xi)$ is, with (8.121),

$$\xi^2 B(\xi) = \frac{pa}{\pi \mu \xi} \left\{ \frac{\sin(\xi a)}{\xi a} - \cos(\xi a) \right\}. \tag{8.131}$$

The potential function ϕ now is, with (8.118),

$$\phi = \frac{pa}{\pi \mu} \int_0^\infty \frac{\exp(-\xi z) J_0(\xi r)}{\xi^2} \left\{ \frac{\sin(\xi a)}{\xi a} - \cos(\xi a) \right\} d\xi. \tag{8.132}$$

One of the most interesting quantities is the normal stress at the surface. This is found to be

$$z = 0 : \sigma_{zz} = -2\mu \frac{\partial^2 \phi}{\partial z^2} = -\frac{2pa}{\pi} \int_0^\infty J_0(\xi r) \left\{ \frac{\sin(\xi a)}{\xi a} - \cos(\xi a) \right\} d\xi. \tag{8.133}$$

Using the Hankel transforms (A.77) and (A.78) this gives

$$z = 0, \ r < a : \sigma_{zz} = -p, \tag{8.134}$$

$$z = 0, \ r > a : \sigma_{zz} = -\frac{2p}{\pi} \left[\arcsin(a/r) - \frac{a}{(r^2 - a^2)^{1/2}} \right]. \tag{8.135}$$

Equation (8.134) confirms the boundary condition (8.117), and (8.135) is a well known result (Sneddon, 1951, p. 495).

Alternative Solution

An alternative for the second step of the derivation is as follows. In this alternative method the Bessel function $J_0(\xi r)$ is eliminated from the boundary condition (8.120) by using the integral (8.110),

$$\int_0^s \frac{r}{(s^2 - r^2)^{1/2}} J_0(\xi r) \, dr = \frac{\sin(\xi s)}{\xi}. \tag{8.136}$$

This boundary condition (8.120) then is transformed into the form

$$\int_0^\infty \xi^2 B(\xi) \sin(\xi s) \, d\xi = \int_0^s \frac{rg(r)}{(s^2 - r^2)^{1/2}} \, dr, \quad 0 \le s < a. \tag{8.137}$$

The function $B(\xi)$ can be written in the form of (8.124), which was obtained by partial integration from the actual definition (8.121), and assuming that $V(0) = 0$,

$$\xi^3 B(\xi) = V(a) \cos(\xi a) + \int_0^a V'(t) \cos(\xi t) \, dt. \tag{8.138}$$

Substitution into (8.137) gives

$$V(a) \int_0^\infty \frac{\sin(\xi s) \cos(\xi a)}{\xi} \, d\xi + \int_0^a V'(t) \left\{ \int_0^\infty \frac{\sin(\xi s) \cos(\xi t)}{\xi} \, d\xi \right\} dt$$

$$= \int_0^s \frac{rg(r)}{(s^2 - r^2)^{1/2}} \, dr, \quad 0 \le s < a, \tag{8.139}$$

where it should be noted that in the second integral $0 < t < a$. The integrals are of the form of (8.113),

$$\int_0^\infty \frac{\sin(\xi t) \cos(\xi s)}{\xi} \, d\xi = \begin{cases} \frac{\pi}{2}, & s < t, \\ 0, & s > t. \end{cases} \tag{8.140}$$

This means that the first integral of (8.139) is zero, and that in the second integral the integration can be restricted to the interval $0 < t < s$. The result is

$$\int_0^s V'(t) \, dt = V(s) = \frac{2}{\pi} \int_0^s \frac{rg(r)}{(s^2 - r^2)^{1/2}} \, dr, \quad 0 \le s < a. \tag{8.141}$$

This is the same solution as obtained before, see (8.129). The present derivation seems to be simpler, as it avoids the Abel integral equation.

8.6 Confined Elastostatics

Although several elastic problems have been successfully solved analytically in the preceding sections, and many more solutions can be found in the literature, the solution methods are relatively complex, and it seems attractive to attempt to develop a simplified approximate method of solution. This may be especially useful as a preparation for the more difficult problems of elastodynamics, which will be considered in the next chapter.

For problems of an elastic half space in which the load consists of vertical normal stresses on the surface only, it can be expected that the vertical displacements are considerably larger than the lateral displacements. This suggests to develop an approximate method of solution by assuming that the horizontal displacements are zero, so that the only remaining displacement is the vertical displacement. Problems solved under these assumptions will be referred to as *confined* elastic problems here. The approximation was introduced by Westergaard (1938), by considering the vanishing of the horizontal deformations as a consequence of a reinforcement of the material by inextensible horizontal sheets.

The basic assumptions are

$$u_x = 0, \tag{8.142}$$

$$u_y = 0, \tag{8.143}$$

$$u_z = w(x, y, z). \tag{8.144}$$

The vertical displacement will be denoted by w, for simplicity.

Using these assumptions the only relevant basic equation is the equation of vertical equilibrium, which now requires that

$$\mu \frac{\partial^2 w}{\partial x^2} + \mu \frac{\partial^2 w}{\partial y^2} + (\lambda + 2\mu) \frac{\partial^2 w}{\partial z^2} = 0. \tag{8.145}$$

The remaining relevant stress components are the stresses on horizontal planes. They are related to the vertical displacements by the equations

$$\sigma_{zz} = (\lambda + 2\mu) \frac{\partial w}{\partial z}, \tag{8.146}$$

$$\sigma_{zx} = \mu \frac{\partial w}{\partial x}, \tag{8.147}$$

$$\sigma_{zy} = \mu \frac{\partial w}{\partial y}. \tag{8.148}$$

8.6.1 An Axially Symmetric Problem

In case of an axially symmetric surface load the differential equation (8.145) can be formulated, using polar coordinates, as

$$\eta^2 \left\{ \frac{\partial^2 w}{\partial r^2} + \frac{1}{r} \frac{\partial w}{\partial r} \right\} + \frac{\partial^2 w}{\partial z^2} = 0, \tag{8.149}$$

where

$$\eta^2 = \frac{\mu}{\lambda + 2\mu} = \frac{1 - 2v}{2(1 - v)}. \tag{8.150}$$

If the load is a uniform load on a circular area, the boundary condition is, with (8.146),

$$z = 0 : \quad (\lambda + 2\mu)\frac{\partial w}{\partial z} = \begin{cases} -p, & r < a, \\ 0, & r > a. \end{cases} \tag{8.151}$$

For the solution of this problem the Hankel transform method seems particularly suited, as in other axially symmetric cases. The Hankel transform of the vertical displacement w is defined by

$$W(\xi, z) = \int_0^\infty r w(r, z) J_0(r\xi) \, dr, \tag{8.152}$$

where $J_0(x)$ is the Bessel function of the first kind and order zero. The inverse transformation is (Sneddon, 1951)

$$w(r, z) = \int_0^\infty \xi W(\xi, z) J_0(\xi r) \, d\xi. \tag{8.153}$$

The differential equation (8.149) becomes, after application of the Hankel transform,

$$\frac{d^2 W}{dz^2} - \xi^2 \eta^2 W = 0, \tag{8.154}$$

which is an ordinary differential equation. The general solution of this equation is

$$W = A \exp(\xi \eta z) + B \exp(-\xi \eta z), \tag{8.155}$$

where the integration constants A and B may depend upon the transformation parameter ξ. In the half space $z > 0$ the constant A can be assumed to vanish, because of the boundary condition at infinity. With the boundary condition (8.151) the value of the constant B is found to be

$$B = \frac{p}{\eta(\lambda + 2\mu)\xi} \int_0^a r J_0(\xi r) \, dr. \tag{8.156}$$

This is a well known integral (Abramowitz and Stegun, 1964, 11.3.20). The result is

$$B = \frac{pa}{\eta(\lambda + 2\mu)\xi^2} J_1(\xi a), \tag{8.157}$$

where $J_1(x)$ is the Bessel function of the first kind and order one.

The vertical displacement w now is

$$w = \frac{pa}{\eta(\lambda + 2\mu)} \int_0^\infty \frac{J_1(\xi a) \exp(-\xi \eta z) J_0(\xi r)}{\xi} \, d\xi. \tag{8.158}$$

The standard tables of integral transforms do not give closed form expressions of this integral. However, for the displacements of the surface one obtains, with $z = 0$, and using (8.150) in order to express the coefficient in terms of the shear modulus μ,

$$z = 0 : w = \frac{pa\eta}{\mu} \int_0^\infty \frac{J_1(\xi a) J_0(\xi r)}{\xi} \, d\xi. \qquad (8.159)$$

This happens to be the same integral as in the exact solution, (8.78). Hence the result is

$$z = 0 : w = \frac{2pa\eta}{\pi\mu} \begin{cases} E(r^2/a^2), & r < a, \\ F(r^2/a^2), & r > a, \end{cases} \qquad (8.160)$$

where

$$F(x) = \sqrt{x}[E(1/x) - (1 - 1/x)K(1/x)], \qquad (8.161)$$

and where $K(x)$ and $E(x)$ are complete elliptic integrals of the first and second kind, respectively.

It is perhaps remarkable that the approximate solution and the exact solution are of precisely the same form, even though the coefficient is slightly different, in its dependence upon Poisson's ratio v. Only in the completely incompressible case, $v = 0.5$, the approximate solution degenerates. This could have been expected, because the only possible deformation is a vertical displacement, which is suppressed in an impermeable material if there are no horizontal displacements. The agreement between the exact elastic solution and the approximate solution for a confined elastic medium may provide support for a similar approach to problems of elastodynamics.

The only difference between the exact solution, as given by (8.79), and the approximate solution (8.160) derived here, is in the coefficient of the solution. In the exact case this coefficient is $1 - v$, and here it is found to be η, where η is defined by (8.150). These two coefficients are compared in Table 8.1. The exact solution appears to give somewhat larger displacements than the approximate solution. This is a general property of approximate solutions obtained by a constraint on the displacement field. The material appears to be somewhat stiffer because of the constraint that there can be no horizontal displacements. Or, as Westergaard stated in his original publication (Westergaard, 1938), because the material has been reinforced

Table 8.1 Comparison of coefficients

v	$1 - v$	η
0.0	1.000	0.707
0.1	0.900	0.667
0.2	0.800	0.612
0.3	0.700	0.534
0.4	0.600	0.408
0.5	0.500	0.000

Fig. 8.13 σ_{zz} for $r = 0$
($\nu = 0$)

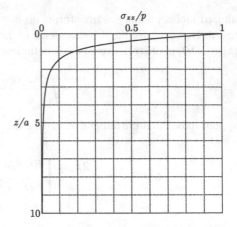

by horizontal inextensible sheets. The vertical normal stress σ_{zz} is, with (8.146) and (8.158),

$$\frac{\sigma_{zz}}{p} = -\int_0^\infty a J_1(\xi a) \exp(-\xi \eta z) J_0(\xi r) \, d\xi. \tag{8.162}$$

For $r = 0$, that is along the vertical axis, this reduces to

$$r = 0 \; : \; \frac{\sigma_{zz}}{p} = -\int_0^\infty a J_1(\xi a) \exp(-\xi \eta z) \, d\xi. \tag{8.163}$$

This integral can be found in a table of Laplace transforms (Churchill, 1972). The result is

$$r = 0 \; : \; \frac{\sigma_{zz}}{p} = -1 + \frac{\eta z}{\sqrt{a^2 + \eta^2 z^2}}. \tag{8.164}$$

This function, illustrated in Fig. 8.13, has the same properties as the exact solution given in (8.84), see also Fig. 8.9. It is not the same, however. One of the major differences is that the present solution depends upon Poisson's ratio. Another difference is that in this approximate solution the stresses tend to zero much faster than in the complete elastic solution.

8.6.2 A Plane Strain Problem

For a case of plane strain deformation in the x, z-plane the basic differential equation is

$$\eta^2 \frac{\partial^2 w}{\partial x^2} + \frac{\partial^2 w}{\partial z^2} = 0, \tag{8.165}$$

where η is a parameter depending upon Poisson's ration, see (8.150).

If the load is a uniform load on a strip of width $2a$, the boundary condition is, with (8.146),

$$z = 0 : (\lambda + 2\mu)\frac{\partial w}{\partial z} = \begin{cases} -p, & |x| < a, \\ 0, & |x| > a. \end{cases} \tag{8.166}$$

Because of the symmetry of the load, the Fourier cosine transform seems to be appropriate in this case,

$$W(\alpha, z) = \int_0^\infty w(x, z) \cos(\alpha x)\, dx. \tag{8.167}$$

The differential equation (8.165) now can be transformed into

$$\frac{d^2 W}{dz^2} - \alpha^2 \eta^2 W = 0. \tag{8.168}$$

The solution vanishing at infinity is

$$W = A \exp(-\alpha \eta z). \tag{8.169}$$

The constant A can be determined using the boundary condition (8.166). The final result is

$$W = \frac{p}{\eta \lambda + 2\mu} \frac{\sin(\alpha a)}{\alpha^2} \exp(-\alpha \eta z). \tag{8.170}$$

Inverse transformation gives

$$w = \frac{2p}{\pi \eta (\lambda + 2\mu)} \int_0^\infty \frac{\sin(\alpha a)\cos(\alpha x)}{\alpha^2} \exp(-\alpha \eta z)\, d\alpha. \tag{8.171}$$

The vertical normal stress σ_{zz} is of particular importance. This is found to be

$$\sigma_{zz} = -\frac{2p}{\pi} \int_0^\infty \frac{\sin(\alpha a)\cos(\alpha x)\exp(-\alpha \eta z)}{\alpha}\, d\alpha, \tag{8.172}$$

or

$$\sigma_{zz} = -\frac{p}{\pi} \int_0^\infty \frac{\sin[\alpha(x + a)] - \sin(\alpha(x - a))}{\alpha} \exp(-\alpha \eta z)\, d\alpha. \tag{8.173}$$

The integrals have the form of Laplace transforms, with t replaced by α. They can be evaluated using a table of standard Laplace transforms. This gives

$$\sigma_{zz} = -\frac{p}{\pi}\left\{ \arctan\left(\frac{x + a}{\eta z}\right) - \arctan\left(\frac{x - a}{\eta z}\right) \right\}. \tag{8.174}$$

Comparison of this result with the solution of the complete elastic problem, see (8.62), shows that there is a certain similarity of the solutions. Again, the approximate solution using Westergaard's approximation appears to depend upon the value of Poisson's ratio, whereas the solution of the complete elastic problem is independent of Poisson's ratio, at least as far as the stresses are concerned. The boundary condition (8.166) is exactly satisfied, of course.

The case of a line load can be obtained by taking the width of the load $a \to 0$, with $F = 2pa$. The simplest way to derive the vertical stress σ_{zz} for this case is by starting from (8.172), which then becomes

$$\sigma_{zz} = -\frac{F}{\pi} \int_0^\infty \cos(\alpha x) \exp(-\alpha \eta z)\, d\alpha. \tag{8.175}$$

This is an elementary Laplace transform, see Table A.1 in Appendix A. It follows that

$$\sigma_{zz} = -\frac{F \eta z}{\pi (x^2 + \eta^2 z^2)}. \tag{8.176}$$

This can be compared with the exact result given in (8.50).

Chapter 9
Elastodynamics of a Half Space

An important and useful basic problem for the analysis of the propagation of waves in soils is the problem of an elastic half space loaded at its surface by a time-dependent load, see Fig. 9.1. The load may be fluctuating sinusoidally with time, or it may be applied in a very short time, and then remain constant. For the case of a concentrated pulse load the solution has first been given by Lamb (1904), and later by others, such as Pekeris (1955) and De Hoop (1960). All these solutions are mathematically rather complex, however. Therefore in the next chapter a simplified approach will be followed, in which the elastic problem is approximated by disregarding the horizontal displacements, and thus considering vertical displacements only. This approximation was first suggested by Westergaard (1938), and is denoted as *confined* elasticity in this book. It has been shown in the previous chapter that this approximate method gives very good results for the elastostatic problems of the same type. The extension to problems of elastodynamics was first suggested by Barends (1980). It will appear in the next chapter that in the case of elastodynamics the most important aspects of the solutions, such as the magnitude of the vertical displacements, and the effect of damping, can be approximated reasonably well. The solution of these problems will be used as basic elements for the analysis of foundation vibrations in Chap. 15.

In later chapters the complete solution of some problems of elastodynamics of a half space (or a half plane) will be presented, using methods developed by Pekeris

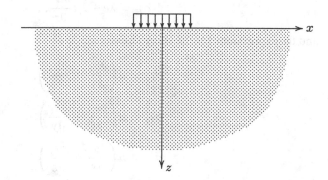

Fig. 9.1 Half space

A. Verruijt, *An Introduction to Soil Dynamics*,
Theory and Applications of Transport in Porous Media 24,
© Springer Science+Business Media B.V. 2010

(1955) and De Hoop (1960). These include the solutions for a line load and a point load on the surface of an elastic half space.

As an introduction to the chapters in which the solutions of particular problems are presented, this chapter presents some general aspects of the propagation of waves in homogeneous elastic media. A brief introduction is given of compression waves and shear waves, which are important waves that appear in the solution of many problems. Also a general description is given of Rayleigh waves, which appear in problems for a half space, and which are mainly responsible for the damage caused by earthquakes.

9.1 Basic Equations of Elastodynamics

The basic equations of elastodynamics are the Navier equations, extended with an inertia term. These equations are

$$(\lambda + \mu)\frac{\partial e}{\partial x} + \mu\nabla^2 u_x = \rho\frac{\partial^2 u_x}{\partial t^2}, \tag{9.1}$$

$$(\lambda + \mu)\frac{\partial e}{\partial y} + \mu\nabla^2 u_y = \rho\frac{\partial^2 u_y}{\partial t^2}, \tag{9.2}$$

$$(\lambda + \mu)\frac{\partial e}{\partial z} + \mu\nabla^2 u_z = \rho\frac{\partial^2 u_z}{\partial t^2}, \tag{9.3}$$

where ρ is the density of the material, and t is the time. The static versions of these equations have been derived in Chap. 8.

The stresses can be expressed into the displacement components by the generalized form of Hooke's law. For an isotropic material the expressions for the normal stresses are

$$\sigma_{xx} = \lambda e + 2\mu\frac{\partial u_x}{\partial x}, \tag{9.4}$$

$$\sigma_{yy} = \lambda e + 2\mu\frac{\partial u_y}{\partial y}, \tag{9.5}$$

$$\sigma_{zz} = \lambda e + 2\mu\frac{\partial u_z}{\partial z}, \tag{9.6}$$

and the expressions for the shear stresses are

$$\sigma_{xy} = \mu\left(\frac{\partial u_x}{\partial y} + \frac{\partial u_y}{\partial x}\right), \tag{9.7}$$

$$\sigma_{yz} = \mu\left(\frac{\partial u_y}{\partial z} + \frac{\partial u_z}{\partial y}\right), \tag{9.8}$$

$$\sigma_{zx} = \mu\left(\frac{\partial u_z}{\partial x} + \frac{\partial u_x}{\partial z}\right). \tag{9.9}$$

Here λ and μ are the Lamé constants, and e is the volume strain,

$$e = \frac{\partial u_x}{\partial x} + \frac{\partial u_y}{\partial y} + \frac{\partial u_z}{\partial z}. \tag{9.10}$$

9.2 Compression Waves

A special solution of the basic equations of elastodynamics can be obtained by differentiating the first equation of motion, (9.1) with respect to x, the second one with respect to y, the third one with respect to z, and then adding the result. This gives

$$(\lambda + 2\mu)\nabla^2 e = \rho \frac{\partial^2 e}{\partial t^2}. \tag{9.11}$$

This is the classical form of the wave equation. It has solutions of the form

$$e = f_1(r - c_p t) + f_2(r + c_p t), \tag{9.12}$$

where r is the direction of the wave, and c_p is the velocity of the wave,

$$c_p = \sqrt{(\lambda + 2\mu)/\rho}. \tag{9.13}$$

These waves are called compression waves, or simply P-waves.

9.3 Shear Waves

Another special solution of the basic equations of elastodynamics can be obtained by differentiating the first equation of motion, (9.1) with respect to y, the second one with respect to x, and then subtracting the result. This gives

$$\mu\nabla^2\omega_{xy} = \rho\frac{\partial^2\omega_{xy}}{\partial t^2}, \tag{9.14}$$

where ω_{xy} is the rotation about the z-axis,

$$\omega_{xy} = \left(\frac{\partial u_x}{\partial y} - \frac{\partial u_y}{\partial x}\right). \tag{9.15}$$

Similar equations can be obtained from other combinations, namely

$$\mu\nabla^2\omega_{yz} = \rho\frac{\partial^2\omega_{yz}}{\partial t^2}, \tag{9.16}$$

$$\mu\nabla^2\omega_{zx} = \rho\frac{\partial^2\omega_{zx}}{\partial t^2}. \tag{9.17}$$

Again equations of the form of the wave equation are obtained. For these rotational waves, or shear waves, or simply S-waves, the propagation velocity is

$$c_s = \sqrt{\mu/\rho}. \tag{9.18}$$

Comparison with (9.13) shows that the velocity of the shear waves in general will be smaller than the velocity of the compression waves. The P-waves and S-waves play an important part in seismology. From the arrival time of these waves the dynamic properties of the material may be derived.

9.4 Rayleigh Waves

The possibility of elastodynamic waves propagating along the surface of an elastic half space was first considered by Lord Rayleigh (1885). This is a wave that propagates near the free surface of an elastic half space, and is strongly decreases exponentially with depth. Derivations of the Rayleigh wave solution can be found in many textbooks on soil dynamics and earthquake engineering (e.g. Kolsky, 1963; Richart et al., 1970; Das, 1993; Kramer, 1996). In this chapter the derivation mainly follows the method used by Achenbach (1975).

It is assumed that a solution of the basic equations of elastodynamics can be represented by the following expressions for the displacements of a wave in the x, z-plane (see Fig. 9.1),

$$u_x = A \exp(-bz) \sin[k(x - c_r t)], \tag{9.19}$$

$$u_z = B \exp(-bz) \cos[k(x - c_r t)], \tag{9.20}$$

where k is a given constant, and b and c_r are as yet unknown parameters. It is assumed that $b > 0$, so that the displacements tend towards zero for $z \to \infty$. The constants A and B are also unknown at this stage. The displacements in the direction perpendicular to the x, z-plane are assumed to vanish, and the other two components are assumed to be independent of y. It should be noted that in this solution, if it is found to exist, the amplitudes of the displacement components are independent of the lateral distance x.

Substitution of (9.19) and (9.20) into the basic equations in x- and z-direction, see (9.1) and (9.3), gives

$$\left[c_s^2 b^2 - (c_p^2 - c_r^2)k^2\right]A + (c_p^2 - c_s^2)kbB = 0, \tag{9.21}$$

$$-(c_p^2 - c_s^2)kbA + \left[c_p^2 b^2 - (c_s^2 - c_r^2)k^2\right]B = 0, \tag{9.22}$$

where c_p and c_s are the velocities of compression waves and shear waves, respectively, as defined by (9.13) and (9.18). A solution of this system of equations is possible only if the determinant of the system is zero. This leads to the condition

$$\left[\left(\frac{b}{k}\right)^2 - \left(\frac{c_p^2 - c_r^2}{c_p^2}\right)\right]\left[\left(\frac{b}{k}\right)^2 - \left(\frac{c_s^2 - c_r^2}{c_s^2}\right)\right] = 0. \tag{9.23}$$

If it is assumed that $c_r < c_s < c_p$ there are two real and positive solutions,

$$b_1/k = \sqrt{1 - c_r^2/c_p^2}, \tag{9.24}$$

$$b_2/k = \sqrt{1 - c_r^2/c_s^2}. \tag{9.25}$$

It now follows from (9.21) and (9.22) that for these two solutions

$$B_1/A_1 = b_1/k, \tag{9.26}$$

$$A_2/B_2 = b_2/k. \tag{9.27}$$

These relations can most conveniently be satisfied by writing $A_1 = kC_1$, $B_1 = b_1C_1$, $A_2 = b_2C_2$ and $B_2 = kC_2$, where C_1 and C_2 now are the unknown constants of the two solutions. The total solution can then be written as

$$u_x = \left[kC_1 \exp(-b_1 z) + b_2 C_2 \exp(-b_2 z)\right] \sin[k(x - c_r t)], \tag{9.28}$$
$$u_z = \left[b_1 C_1 \exp(-b_1 z) + kC_2 \exp(-b_2 z)\right] \cos[k(x - c_r t)]. \tag{9.29}$$

The solution is supposed to be applicable to the region near the free surface of a half space. Thus, the boundary conditions are

$$z = 0 : \sigma_{zz} = 0, \tag{9.30}$$

$$z = 0 : \sigma_{zx} = 0. \tag{9.31}$$

Using the relations (9.6) and (9.9) these boundary conditions lead to the equations

$$(2 - c_r^2/c_s^2)C_1 + 2\sqrt{1 - c_r^2/c_s^2}C_2 = 0, \tag{9.32}$$

$$2\sqrt{1 - c_r^2/c_p^2}\,C_1 + (2 - c_r^2/c_s^2)C_2 = 0. \tag{9.33}$$

This system of equations will have a non-zero solution only if the determinant of the system is zero. This gives

$$(2 - c_r^2/c_s^2)^2 - 4\sqrt{1 - \eta^2 c_r^2/c_s^2}\sqrt{1 - c_r^2/c_s^2} = 0, \tag{9.34}$$

where

$$\eta^2 = c_s^2/c_p^2 = (1 - 2\nu)/[2(1 - \nu)]. \tag{9.35}$$

The Rayleigh wave velocity c_r can be determined from the condition (9.34).

A simple way to determine this value is to write $p = c_s^2/c_r^2$. It then follows from (9.34) that

$$(2p - 1)^4 = 16p^2(p - \eta^2)(p - 1), \tag{9.36}$$

or

$$16(1 - \eta^2)p^3 - 8(3 - 2\eta^2)p^2 + 8p - 1 = 0. \tag{9.37}$$

This is a cubic equation, for which an analytical method of solution is available (see e.g. Abramowitz and Stegun, 1964, p. 17). This will show that there is only one real solution. This solution can also be derived in an approximate way by noting that it can be expected (and follows from the analytical solution) that $p = 1 + a$, where $a \ll 1$. Equation (9.37) then gives

$$a(a + 1)[2(1 - \eta^2)a + (1 - 2\eta^2)] = 1/8, \tag{9.38}$$

or, using the definition (9.35) of the parameter η^2,

$$a = \frac{1 - v}{8(1 + a)(v + a)}. \tag{9.39}$$

If v is sufficiently large, the value of a can be determined iteratively, from this equation, starting from an initial small value, say $a = 0.01$. For small values of v, say $v < 0.1$, it may be more effective to write $v + a = a(1 + v/a)$, so that

$$a^2 = \frac{1 - v}{8(1 + a)(1 + v/a)}, \tag{9.40}$$

which can be used to determine the value of a iteratively, starting from the same initial estimate, $a = 0.01$.

Once that the value of a has been determined, it follows that $p = 1 + a$, and thus

$$\frac{c_r}{c_s} = \frac{1}{\sqrt{1 + a}}. \tag{9.41}$$

A function (in C) to calculate the value of c_r/c_s as a function of Poisson's ratio is shown below.

```
double crcs(double nu)
{
double a,b,e,f;
e=0.000001;e*=e;f=1;b=0.01;
if (nu>0.1) {while (f>e) {a=b;b=(1-nu)/(8*(1+a)*(nu+a));f=fabs(b-a);}}
else {while (f>e) {a=b;b=sqrt((1-nu)/(8*(1+a)*(1+nu/a)));f=fabs(b-a);}}
return(1/sqrt(1+a));
}
```

Some numerical values are shown in Table 9.1. The table also gives the values of c_p/c_s, the ratio of the velocities of compression waves and shear waves.

A graphical representation of the ratio of the velocities of Rayleigh waves and shear waves is shown in Fig. 9.2. It appears that the Rayleigh wave is always somewhat slower than the shear wave.

The relation between the two coefficients C_1 and C_2 in the solution can be obtained from either of (9.32) and (9.33). The two components of the displacements in a Rayleigh wave can then be determined from (9.28) and (9.29). This results in

Table 9.1 Velocity of Rayleigh waves

v	c_r/c_s	c_p/c_s
0.00	0.874032	1.414214
0.05	0.883695	1.452966
0.10	0.893106	1.500000
0.15	0.902220	1.558387
0.20	0.910996	1.632993
0.25	0.919402	1.732051
0.30	0.927413	1.870829
0.35	0.935013	2.081667
0.40	0.942195	2.449490
0.45	0.948960	3.316625
0.50	0.955313	∞

Fig. 9.2 Velocity of Rayleigh waves

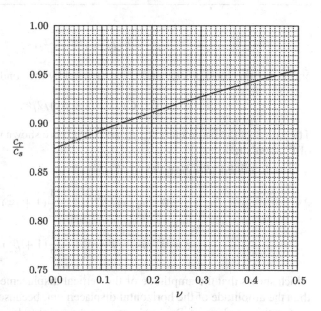

the following expressions for the two displacement components

$$u_x = kC_1\left[\exp(-\beta_1 kz) - \frac{1}{2}(1+\beta_2^2)\exp(-\beta_2 kz)\right]\sin[k(x-c_r t)], \quad (9.42)$$

$$u_z = kC_2\left[\exp(-\beta_2 kz) - \frac{1}{2}(1+\beta_2^2)\exp(-\beta_1 kz)\right]\cos[k(x-c_r t)], \quad (9.43)$$

where

$$\beta_1 = b_1/k = \sqrt{1 - c_r^2/c_p^2}, \quad (9.44)$$

Table 9.2 Table of Rayleigh wave velocities and some parameters

ν	c_r/c_s	c_p/c_s	β_1	β_2	C_2/C_1
0.00	0.874032	1.414214	0.786151	0.485868	−1.272020
0.05	0.883695	1.452966	0.793783	0.468064	−1.302263
0.10	0.893106	1.500000	0.803426	0.449846	−1.336414
0.15	0.902220	1.558387	0.815367	0.431277	−1.374987
0.20	0.910996	1.632993	0.829929	0.412415	−1.418577
0.25	0.919402	1.732051	0.847487	0.393320	−1.467890
0.30	0.927413	1.870829	0.868481	0.374040	−1.523776
0.35	0.935013	2.081666	0.893448	0.354613	−1.587294
0.40	0.942195	2.449490	0.923063	0.335064	−1.659785
0.45	0.948960	3.316625	0.958193	0.315397	−1.743000
0.50	0.955313	∞	1.000000	0.295598	−1.839287

$$\beta_2 = b_2/k = \sqrt{1 - c_r^2/c_s^2}, \tag{9.45}$$

and where the constants C_1 and C_2 are related by the condition

$$C_2/C_1 = -(1 + \beta_2^2)/2\beta_2. \tag{9.46}$$

The values of the parameters β_1, β_2 and C_2/C_1 are shown in Table 9.2, as a function of Poisson's ratio ν.

The amplitudes at the surface $z = 0$ are

$$z = 0 : |u_x| = k|C_1|\left[1 - \frac{1}{2}(1 + \beta_2^2)\right], \tag{9.47}$$

$$z = 0 : |u_z| = k|C_2|\left[1 - \frac{1}{2}(1 + \beta_2^2)\right], \tag{9.48}$$

which shows that the amplitude of the vertical displacement at the surface is larger than the amplitude of the horizontal displacement, because $|C_2| > |C_1|$.

For three values of Poisson's ratio ν the amplitudes of the displacements are shown, as a function of z/L, where L is the wave length, $L = 2\pi/k$. The vertical displacement at the surface, indicated by u_0, is used as a scaling factor.

9.5 Love Waves

In a non-homogenous elastic material, such as a material consisting of various horizontal layers (a common occurrence in nature), compression waves and shear waves may be reflected and partly transmitted on the interfaces, as was illustrated for the one-dimensional case in Chap. 3. Successive reflections on the two sides of a thin

Fig. 9.3 Displacements for
Rayleigh wave

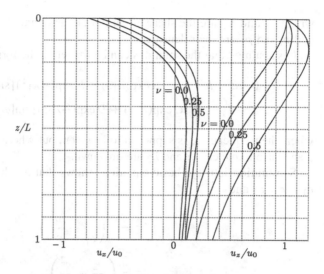

Fig. 9.4 Soft layer on a stiff
half space

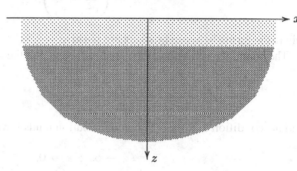

soft layer on top of a stiffer subsoil may lead to a special type of wave, the Love
wave. At the interface of two solids with certain properties a Stoneley wave may
be generated. This resembles a Rayleigh wave in the sense that it is confined to the
vicinity of the interface. For soil mechanics practice the Love wave is especially relevant. Therefore this wave will be considered in some detail here. For the Stonely
wave see e.g. Ewing et al. (1957), Cagniard et al. (1962), Achenbach (1975).

The simplest case of a Love wave occurs in a thin soft layer on a relatively stiff
half space, see Fig. 9.4. It is assumed that the only non-vanishing displacement is a
displacement $v = v(x, z, t)$ in the y-direction, that is the direction perpendicular to
the plane in which the wave propagates.

The basic equations are

$$0 < z < h \ : \ \frac{\partial^2 v}{\partial t^2} = \frac{\mu_1}{\rho_1}\left(\frac{\partial^2 v}{\partial x^2} + \frac{\partial^2 v}{\partial z^2}\right), \tag{9.49}$$

$$z > h \ : \ \frac{\partial^2 v}{\partial t^2} = \frac{\mu_2}{\rho_2}\left(\frac{\partial^2 v}{\partial x^2} + \frac{\partial^2 v}{\partial z^2}\right), \tag{9.50}$$

where μ_1 and μ_2 are the shear moduli of the layer and the base rock, respectively, and ρ_1 and ρ_2 are their densities.

It is assumed that the solutions can be written in the form

$$0 < z < h \ : \ v = [A \exp(\lambda_1 z) + B \exp(-\lambda_1 z)] \sin[\omega(t - x/c)], \quad (9.51)$$

$$z > h \ : \ v = [C \exp(\lambda_2 z) + D \exp(-\lambda_2 z)] \sin[\omega(t - x/c)], \quad (9.52)$$

where the frequency ω is supposed to be given, but where the propagation velocity c and the parameters λ_1 and λ_2 are unknown.

Substitution into the basic equations shows that a solution may be obtained if $c_1 < c < c_2$ and

$$\lambda_1^2 = -\left(\frac{\omega^2}{c_1^2} - \frac{\omega^2}{c^2}\right), \quad (9.53)$$

$$\lambda_2^2 = \left(\frac{\omega^2}{c^2} - \frac{\omega^2}{c_2^2}\right), \quad (9.54)$$

where the terms between brackets are positive.

The boundary condition at the free surface is that the shear stress is zero, so that

$$z = 0 \ : \ \frac{\partial v}{\partial z} = 0, \quad (9.55)$$

and the condition at infinity is that the solution tends towards zero,

$$z \to \infty \ : \ v \to 0. \quad (9.56)$$

Using these conditions it follows that the solution reduces to

$$0 < z < h \ : \ v = A \cos\left(\omega z \sqrt{1/c_1^2 - 1/c^2}\right) \sin[\omega(t - x/c)], \quad (9.57)$$

$$z > h \ : \ v = D \exp\left(-\omega z \sqrt{1/c^2 - 1/c_2^2}\right) \sin[\omega(t - x/c)]. \quad (9.58)$$

The conditions at the interface $z = h$ are that the displacement and the shear stress are continuous. The first condition leads to the equation

$$A \cos\left(\omega h \sqrt{1/c_1^2 - 1/c^2}\right) = D \exp\left(-\omega h \sqrt{1/c^2 - 1/c_2^2}\right). \quad (9.59)$$

The second condition leads to the equation

$$-\mu_1 A \sqrt{1/c_1^2 - 1/c^2} \sin\left(\omega h \sqrt{1/c_1^2 - 1/c^2}\right)$$

$$= -\mu_2 D \sqrt{1/c^2 - 1/c_2^2} \exp\left(-\omega h \sqrt{1/c^2 - 1/c_2^2}\right). \quad (9.60)$$

Fig. 9.5 Determination of the velocity of a Love wave, $c_2/c_1 = 5$, $\omega h/c_1 = 1$

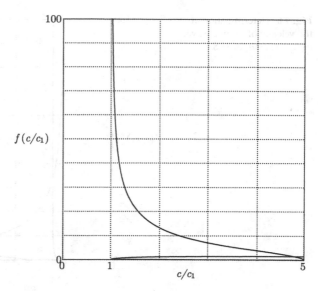

This system of equations has a non-zero solution only if the determinant of the system of equations is zero. This gives

$$\tan\left(\frac{\omega h}{c_1}\sqrt{1 - c_1^2/c^2}\right) = \frac{\rho_2 c_2}{\rho_1 c_1}\sqrt{\frac{c_2^2/c_1^2 - c^2/c_1^2}{c^2/c_1^2 - 1}}. \tag{9.61}$$

This equation contains two given parameters: c_2/c_1 and $\omega h/c_1$. The unknown value of c/c_1 can be determined by determining the intersection point of the two functions in the left and right hand side, respectively. The procedure is illustrated in Fig. 9.5, for the case $c_2/c_1 = 5$ and $\omega h/c_1 = 1$. It has been assumed that the densities of the two layers are equal. The location of the intersection point of the two curves indicates that the value of the Love wave velocity c in this case is very close to c_2, the shear wave velocity in the deep layer. This will be found for all values of $\omega h/c_1 < 1$. It means that for slow vibrations the velocity of the shear waves in the deep layer dominates the velocity of the Love wave in the upper layer.

For high frequencies, say $\omega h/c_1 > 4$, the first zero is close to $c = c_1$, but there may be several possible solutions in the range $c_1 < c < c_2$. In the case $\omega h/c_1 = 8$, shown in Fig. 9.6, there appear to be three solutions. For larger values of the frequency the number of zeroes further increases.

The velocity of the (first) Love wave is shown as a function of the frequency ω in Fig. 9.7, for three values of the parameter c_2/c_1. For high frequencies the value of c approaches c_1, and for small frequencies it approaches c_2, as mentioned before.

Fig. 9.6 Determination of
the velocity of a Love wave,
$c_2/c_1 = 5$, $\omega h/c_1 = 8$

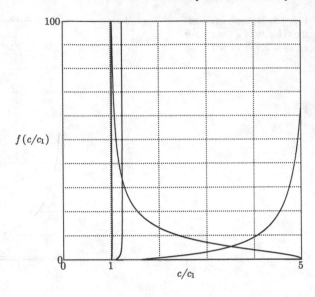

Fig. 9.7 Velocity of Love
wave

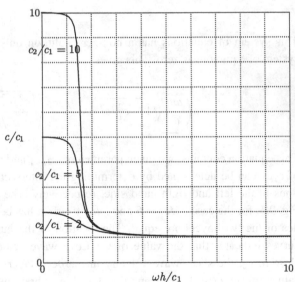

9.5.1 A Practical Implication

For geotechnical engineering an interesting situation is a soft layer of limited thickness on top of a hard rock of great thickness, in which an earthquake wave is generated. A normal value for the density of the rock is $\rho_2 = 2500$ kg/m³ and a normal value for the density of the soft soil is $\rho_1 = 2000$ kg/m³. The shear modulus of the rock may be as large as $\mu_2 = 10$ MPa $= 10 \times 10^9$ kg/ms². This means that the velocity of shear waves in the rock is about $c_2 = 2000$ m/s. The shear modulus of the soft soil is of the order of magnitude $\mu_1 = 20$ kPa $= 20 \times 10^6$ kg/ms^2, so that the velocity of shear waves in the top layer is about $c_1 = 100$ m/s. Thus the ratio of the shear waves is about $c_2/c_1 = 20$.

The frequency of earthquake vibrations is of the order of magnitude $\omega = 30$ s^{-1}, indicating a period of about $T = 2\pi/\omega \approx 0.2$ s. For a layer of 20 m thickness the parameter $\omega h/c_1$ now is about 6, which is large enough to conclude that several modes of Love waves will be possible, one with $c \approx c_1$ and one approaching c_2. Considering a Love wave for which $c \approx c_2$, the solution for the displacements in the top layer is, with (9.57),

$$0 < z < h \; : \; v = A\cos\left(\omega z\sqrt{1/c_1^2 - 1/c_2^2}\right)\sin[\omega(t - x/c_2)], \qquad (9.62)$$

or, because $c_2 \gg c_1$,

$$0 < z < h \; : \; v = A\cos(\omega z/c_1)\sin[\omega(t - x/c_2)]. \qquad (9.63)$$

This is precisely the expression (3.16) used in Chap. 3. It appears that the approximate solutions considered in that chapter can be considered as approximations of a Love wave.

Chapter 10
Confined Elastodynamics

For a particular problem of elastodynamics, characterized by its boundary conditions, the basic equations are often very difficult to solve, both analytically or numerically. Some insight can be obtained by studying special solutions, such as those describing compression waves and shear waves. Another way of gaining some insight into the dynamic behaviour of an elastic continuum is to simplify the problem by an appropriate restriction on the displacement field. For this purpose it will be assumed here that the two horizontal displacements are so small compared to the vertical displacement that they may be neglected. This assumption was first proposed by Westergaard (1938) for problems of elastostatics, and has been used in Chap. 8. The generalization to problems of elastodynamics was first made by Barends (1980). Problems solved under these assumptions will be referred to as *confined elastodynamic* problems in this chapter.

The basic assumptions are

$$u_x = 0, \tag{10.1}$$

$$u_y = 0, \tag{10.2}$$

$$u_z = w(x, y, x). \tag{10.3}$$

The vertical displacement will be denoted by w, for simplicity.

Using these assumptions the only remaining basic equation is the equation of vertical equilibrium, which now requires that

$$\mu \frac{\partial^2 w}{\partial x^2} + \mu \frac{\partial^2 w}{\partial y^2} + (\lambda + 2\mu) \frac{\partial^2 w}{\partial z^2} = \rho \frac{\partial^2 w}{\partial t^2}. \tag{10.4}$$

The remaining relevant stress components are the stresses on horizontal planes. They are related to the vertical displacements by the equations

$$\sigma_{zz} = (\lambda + 2\mu) \frac{\partial w}{\partial z}, \tag{10.5}$$

$$\sigma_{zx} = \mu \frac{\partial w}{\partial x}, \tag{10.6}$$

A. Verruijt, *An Introduction to Soil Dynamics*,
Theory and Applications of Transport in Porous Media 24,
© Springer Science+Business Media B.V. 2010

$$\sigma_{zy} = \mu \frac{\partial w}{\partial y}. \tag{10.7}$$

10.1 Line Load on Half Space

As a first example consider the problem of a line load as a step in time. The load is applied in a very short time, at time $t = 0$, and then remains constant, see Fig. 10.1. In this case of a line load, with the line following the y-axis, the vertical displacement w can be assumed to be independent of y, so that the basic equation (10.4) reduces to

$$\mu \frac{\partial^2 w}{\partial x^2} + (\lambda + 2\mu) \frac{\partial^2 w}{\partial z^2} = \rho \frac{\partial^2 w}{\partial t^2}. \tag{10.8}$$

The boundary condition is

$$z = 0 : (\lambda + 2\mu) \frac{\partial w}{\partial z} = \begin{cases} 0, & \text{if } t < 0, \\ -P \, \delta(x), & \text{if } t > 0, \end{cases} \tag{10.9}$$

where $\delta(x)$ is a function that is everywhere zero, except in the origin, where it is infinitely large, such that the integral over x is 1, whenever the origin is included, i.e. for all positive values of a,

$$\int_{-a}^{+a} \delta(x) \, dx = 1. \tag{10.10}$$

The dimension of P is [F]/[L], i.e. kN/m in SI-units.

The initial condition is supposed to be that before $t = 0$ the displacement w and its derivative (the velocity) are zero,

$$t = 0 : w = 0, \tag{10.11}$$

$$t = 0 : \frac{\partial w}{\partial t} = 0. \tag{10.12}$$

The problem can be solved by using the Laplace transform method (see e.g. Churchill, 1972). The Laplace transform of the displacement w is defined by

$$\overline{w} = \int_0^\infty w \exp(-st) \, dt, \tag{10.13}$$

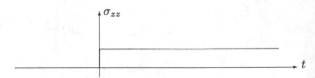

Fig. 10.1 Step load

where now $\overline{w}$ is a function of the Laplace transform parameter s as well as the spatial variables x and z.

Applying the Laplace transformation to the differential equation (10.8) gives

$$\mu \frac{\partial^2 \overline{w}}{\partial x^2} + (\lambda + 2\mu) \frac{\partial^2 \overline{w}}{\partial z^2} = \rho s^2 \overline{w}, \tag{10.14}$$

and the transformed boundary condition is

$$z = 0 : (\lambda + 2\mu) \frac{\partial \overline{w}}{\partial z} = -\frac{P}{s} \delta(x). \tag{10.15}$$

The partial differential equation (10.14) can be solved by the Fourier transform method (see e.g. Sneddon, 1951). The Fourier transform is defined by

$$W = \int_{-\infty}^{+\infty} \overline{w} \exp(i\alpha x) \, dx, \tag{10.16}$$

and the general inversion formula is given by the fundamental theorem of the theory of Fourier transforms (Sneddon, 1951),

$$\overline{w} = \frac{1}{2\pi} \int_{-\infty}^{-\infty} W \exp(-i\alpha x) \, d\alpha. \tag{10.17}$$

Applying the Fourier transform to the differential equation (10.14) gives

$$-\mu \alpha^2 \overline{W} + (\lambda + 2\mu) \frac{d^2 \overline{W}}{dz^2} = \rho s^2 \overline{W}, \tag{10.18}$$

which is an ordinary differential equation. After some rearranging it can also be written as

$$\frac{d^2 \overline{W}}{dz^2} = \eta^2 \gamma^2 \overline{W}, \tag{10.19}$$

where

$$\gamma^2 = \alpha^2 + s^2/c_s^2, \tag{10.20}$$

and

$$\eta^2 = \frac{\mu}{\lambda + 2\mu} = \frac{1 - 2\nu}{2(1 - \nu)}. \tag{10.21}$$

The parameter c_s is the velocity of shear waves in the medium,

$$c_s^2 = \frac{\mu}{\rho}. \tag{10.22}$$

The solution of (10.19) vanishing at infinity is

$$\overline{W} = A \exp(-\gamma \eta z). \tag{10.23}$$

The Fourier transform of the boundary condition (10.15) is

$$z = 0 : \quad (\lambda + 2\mu)\frac{d\overline{W}}{dz} = -\frac{P}{2\varepsilon s}\int_{-\varepsilon}^{+\varepsilon}\exp(i\alpha x)dx = -\frac{P}{s}. \tag{10.24}$$

From this condition the constant A can be determined,

$$A = \frac{P}{\eta(\lambda + 2\mu)s\gamma}, \tag{10.25}$$

so that the final solution of the transformed problem is

$$\overline{W} = \frac{P}{\eta(\lambda + 2\mu)s\gamma}\exp(-\gamma\eta z). \tag{10.26}$$

In principle the problem is solved now. What remains is to evaluate the inverse Fourier and Laplace transforms, which in general may be a formidable mathematical problem.

In this case the Fourier inverse of the expression (10.26) can formally be written as

$$\overline{w} = \frac{P}{2\pi\eta(\lambda + 2\mu)s}\int_{-\infty}^{+\infty}\frac{\exp[-\eta z\sqrt{\alpha^2 + s^2/c_s^2}]}{\sqrt{\alpha^2 + s^2/c_s^2}}\exp(-i\alpha x)d\alpha, \tag{10.27}$$

or, because the integrand is even,

$$\overline{w} = \frac{P}{\pi\eta(\lambda + 2\mu)s}\int_{0}^{\infty}\frac{\exp[-\eta z\sqrt{\alpha^2 + s^2/c_s^2}]}{\sqrt{\alpha^2 + s^2/c_s^2}}\cos(\alpha x)d\alpha. \tag{10.28}$$

It remains to evaluate this integral, and then to perform the inverse Laplace transformation.

The integral (10.28) happens to be a well known Fourier transform (Erdélyi et al., 1954, 1.4.27). The result is

$$\overline{w} = \frac{P}{\pi\eta(\lambda + 2\mu)s}K_0\left(\frac{s}{c_s}\sqrt{x^2 + \eta^2 z^2}\right), \tag{10.29}$$

where $K_0(x)$ is the modified Bessel function of the second kind and order zero.

The inverse Laplace transform of the function (10.29) is also well known (Erdélyi et al., 1954, 5.15.9). Thus the final expression for the vertical displacement is

$$w = \frac{P}{\pi\eta(\lambda + 2\mu)}\operatorname{arccosh}(t/t_0)\,\mathrm{H}(t - t_0), \tag{10.30}$$

where t_0 is the arrival time of the wave, taking into account the apparent scale transformation of the vertical coordinate,

$$t_0 = \frac{\sqrt{x^2 + \eta^2 z^2}}{c_s}, \tag{10.31}$$

and $H(t - t_0)$ is Heaviside's unit step function,

$$H(t - t_0) = \begin{cases} 0, & \text{if } t < t_0, \\ 1, & \text{if } t > t_0. \end{cases} \tag{10.32}$$

In view of the complexity of the original function (10.28) the simplicity of the final result (10.30) is perhaps surprising.

If the Laplace transform of the vertical normal stress σ_{zz} is defined as

$$\overline{\sigma}_{zz} = \int_0^\infty \sigma_{zz} \exp(-st)\, dt, \tag{10.33}$$

then the solution for $\overline{\sigma}_{zz}$ is, because $\overline{\sigma}_{zz} = (\lambda + 2\mu) d\overline{w}/dz$,

$$\overline{\sigma}_{zz} = -\frac{P}{\pi s} \int_0^\infty \exp[-\eta z \sqrt{\alpha^2 + s^2/c_s^2}] \cos(\alpha x)\, d\alpha. \tag{10.34}$$

Again this integral is a well known Fourier transform (Erdélyi et al., 1954, 1.4.26). The result is

$$\overline{\sigma}_{zz} = -\frac{P}{\pi c_s} \frac{\eta z}{\sqrt{x^2 + \eta^2 z^2}} K_1 \left(\frac{s}{c_s} \sqrt{x^2 + \eta^2 z^2} \right), \tag{10.35}$$

where $K_1(x)$ is the modified Bessel function of the second kind and order one.

The inverse Laplace transform of the expression (10.35) is, with (5.15.10) from Erdélyi et al. (1954),

$$\sigma_{zz} = -\frac{P}{\pi} \frac{\eta z}{x^2 + \eta^2 z^2} \frac{t}{\sqrt{t^2 - t_0^2}} H(t - t_0). \tag{10.36}$$

This is the final expression for the normal stresses in the half space. Of course, this formula can also be obtained from (10.30) by direct differentiation, using the simplified form of Hooke's law, (10.5). The value of t_0, the arrival time of the wave, is defined by (10.31).

A quantity of great practical interest is the vertical velocity, $\partial w/\partial t$. This is found to be, after differentiation of (10.30),

$$\frac{\partial w}{\partial t} = \frac{P}{\pi \eta (\lambda + 2\mu)} \frac{1}{\sqrt{t^2 - t_0^2}} H(t - t_0). \tag{10.37}$$

At the moment of arrival of the wave this is infinitely large, indicating the passage of a shock. A certain time after this passage, say at $t = t_0 + \Delta t$, the velocity is, approximately, assuming that $\Delta t \ll t_0$,

$$t = t_0 + \Delta t : \quad \frac{\partial w}{\partial t} \approx \frac{P}{\pi \eta (\lambda + 2\mu)} \frac{1}{\sqrt{2t_0 \Delta t}}. \tag{10.38}$$

As the travel time t_0 is a linear function of the distance from the source of the disturbance, see (10.31), this means that the velocities after the passage of the shock are smaller at greater distance from the source, inversely proportional to the square root of the distance.

It should be noted that, although certain characteristics of the complete elastodynamic solution are obtained, the solution of the present problem is rather different from the true solution of the elastodynamic problem for the line load on a half space. In this complete solution (which is presented in Chap. 11), three waves can be distinguished: a compression wave arriving first, then a shear wave, and slightly later the Rayleigh wave. This is the most important wave, because in two dimensions its magnitude remains constant, without any attenuation. After an earthquake the main damage away from the source of the disturbance is caused by the Rayleigh wave.

10.2 Line Pulse on Half Space

The solution for the case of a line pulse, that is a line load of very short duration, see Fig. 10.2, can be derived from the solution for the previous case by replacing the boundary condition (10.9) by the condition

$$z = 0 : (\lambda + 2\mu)\frac{\partial w}{\partial z} = -Q\,\delta(t)\delta(x) \tag{10.39}$$

which may be considered as the time derivative of the boundary condition in the previous problem. In order for the formulas to be dimensionally correct, the dimension of Q should be [F][T]/[L], i.e. kNs/m in SI-units.

Because differentiation with respect to time corresponds to multiplication of the Laplace transform space by s, the solution of the present problem in Laplace transform space can be obtained from the previous solution by multiplication by the parameter s.

In particular, the solution for the Laplace transform of the vertical displacement now will be, multiplying the solution (10.29) by s,

$$\overline{w} = \frac{Q}{\pi\eta(\lambda + 2\mu)} K_0\left(\frac{s}{c_s}\sqrt{x^2 + \eta^2 z^2}\right), \tag{10.40}$$

Fig. 10.2 Line pulse

It now remains to perform the inverse Laplace transformation. Using the formula (5.15.8) from Erdélyi et al. (1954) one obtains

$$w = \frac{Q}{\pi\eta(\lambda+2\mu)}\frac{1}{\sqrt{t^2-t_0^2}}H(t-t_0),\qquad(10.41)$$

where t_0 is the arrival time of the wave, as given by (10.31). This solution can also be obtained from the solution of the previous problem, (10.30), by differentiation with respect to t. The solution (10.41) was first given by Barends (1980). The derivation of expressions for the stresses and the velocity is left as an exercise for the reader.

10.3 Strip Load on Half Space

10.3.1 Strip Pulse

The next problem to be considered in this chapter is the case of a strip load on an elastic half plane, i.e. a constant load over a strip on the surface of the half plane. As a function of time the load may be a pulse of short duration, or a load constant in time. The pulse load will be considered first.

The elastodynamic solution to be derived in this chapter should reduce to the solution for a line load obtained in the previous section, if the width of the loaded strip ($2a$) becomes very small.

In this case the boundary condition for the vertical normal stress is

$$z=0 : \sigma_{zz} = (\lambda+2\mu)\frac{\partial w}{\partial z} = \begin{cases} -q\,\delta(t), & \text{if } |x| < a, \\ 0, & \text{if } |x| > a. \end{cases}\qquad(10.42)$$

The Laplace transform of this condition is

$$z=0 : (\lambda+2\mu)\frac{\partial\overline{w}}{\partial z} = \begin{cases} -q, & \text{if } |x| < a, \\ 0, & \text{if } |x| > a. \end{cases}\qquad(10.43)$$

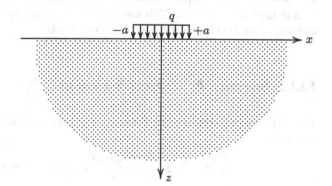

Fig. 10.3 Half plane with strip load

The Fourier transform of this condition is

$$z = 0 \; : \; (\lambda + 2\mu)\frac{d\overline{W}}{dz} = -q\int_{-a}^{a} \exp(i\alpha x)\,dx = -\frac{2q}{\alpha}\sin(\alpha a). \tag{10.44}$$

It is recalled from (10.23) that the general solution of the problem for the half plane $z > 0$ is

$$\overline{W} = A\exp(-\eta z\sqrt{\alpha^2 + s^2/c_s^2}), \tag{10.45}$$

where

$$\eta^2 = \frac{\mu}{\lambda + 2\mu} = \frac{1 - 2\nu}{2(1 - \nu)}. \tag{10.46}$$

It follows from (10.44) and (10.45) that

$$A = \frac{2q\sin(\alpha a)}{(\lambda + 2\mu)\eta\alpha\sqrt{\alpha^2 + s^2/c_s^2}}. \tag{10.47}$$

The Fourier transform of the solution can be obtained by substituting (10.47) into (10.45). This gives

$$\overline{W} = \frac{2q}{(\lambda + 2\mu)}\frac{\sin(\alpha a)\exp(-\eta z\sqrt{\alpha^2 + s^2/c_s^2})}{\eta\alpha\sqrt{\alpha^2 + s^2/c_s^2}}. \tag{10.48}$$

Inverse Fourier transformation, using (10.17), gives

$$\overline{w} = \frac{q}{\pi(\lambda + 2\mu)}\int_{-\infty}^{+\infty}\frac{\sin(\alpha a)\exp(-\eta z\sqrt{\alpha^2 + s^2/c_s^2})}{\eta\alpha\sqrt{\alpha^2 + s^2/c_s^2}}\exp(-i\alpha x)\,d\alpha. \tag{10.49}$$

The vertical normal stress is of particular importance, and probably easier to determine. Its Laplace transform is, because $\sigma_{zz} = (\lambda + 2\mu)\partial w/\partial z$,

$$\overline{\sigma}_{zz} = -\frac{q}{\pi}\int_{-\infty}^{+\infty}\frac{\sin(\alpha a)\exp(-\eta z\sqrt{\alpha^2 + s^2/c_s^2})}{\alpha}\exp(-i\alpha x)\,d\alpha. \tag{10.50}$$

The final mathematical problem now is to determine this integral, and then find the inverse Laplace transform. This can be accomplished by a transformation of the integral, using De Hoop's method (De Hoop, 1960).

10.3.2 Inversion by De Hoop's Method

The first step is to replace the Fourier parameter α by $s\alpha$, so that (10.50) can be written as

$$\overline{\sigma}_{zz} = -\frac{q}{\pi}\int_{-\infty}^{+\infty}\frac{\sin(\alpha as)}{\alpha}\exp[-s(i\alpha x + k\eta z)]\,d\alpha, \tag{10.51}$$

where k is defined by

$$k = \sqrt{1/c_s^2 + \alpha^2}. \tag{10.52}$$

Following a suggestion by Stam (1990), the function $\sin(\alpha a s)$ can be brought into the exponential function by using the relation

$$\sin(\alpha a s) = \frac{\exp(i\alpha a s) - \exp(-i\alpha a s)}{2i}. \tag{10.53}$$

This gives

$$\frac{\overline{\sigma}_{zz}}{q} = \overline{g}(x + a, z, s) - \overline{g}(x - a, z, s), \tag{10.54}$$

where

$$\overline{g}(x, z, s) = \frac{1}{2\pi i} \int_{-\infty}^{+\infty} \frac{\exp[-s(i\alpha x + k\eta z)]}{\alpha} d\alpha. \tag{10.55}$$

This integral will be evaluated, for positive or negative values of x.

It may be noted that the Laplace transform parameter s occurs only once in (10.55), as a factor in an exponential function. It will be attempted to transform the integrand so that the factor $(i\alpha x + k\eta z)$ is replaced by t, which then indicates a Laplace transform.

The integration parameter α is now replaced by p, such that $p = i\alpha$. Equation (10.55) is then transformed into

$$\overline{g}(x, z, s) = \frac{1}{2\pi i} \int_{-i\infty}^{+i\infty} \frac{\exp[-s(px + k\eta z)]}{p} dp, \tag{10.56}$$

where now

$$k = \sqrt{1/c_s^2 - p^2}. \tag{10.57}$$

The next step is to transform the integration path in the complex p-plane, see Fig. 10.4. Two branch cuts are needed, to avoid multiple values for the parameter k. The branch points are located at the points $p \pm 1/c_s$. It is most convenient to let the branch cuts follow the real axis, as indicated in the figure. The case $x > 0$ is considered first.

It is assumed that along the transformed integration path a real positive parameter t (time) can be defined such that

$$t = px + k\eta z. \tag{10.58}$$

It follows from (10.57) and (10.58) that

$$k^2 = 1/c_s^2 - p^2 = (t^2 - 2tpx + p^2 x^2)/\eta^2 z^2, \tag{10.59}$$

or

$$r^2 p^2 - 2txp - \eta^2 z^2/c_s^2, \tag{10.60}$$

Fig. 10.4 Transformed integration paths, for $x < 0$ and $x > 0$

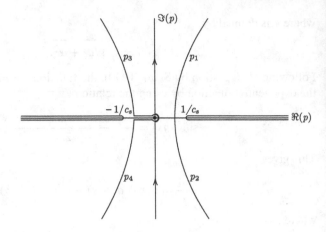

where

$$r^2 = x^2 + \eta^2 z^2. \tag{10.61}$$

Equation (10.60) is a quadratic equation in p, with the two solutions

$$p_1 = \frac{tx}{r^2} + \frac{i\eta z}{r^2}\sqrt{t^2 - r^2/c_s^2}, \qquad p_2 = \frac{tx}{r^2} - \frac{i\eta z}{r^2}\sqrt{t^2 - r^2/c_s^2}. \tag{10.62}$$

If it is assumed that along the two parts of the transformed integration path $r/c_s < t < \infty$, it follows that on the curve p_1, the upper half of the integration path, the value of p varies from the real value $p = x/rc_s$, if $t = r/c_s$, to a complex value $p = (x + i\eta z)t/r^2$, if $t \to \infty$. The point $p = x/rc_s$ is always located between the origin and the branch point $p = 1/c_s$, if $x > 0$, which is assumed here.

The transformation of the integration path from the original path along the imaginary axis to the path consisting of the curves p_2 and p_1 in Fig. 10.4 is permissible if the contributions of the parts along a closing contour at infinity vanish. This will indeed be the case if $x > 0$, because in the factor $\exp(-spx)$ in (10.56) the parameters s, $\Re(p)$ and x are all positive and $p \to \infty$ at infinity. The transformation of the integration path also requires that there are no singularities between the two paths. This means that it must be assumed that the pole at $p = 0$ in the integrand of (10.56) is located just to the left of the original integration path.

It follows from (10.62) that along the paths p_1 and p_2

$$\frac{dp_1}{dt} = \frac{x}{r^2} + \frac{i\eta z}{r^2}\frac{t}{\sqrt{t^2 - r^2/c_s^2}}, \qquad \frac{dp_2}{dt} = \frac{x}{r^2} - \frac{i\eta z}{r^2}\frac{t}{\sqrt{t^2 - r^2/c_s^2}}, \tag{10.63}$$

so that, after some elementary operations,

$$\frac{dp_1/dt}{p_1} = \frac{t\sqrt{t^2 - r^2/c_s^2} + i\eta xz/c_s^2}{(t^2 - \eta^2 z^2/c_s^2)\sqrt{t^2 - r^2/c_s^2}},$$

$$\frac{dp_2/dt}{p_2} = \frac{t\sqrt{t^2 - r^2/c_s^2} - i\eta xz/c_s^2}{(t^2 - \eta^2 z^2/c_s^2)\sqrt{t^2 - r^2/c_s^2}}, \tag{10.64}$$

which are complex conjugates.

Substitution into the integral (10.56) now gives, taking into account that the transformed integration path consists of the two branches p_1 and p_2, with the integration path on p_1 from $t = r/c_s$ to $t = \infty$, and on p_2 from $t = \infty$ to $t = r/c_s$,

$$x > 0 : \bar{g}(x, z, s) = \frac{1}{\pi} \int_{r/c_s}^{\infty} \frac{\eta x z / c_s^2}{(t^2 - \eta^2 z^2 / c_s^2)\sqrt{t^2 - r^2/c_s^2}} \exp(-st)\, dt, \quad (10.65)$$

or

$$x > 0 : \bar{g}(x, z, s) = \frac{1}{\pi} \int_0^{\infty} \frac{\eta x z / c_s^2}{(t^2 - \eta^2 z^2 / c_s^2)\sqrt{t^2 - r^2/c_s^2}} H(t - r/c_s) \exp(-st)\, dt, \quad (10.66)$$

where $H(t - r/c_s)$ is Heaviside's unit step function.

The integral (10.66) has the form of a Laplace transform, which was the purpose of the transformation of the original Fourier integral (10.55). Inverse Laplace transformation now leads to

$$x > 0 : g(x, z, t) = \frac{1}{\pi} \frac{\eta x z / c_s^2}{(t^2 - \eta^2 z^2 / c_s^2)\sqrt{t^2 - r^2/c_s^2}} H(t - r/c_s). \quad (10.67)$$

If $x < 0$ the integration path must be transformed by moving the integration path to the left, see Fig. 10.4, in order that the contributions by the arcs at infinity vanish. This means that the pole at $p = 0$ will be passed, resulting in a contribution to the integral. In the figure the transformed integration path is indicated by the curves p_3 and p_4, with a loop around the pole. It can be shown that the result of the integration along p_3 and p_4 will be the same as before, see (10.67). However, to this expression the contribution by integrating around the pole must be added. Along this path the integration variable p is

$$p = \varepsilon \exp(i\theta), \quad (10.68)$$

where $\varepsilon \to 0$, and the angle θ runs from $\theta = -\pi$ to $\theta = +\pi$ along the small circle around the pole. This contribution can be determined by considering the limiting value of the Laplace transform $\bar{g}(x, z, s)$, as defined in (10.56), for $p \to 0$. This leads to an additional contribution

$$\Delta \bar{g}(x, z, s) = \exp(-s\eta z/c_s)\{1 - H(x)\}, \quad (10.69)$$

where the factor $1 - H(x)$ has been added to indicate that this contribution applies only if $x < 0$. Inverse Laplace transformation of (10.69) gives

$$\Delta g(x, z, t) = \delta(t - \eta z/c_s)\{1 - H(x)\}. \quad (10.70)$$

The results for $\xi > 0$ and $\xi < 0$ can be combined in the single formula

$$g(x, z, t) = \frac{1}{\pi} \frac{\eta x z / c_s^2}{(t^2 - \eta^2 z^2 / c_s^2)\sqrt{t^2 - r^2/c_s^2}} H(t - r/c_s) + \delta(t - \eta z/c_s)\{1 - H(x)\}.$$

$$(10.71)$$

For the calculation of numerical values it is convenient to introduce the dimensionless parameters

$$\xi = x/a, \qquad \zeta = z/a, \qquad \tau = c_s t/a, \qquad \rho = \sqrt{\xi^2 + \zeta^2}. \tag{10.72}$$

Using these parameters, (10.71) can be written as

$$g(x, z, t) = h(\xi, \zeta, \tau)/t_a + \Delta h(\xi, \zeta, \tau)/t_a, \tag{10.73}$$

with

$$h(\xi, \zeta, \tau) = \frac{1}{\pi} \frac{\eta \xi \zeta}{(\tau^2 - \eta^2 \zeta^2)\sqrt{\tau^2 - \rho^2}} H(\tau - \rho), \tag{10.74}$$

$$\Delta h(\xi, \zeta, \tau) = \delta(\tau - \eta \zeta)\{1 - H(\xi)\}. \tag{10.75}$$

In (10.73) t_a is a reference time, defined as

$$t_a = a/c_s. \tag{10.76}$$

It has been assumed that $\delta(t - \eta z/c_s) = (1/t_a)\delta(\tau - \eta \zeta)$, because both delta functions should have an area equal to 1,

$$\int_{-\infty}^{+\infty} \delta(t - \eta z/c_s)\, dt = t_a \int_{-\infty}^{+\infty} \delta(t - \eta z/c_s)\, d\tau = \int_{-\infty}^{+\infty} \delta(\tau - \eta \zeta)\, d\tau = 1. \tag{10.77}$$

Using (10.73) the expression for the vertical normal stress, (10.54) becomes, after inverse Laplace transformation,

$$\frac{\sigma_{zz}}{\sigma_a} = h(\xi + 1, \zeta, \tau) + \Delta h(\xi + 1, \zeta, \tau) - h(\xi - 1, \zeta, \tau) - \Delta h(\xi - 1, \zeta, \tau), \tag{10.78}$$

where σ_a is a reference stress, defined as

$$\sigma_a = q/t_a = qc_s/a. \tag{10.79}$$

It may be noted that the physical dimension of q is a stress multiplied by time, because the physical dimension of the delta function $\delta(t)$ in the boundary condition (10.42) is the inverse of time, to ensure that its integral over time is 1. Thus, the physical dimension of σ_a is indeed a stress, and σ_{zz}/σ_a is dimensionless.

10.3.3 Constant Strip Load

The second problem to be considered in this section is the case of a strip load on an elastic half plane, i.e. a load that is applied at time $t = 0$, and then remains constant in time, see Fig. 10.5. The solution will be obtained by an integration of

Fig. 10.5 Half plane with
strip load

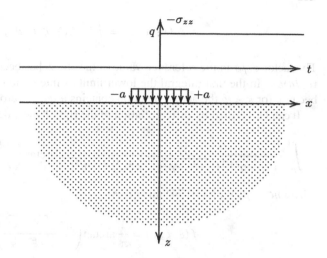

the solution of the problem of a strip pulse, considered above, with respect to the
time parameter t.

The elastostatic equivalent of this problem is a classical problem of applied mechanics (Timoshenko and Goodier, 1970; Sneddon, 1951). This means that the confined elastodynamic solutions to be derived in this section should reduce to the
confined elastostatic limit if $t \rightarrow \infty$. Also, the solution should reduce to the one
obtained for a line load in an earlier section, if the width of the loaded strip $(2a)$
becomes very small.

In this case the boundary condition is

$$z = 0 : \sigma_{zz} = (\lambda + 2\mu)\frac{\partial w}{\partial z} = \begin{cases} -q\, H(t), & \text{if } |x| < a, \\ 0, & \text{if } |x| > a. \end{cases} \tag{10.80}$$

The Laplace transform of this condition is

$$z = 0 : (\lambda + 2\mu)\frac{\partial \overline{w}}{\partial z} = \begin{cases} -q/s, & \text{if } |x| < a, \\ 0, & \text{if } |x| > a. \end{cases} \tag{10.81}$$

Compared to the boundary condition in case of a strip pulse, see (10.43), the difference is a division by s. In the time domain this corresponds to integration with
respect to time t. The stresses will be evaluated for this case, taking the solution for
the strip impulse as the starting point.

The vertical normal stress is, on the basis of a time integration of (10.78),

$$\frac{\sigma_{zz}}{q} = f(\xi + 1, \zeta, \tau) + \Delta f(\xi + 1, \zeta, \tau) - f(\xi - 1, \zeta, \tau) - \Delta f(\xi - 1, \zeta, \tau), \tag{10.82}$$

where

$$f(\xi, \zeta, \tau) = \int_{\rho}^{\tau} h(\xi, \zeta, \kappa)\, d\kappa, \tag{10.83}$$

$$\Delta f(\xi, \zeta, \tau) = \int_0^\tau \Delta h(\xi, \zeta, \kappa)\, d\kappa. \tag{10.84}$$

The factor c_s/a in the reference value of the stress has been omitted, because $dt = (c_s/a)d\kappa$. In the first integral the lower limit of integration has been set equal to ρ, because for $\kappa < \rho$ the actual function contains a factor zero.

It can be verified by differentiation of the right hand side with respect to τ that

$$\int_\rho^\tau \frac{d\kappa}{(\kappa^2 - \eta^2\zeta^2)\sqrt{\kappa^2 - \rho^2}} = \frac{1}{\eta\zeta\sqrt{\rho^2 - \eta^2\zeta^2}} \arctan\left(\frac{\eta\zeta\sqrt{\tau^2 - \rho^2}}{\tau\sqrt{\rho^2 - \eta^2\zeta^2}}\right), \tag{10.85}$$

where $\eta\zeta < \rho$. With (10.74) this gives

$$f(\xi, \zeta, \tau) = \frac{1}{\pi} \arctan\left(\frac{\eta\zeta\sqrt{\tau^2 - \rho^2}}{\tau\xi}\right), \tag{10.86}$$

where it has been used that $\rho^2 = \xi^2 + \eta^2\zeta^2$.

Furthermore, with (10.75) and (10.84) it follows that

$$\Delta f(\xi, \zeta, \tau) = H(\tau - \eta\zeta)\{1 - H(\xi)\}. \tag{10.87}$$

All elements of the expression (10.82) now have been evaluated. This can now be written as

$$\frac{\sigma_{zz}}{q} = \frac{1}{\pi} \arctan\left\{\frac{\eta\zeta\sqrt{\tau^2 - \tau_1^2}}{\tau(\xi + 1)}\right\} H(\tau - \tau_1) - \frac{1}{\pi} \arctan\left\{\frac{\eta\zeta\sqrt{\tau^2 - \tau_2^2}}{\tau(\xi - 1)}\right\} H(\tau - \tau_2)$$
$$+ H(\tau - \eta\zeta)\{H(\xi - 1) - H(\xi + 1)\}, \tag{10.88}$$

where

$$\tau_1^2 = (\xi + 1)^2 + \eta^2\zeta^2, \qquad \tau_2^2 = (\xi - 1)^2 + \eta^2\zeta^2. \tag{10.89}$$

The last term in (10.88) represents a block wave just below the load, travelling with the compression wave velocity in vertical direction.

To avoid passages through infinity at $\xi \pm 1$ it is convenient to transform this equation, using the property that $\arctan(x) = \pi/2 - \arctan(1/x)$ for all positive values of x, and the property that $\arctan(-x) = -\arctan(x)$ for all values of x. This finally gives

$$\frac{\sigma_{zz}}{q} = -\frac{1}{\pi} \arctan\left\{\frac{\tau(\xi + 1)/\eta\zeta}{\sqrt{\tau^2 - \tau_1^2}}\right\} H(\tau - \tau_1) + \frac{1}{\pi} \arctan\left\{\frac{\tau(\xi - 1)/\eta\zeta}{\sqrt{\tau^2 - \tau_2^2}}\right\} H(\tau - \tau_2)$$

$$+ \{H(\xi - 1) - H(\xi + 1)\}H(\tau - \eta\zeta) + \left\{H(\xi + 1) - \frac{1}{2}\right\}H(\tau - \tau_1)$$

$$- \left\{H(\xi - 1) - \frac{1}{2}\right\}H(\tau - \tau_2). \tag{10.90}$$

In terms of the original variables this solution can be written as

$$\frac{\sigma_{zz}}{q} = -\frac{1}{\pi}\arctan\left\{\frac{t(x+a)/\eta z}{\sqrt{t^2-t_1^2}}\right\}H(t-t_1) + \frac{1}{\pi}\arctan\left\{\frac{t(x-a)/\eta z}{\sqrt{t^2-t_2^2}}\right\}H(t-t_2)$$

$$+ \{H(x-a) - H(x+a)\}H(t-\eta z) + \left\{H(x+a) - \frac{1}{2}\right\}H(t-t_1)$$

$$- \left\{H(x-a) - \frac{1}{2}\right\}H(t-t_2), \tag{10.91}$$

where

$$t_1^2 = \{(x+a)^2 + \eta^2 z^2\}/c_s^2, \qquad t_2^2 = \{(x-a)^2 + \eta^2 z^2\}/c_s^2. \tag{10.92}$$

For very large values of time $t \to \infty$ and the solution reduces to

$$\tau \to \infty : \frac{\sigma_{zz}}{q} = -\frac{1}{\pi}\arctan\left\{\frac{x+a}{\eta z}\right\} + \frac{1}{\pi}\arctan\left\{\frac{x-a}{\eta z}\right\}. \tag{10.93}$$

This is indeed the solution of the elastostatic (and confined) problem, see (8.174).

A function to calculate the value of σ_{zz}/q for given values of ξ, ζ, τ and Poisson's ratio ν, is given below, in C. In this program the variables ξ, ζ, τ and ν are denoted by x, z, t and nu.

```
double stress(double x,double z,double t,double nu)
{
double s,eta,eta2,t1,t2;
eta2=(1-2*nu)/(2*(1-nu));eta=sqrt(eta2);s=0;
t1=sqrt((x+1)*(x+1)+eta2*z*z);t2=sqrt((x-1)*(x-1)+eta2*z*z);
if (t>t1) {s-=0.5+atan((x+1)/((eta*z)*sqrt(1-t1*t1/(t*t))))/PI;if
(x+1>0) s+=1;}
if (t>t2) {s+=0.5+atan((x-1)/((eta*z)*sqrt(1-t2*t2/(t*t))))/PI;if
(x-1>0) s-=1;}
if (t>=eta*z) {if (x-1>0) s+=1;if (x+1>0) s-=1;}
return(s);
}
```

Some results have been calculated by a computer program using this function. The stresses below the strip load, as calculated using the present confined solution are shown in the left half of Fig. 10.6, for a region of depth $10\,a$. It has been assumed that the value of time is such that the compression wave has just reached that depth. The value of Poisson's ratio has been assumed to be $\nu = 0$. At a small depth ($z = a$) the stresses are practically equal to the static values. The right half of the figure shows the solution for the full elastodynamic problem, which is considered in Chap. 12 of this book.

For a different value of Poisson's ratio, $\nu = 0.499$, the stresses for the confined solution are shown in the left half of Fig. 10.7. In this practically incompressible case the compression wave travels down very fast, and it will take (relatively) longer for the stresses to approach the static values. It may be noted that the material is

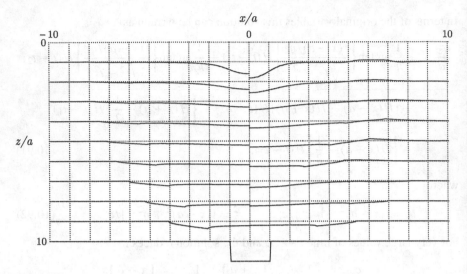

Fig. 10.6 Stresses below a strip load, $\nu = 0$, $c_p t/a = 10$

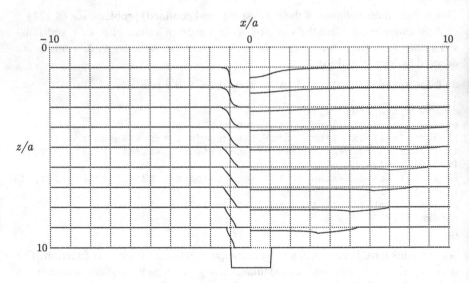

Fig. 10.7 Stresses below a strip load, $\nu = 0.499$, $c_p t/a = 10$

practically undeformable for $\nu \to 0.5$. Because it has been assumed that there are no horizontal deformations, the assumption of incompressibility means that there can be no vertical deformations either. The solution clearly degenerates for $\nu \to 0.5$. The exact solution of the full elastic problem is shown in the right half of the figure. The differences with the confined solution appear to be very large.

In the case $\nu = 0$ the differences between the two solutions are not so very large, although the confined solution does not show the effect of the Rayleigh wave, which is present in the full solution. In the case $v = 0.499$ the differences are so large that the conclusion must be that these confined solutions are perhaps interesting from an educational viewpoint, but can not be considered as a serious alternative for the full elastodynamic solution.

10.4 Point Load on Half Space

Another important case is that of the sudden application of a point load, see Fig. 10.8.

In this case the use of polar coordinates is suggested by the axial symmetry of the problem. Thus the differential equation is

$$\mu\left(\frac{\partial^2 w}{\partial r^2} + \frac{1}{r}\frac{\partial w}{\partial r}\right) + (\lambda + 2\mu)\frac{\partial^2 w}{\partial z^2} = \rho\frac{\partial^2 w}{\partial t^2}. \tag{10.94}$$

The boundary condition is supposed to be

$$z = 0 : (\lambda + 2\mu)\frac{\partial w}{\partial z} = \begin{cases} 0, & \text{if } t < 0 \text{ or } r > a, \\ -F/\pi a^2, & \text{if } t > 0 \text{ and } r < a. \end{cases} \tag{10.95}$$

where a is the radius of the loaded area, which is assumed to be very small.

The initial conditions are that before $t = 0$ the displacement w and its derivative (the velocity) are zero,

$$t = 0 : w = 0, \tag{10.96}$$

$$t = 0 : \frac{\partial w}{\partial t} = 0. \tag{10.97}$$

The Laplace transform of the displacement w is defined by

$$\overline{w} = \int_0^\infty w \exp(-st)\, dt. \tag{10.98}$$

Applying the Laplace transformation to the differential equation (10.94) gives

$$\mu\left(\frac{\partial^2 \overline{w}}{\partial r^2} + \frac{1}{r}\frac{\partial \overline{w}}{\partial r}\right) + (\lambda + 2\mu)\frac{\partial^2 \overline{w}}{\partial z^2} = \rho s^2 \overline{w}. \tag{10.99}$$

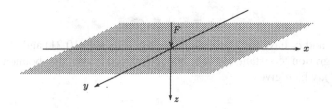

Fig. 10.8 Point load on half space

For radially symmetric problems the Hankel transform is a useful method (Sneddon, 1951). This is defined as

$$\overline{W} = \int_0^\infty \overline{w}\, r\, J_0(\xi r)\, dr,$$

(10.100)

where $J_0(x)$ is the Bessel function of the first kind and order zero. The inverse transform is

$$\overline{w} = \int_0^\infty \overline{W}\, \xi\, J_0(r\xi)\, d\xi.$$

(10.101)

The Hankel transform has the property that the operator

$$\frac{\partial^2}{\partial r^2} + \frac{1}{r}\frac{\partial}{\partial r}$$

is transformed into multiplication by $-\xi^2$. Thus the differential equation (10.99) becomes, after application of the Hankel transformation,

$$-\mu\xi^2\,\overline{W} + (\lambda + 2\mu)\frac{d^2\overline{W}}{dz^2} = \rho s^2\overline{W},$$

(10.102)

which is an ordinary differential equation.

The transformed boundary condition is, applying first the Laplace transform and then the Hankel transform to (10.95),

$$z = 0 \,:\, (\lambda + 2\mu)\frac{d\overline{W}}{dz} = -\frac{F}{\pi a^2 s}\int_0^a r\, J_0(\xi r)\, dr.$$

(10.103)

When a is very small the Bessel function may be approximated by its first term in a series expansion, which is 1, so that one obtains

$$z = 0 \,:\, (\lambda + 2\mu)\frac{d\overline{W}}{dz} = -\frac{F}{2\pi s}.$$

(10.104)

The general solution of (10.102) vanishing for $z \to \infty$ is

$$\overline{W} = A\exp(-\gamma\eta z),$$

(10.105)

with

$$\gamma = \xi^2 + s^2/c_s^2,$$

(10.106)

and where η and c_s have the same meaning as before, see (10.21) and (10.22).

The integration constant A can be determined from the boundary condition (10.104), which gives

$$A = \frac{F}{2\pi\eta(\lambda + 2\mu)\gamma s}.$$

(10.107)

The final solution for the transformed displacement is

$$\overline{W} = \frac{F}{2\pi\eta(\lambda+2\mu)s}\frac{\exp[-\eta z\sqrt{\xi^2+s^2/c_s^2}]}{\sqrt{\xi^2+s^2/c_s^2}}. \tag{10.108}$$

Although this may appear to be a rather complex formula, it happens that its inverse Hankel transform can be found in the literature (Erdélyi et al., 1954, 8.2.24). The result is

$$\overline{w} = \frac{F}{2\pi\eta(\lambda+2\mu)s}\frac{1}{\sqrt{r^2+\eta^2z^2}}\exp\left(-\frac{s}{c_s}\sqrt{r^2+\eta^2z^2}\right). \tag{10.109}$$

The inverse Laplace transform is very simple (Churchill, 1972),

$$w = \frac{F}{2\pi\eta(\lambda+2\mu)}\frac{1}{\sqrt{r^2+\eta^2z^2}}H(t-t_0), \tag{10.110}$$

where

$$t_0 = \frac{\sqrt{r^2+\eta^2z^2}}{c_s}. \tag{10.111}$$

Equation (10.110) is the solution of the problem. Again it may be surprising that such a simple solution has been obtained. In this case there is a downward displacement which occurs at the arrival of the wave. The magnitude of the displacement decreases inversely proportional with the distance from the source of the disturbance. It may be noted that the steady state displacement, for $t \to \infty$, agrees in form with the fully elastic solution given in Chap. 8, see (8.28). The displacement is inversely proportional to the distance from the source in both formulas, and inversely proportional to the modulus of elasticity E (although that is not very surprising in a linear model). The two formulas differ only in their respective dependence upon Poisson's ratio ν.

It deserves to be mentioned that the approximate solution derived here, for the case of horizontally confined displacements, markedly differs from the complete elastic solution (Pekeris, 1955). When considering this complete solution, see Chap. 13, it will appear that shortly after the arrival of the shear wave considered here very large displacements occur, due to the generation of Rayleigh waves.

10.5 Periodic Load on a Half Space

In the previous sections some solutions of problems of wave propagation in a confined elastic half space have been considered, especially for loads that were applied stepwise. Another important class of problems is that of a half space with a periodic load on its surface. A problem of this class will be considered in this section, namely the problem of a uniform periodic load over a circular area, on a confined elastic half space.

Fig. 10.9 Circular load on half space

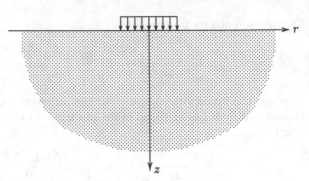

As in the previous sections the problem is simplified by assuming that the only non-vanishing displacement is the vertical displacement w, for which the differential equation then is, in the case of radial symmetry,

$$\mu\left(\frac{\partial^2 w}{\partial r^2} + \frac{1}{r}\frac{\partial w}{\partial r}\right) + (\lambda + 2\mu)\frac{\partial^2 w}{\partial z^2} = \rho\frac{\partial^2 w}{\partial t^2}. \qquad (10.112)$$

In the problem to be considered the load is a periodically varying load on a circular area at the surface, see Fig. 10.9.

The boundary condition is

$$z = 0 : (\lambda + 2\mu)\frac{\partial w}{\partial z} = \begin{cases} 0, & \text{if } t < 0 \text{ or } r > a, \\ -p\sin(\omega t), & \text{if } t > 0 \text{ and } r < a, \end{cases} \qquad (10.113)$$

where a is the radius of the loaded area, and ω is the circular frequency of the periodic load.

The Laplace transform of the vertical displacement w is defined as

$$\overline{w} = \int_0^\infty w \exp(-st)\,dt. \qquad (10.114)$$

Assuming that the initial values of the displacement and velocity are zero, the differential equation (10.112) now becomes

$$\mu\left(\frac{\partial^2 \overline{w}}{\partial r^2} + \frac{1}{r}\frac{\partial \overline{w}}{\partial r}\right) + (\lambda + 2\mu)\frac{\partial^2 \overline{w}}{\partial z^2} = \rho s^2 \overline{w}, \qquad (10.115)$$

and the boundary condition (10.113) is transformed into

$$z = 0 : (\lambda + 2\mu)\frac{\partial \overline{w}}{\partial z} = \begin{cases} 0, & \text{if } r > a, \\ -p\omega/(s^2 + \omega^2), & \text{if } r < a. \end{cases} \qquad (10.116)$$

The radial symmetry of the problem suggests the use of the Hankel transform

$$W = \int_0^\infty \overline{w}\, r\, J_0(r\xi)\,dr. \qquad (10.117)$$

The differential equation (10.115) then is transformed into the ordinary differential equation

$$(\lambda + 2\mu)\frac{d^2\overline{W}}{dz^2} = (\rho s^2 + \mu\xi^2)\overline{W}, \tag{10.118}$$

or

$$\frac{d^2\overline{W}}{dz^2} = \eta^2(s^2/c_s^2 + \xi^2)\overline{W}, \tag{10.119}$$

where c_s is the velocity of shear waves,

$$c_s^2 = \frac{\mu}{\rho}, \tag{10.120}$$

and η is an elastic coefficient, defined by

$$\eta^2 = \frac{\mu}{\lambda + 2\mu} = \frac{1 - 2\nu}{2(1 - \nu)}. \tag{10.121}$$

If a parameter γ is introduced by the definition

$$\gamma^2 = s^2/c_s^2 + \xi^2, \tag{10.122}$$

the solution of the differential equation (10.119) vanishing at infinity can be written as

$$\overline{W} = A\exp(-\gamma\eta z). \tag{10.123}$$

The integration constant A must be determined from the Hankel transform of the boundary condition (10.116). Using the well known integral (Erdélyi et al., 1954, 8.3.18)

$$\int_0^a r\,J_0(r\xi)\,dr = \frac{a}{\xi}J_1(a\xi), \tag{10.124}$$

this gives

$$A = \frac{\rho\omega a}{\gamma\eta(\lambda + 2\mu)(s^2 + \omega^2)\xi}J_1(a\xi), \tag{10.125}$$

so that the solution of the transformed problem is

$$\overline{W} = \frac{\rho\omega a}{\gamma\eta(\lambda + 2\mu)(s^2 + \omega^2)\xi}J_1(a\xi)\exp(-\gamma z/m). \tag{10.126}$$

The inverse Hankel transformation of this result is

$$\overline{w} = \frac{\rho\omega a}{\eta(\lambda + 2\mu)(s^2 + \omega^2)}\int_0^\infty \frac{J_1(a\xi)\,J_0(r\xi)\exp(-\gamma z/m)}{\gamma}\,d\xi. \tag{10.127}$$

It will not be attempted to evaluate this integral. Restriction will be made to two special results: the displacement of the center of the loaded area, $r = 0, z = 0$, and the displacements for a vibrating point load.

Displacement of the Origin

The displacement of the point $r = 0, z = 0$ is, with (10.127),

$$\overline{w}_0 = \frac{p\omega a}{\eta(\lambda + 2\mu)(s^2 + \omega^2)} \int_0^\infty \frac{J_1(a\xi)}{\gamma} \, d\xi, \quad (10.128)$$

or, in terms of the original parameters,

$$\overline{w}_0 = \frac{p\omega a}{\eta(\lambda + 2\mu)(s^2 + \omega^2)} \int_0^\infty \frac{J_1(a\xi)}{\sqrt{\xi^2 + s^2/c_s^2}} \, d\xi. \quad (10.129)$$

This is a well known integral (Erdélyi et al., 1954, 8.4.3). The result is

$$\overline{w}_0 = \frac{p\omega c}{\eta(\lambda + 2\mu) s \, (s^2 + \omega^2)} [1 - \exp(-as/c_s)]. \quad (10.130)$$

This is the Laplace transform of the displacement of the center of the loaded area. Inverse Laplace transformation gives

$$w_0 = \frac{pc_s}{\eta(\lambda + 2\mu)\omega} \{H(t) - H(t - 2t_c) - \cos(\omega t) + \cos[\omega(t - 2t_c)]\}, \quad (10.131)$$

where t_c is a characteristic time,

$$t_c = a/2c_s, \quad (10.132)$$

and $H(t)$ is Heaviside's unit step function,

$$H(t) = \begin{cases} 0, & \text{if } t < 0, \\ 1, & \text{if } t > 0. \end{cases} \quad (10.133)$$

For large values of time the two step functions cancel and the solution reduces to

$$w_0 = -\frac{pc}{\eta(\lambda + 2\mu)\omega} \{\cos(\omega t) - \cos[\omega(t - 2t_c)]\}. \quad (10.134)$$

After some elaboration this can also be written as

$$w_0 = \frac{pa \, \sin(\omega t_c)}{\eta(\lambda + 2\mu)(\omega t_c)} \sin[\omega(t - t_c)]. \quad (10.135)$$

The phase angle turns out to be ωt_c. As in the previous cases the simplicity of the final solution may be noted.

For very small frequencies, $\omega \to 0$, the solution approaches the static result

$$\omega \to 0 : w_0 = w_s = \frac{pa}{\eta(\lambda + 2\mu)}. \quad (10.136)$$

This means that the dynamic amplification factor can be written as

$$\frac{|w_0|}{|w_s|} = \frac{|\sin(\omega t_c)|}{\omega t_c}.$$ (10.137)

This is shown in Fig. 10.10 as a function of a dimensionless frequency ω/ω_c, where ω_c is defined by

$$\omega_c = \frac{1}{t_c} = \frac{2c_s}{a} = \sqrt{\frac{4\mu}{\rho a^2}}.$$ (10.138)

The characteristic frequency ω_c has the character of the square root of the ratio of a spring stiffness and a mass, as usual in dynamic problems. In engineering practice the shear wave velocity c_s usually is of the order of magnitude of 100 m/s, and the physical dimension of the foundation size a is of the order of 1 m, or perhaps as big as 10 m. This means that the characteristic frequency is of the order of magnitude of 20 s^{-1} or 200 s^{-1}. This is a rather large value, and it means that in many cases the value of ω/ω_0 will be rather small. Only in case of very rapid fluctuations the dimensionless frequency may be larger than one. An example of such a phenomenon is pile driving, by hammering or by high frequency vibrating.

Because it can be expected that in engineering practice the value of ω/ω_c will usually be of the order of magnitude of 1, or smaller, the most common values in Fig. 10.10 will be located at the left part of the figure. It may be noted that for certain large values of ω/ω_c the dynamic amplitude may be zero. This can also be seen from (10.137), from which it follows that $w_0 = 0$ for all values of the frequency for which $\omega/\omega_c = k\pi$, where k is any integer. For these frequencies the dynamic amplitude is zero, indicating extremely stiff behaviour. Such a very stiff behaviour will not really be observed in practice, because the assumptions underlying the present theory are only weak reflections of the complex behaviour of real soils. Also, the displacement at the center of the circle may be zero, but this does not mean that the displacements are zero over the entire loaded area.

The phase angle ψ has been found to be ωt_c. Thus there may be a considerable damping, except when the frequency ω is extremely small. This phenomenon is sometimes called *radiation damping*. It is produced by the spreading of the energy over an ever larger area.

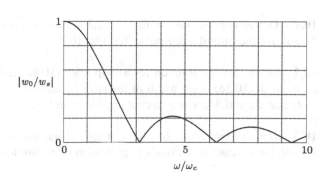

Fig. 10.10 Dynamic amplification

Vibrating Point Load

If the radius of the loaded area a is very small the Bessel function $J_1(a\xi)$ in the solution (10.127) can be approximated by the first term in its series expansion, $J_1(a\xi) \approx \frac{1}{2}a\xi$. This solution then reduces to

$$\overline{w} = \frac{p\omega a^2}{2\eta(\lambda + 2\mu)(s^2 + \omega^2)} \int_0^\infty \frac{\xi\, J_0(r\xi)\exp(-\gamma\eta z)}{\gamma}\, d\xi \qquad (10.139)$$

or, writing $F = p\pi a^2$ for the total load,

$$\overline{w} = \frac{F\omega}{2\pi\eta(\lambda + 2\mu)(s^2 + \omega^2)} \int_0^\infty \frac{\xi\, J_0(r\xi)\exp[-(\eta z)\sqrt{\xi^2 + s^2/c_s^2}]}{\sqrt{\xi^2 + s^2/c_s^2}}\, d\xi. \qquad (10.140)$$

This is a well known inverse Hankel transform (Erdélyi et al., 1954, 8.2.24). The result is

$$\overline{w} = \frac{F\omega}{2\pi\eta(\lambda + 2\mu)(s^2 + \omega^2)} \cdot \frac{\exp[-(s/c_s)\sqrt{r^2 + \eta^2 z^2}]}{\sqrt{r^2 + \eta^2 z^2}}. \qquad (10.141)$$

Inverse Laplace transformation now is simple, using the standard formula for the Laplace transform of the function $\sin(\omega t)$ and the translation theorem,

$$w = \frac{F}{2\pi\eta(\lambda + 2\mu)}\, \frac{\sin[\omega(t - t_0)]}{\sqrt{r^2 + \eta^2 z^2}}\, H(t - t_0), \qquad (10.142)$$

where, as before,

$$t_0 = \frac{\sqrt{r^2 + \eta^2 z^2}}{c_s}. \qquad (10.143)$$

Again a simple result is obtained.

Problems

10.1   Derive expressions for the vertical normal stress σ_{zz} and for the velocity $\partial w/\partial t$, for the case of a line pulse, see Fig. 10.2.

10.2   Verify that the solution for a strip load, (10.91) reduces to the solution for a line load, (10.36) if the width of the strip $2a$ tends towards zero, with $P = 2qa$. Note that the variable x is also contained in t_1 and t_2.

10.3   Derive expressions for the vertical normal stress σ_{zz} and for the velocity $\partial w/\partial t$, for the case of the sudden application of a point load, see Fig. 10.8.

Chapter 11
Line Load on Elastic Half Space

In this chapter some basic problems of an elastic half space are considered, in particular problems for a line pulse or a line load on the surface of the half space. A problem of this type is often denoted as a Lamb problem, because the first solutions for such problems were obtained by Lamb (1904). Lamb's solution, which started from the solution of the problem for a periodic load, can be found, using more modern formulations and techniques, in many textbooks, see e.g. Fung (1965), Achenbach (1975), Graff (1975) and Miklowitz (1978). In the present book the solutions will be obtained by De Hoop's version of the Cagniard method, which uses a combination of Laplace and Fourier transform methods (De Hoop, 1960, 1970; Cagniard et al., 1962), see also Appendix A. An alternative technique has been presented by Eringen and Suhubi (1975), using a *self-similar solution* method, in which the number of independent variables is reduced by one, which is applicable in the case of a concentrated load.

The problems to be considered in this chapter are the displacements due to a line pulse on the surface, and the stresses due to a constant line load on the surface. The solution of the first problem will be given in great detail. For the second problem the solutions are given with only an outline of the derivation, as the solution methods are quite similar. It will be shown that, in the limit for large values of time, the solution of the elastodynamic problem reduces to the known solution of the elastostatic solution. The solutions also appear to be in agreement with general results of theoretical elastodynamics, such as the appearance and the behaviour of Rayleigh waves.

The solutions will be given in the form of analytic expressions, with elementary algorithms to calculate numerical data. A computer program for a constant line load is available as the program LINELOAD.

11.1 Line Pulse

11.1.1 Description of the Problem

The first problem to be considered is the case of a line pulse on an elastic half plane, see Fig. 11.1. This is an important problem in seismology, where the load is caused

Fig. 11.1 Half plane with impulse load

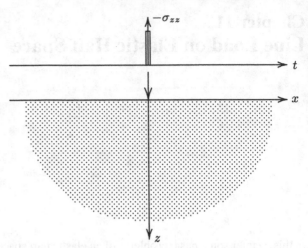

by an explosion of very short duration, and the displacements of the surface are measured, at various distances from the load, as a function of time.

The basic equations are the equations of motion in two dimensions,

$$\frac{\partial \sigma_{xx}}{\partial x} + \frac{\partial \sigma_{zx}}{\partial z} = \rho \frac{\partial^2 u}{\partial t^2},\tag{11.1}$$

$$\frac{\partial \sigma_{xz}}{\partial x} + \frac{\partial \sigma_{zz}}{\partial z} = \rho \frac{\partial^2 w}{\partial t^2},\tag{11.2}$$

where u and w are the displacement components in x-direction and z-direction, respectively, and where ρ is the density of the material.

The material is supposed to be linear elastic, so that the stresses and the strains are related by the generalized form of Hooke's law,

$$\sigma_{xx} = \lambda \left(\frac{\partial u}{\partial x} + \frac{\partial w}{\partial z} \right) + 2\mu \frac{\partial u}{\partial x},\tag{11.3}$$

$$\sigma_{zz} = \lambda \left(\frac{\partial u}{\partial x} + \frac{\partial w}{\partial z} \right) + 2\mu \frac{\partial w}{\partial z},\tag{11.4}$$

$$\sigma_{zx} = \mu \left(\frac{\partial u}{\partial z} + \frac{\partial w}{\partial x} \right),\tag{11.5}$$

where λ and μ are the Lamé constants of the material.

Substitution of (11.3)–(11.5) into (11.1) and (11.2) leads to the basic differential equations

$$(\lambda + \mu) \frac{\partial}{\partial x} \left(\frac{\partial u}{\partial x} + \frac{\partial w}{\partial z} \right) + \mu \left(\frac{\partial^2 u}{\partial x^2} + \frac{\partial^2 u}{\partial z^2} \right) = \rho \frac{\partial^2 u}{\partial t^2},\tag{11.6}$$

$$(\lambda + \mu) \frac{\partial}{\partial z} \left(\frac{\partial u}{\partial x} + \frac{\partial w}{\partial z} \right) + \mu \left(\frac{\partial^2 w}{\partial x^2} + \frac{\partial^2 w}{\partial z^2} \right) = \rho \frac{\partial^2 w}{\partial t^2}.\tag{11.7}$$

These equations can also be written as

$$(\lambda + 2\mu)\frac{\partial^2 u}{\partial x^2} + (\lambda + \mu)\frac{\partial^2 w}{\partial z\partial x} + \mu\frac{\partial^2 u}{\partial z^2} = \rho\frac{\partial^2 u}{\partial t^2}, \tag{11.8}$$

$$(\lambda + 2\mu)\frac{\partial^2 w}{\partial z^2} + (\lambda + \mu)\frac{\partial^2 u}{\partial z\partial x} + \mu\frac{\partial^2 w}{\partial x^2} = \rho\frac{\partial^2 w}{\partial t^2}. \tag{11.9}$$

The boundary conditions for a line pulse on the surface $z = 0$ are

$$z = 0 : \sigma_{zx} = 0, \tag{11.10}$$

$$z = 0 : \sigma_{zz} = -Q\delta(t)\delta(x), \tag{11.11}$$

where Q is the strength of the line pulse, and $\delta(x)$ and $\delta(t)$ are Dirac delta functions (Churchill, 1972), for instance

$$\delta(t) = \begin{cases} 0, & \text{if } |t| > \epsilon, \\ 1/2\epsilon, & \text{if } |t| < \epsilon, \end{cases} \tag{11.12}$$

with $\epsilon \to 0$. The total area below the function is 1, $\int_{-\infty}^{\infty} \delta(t)\,dt = 1$.

11.1.2 Solution by Integral Transform Method

The solution of the problem is sought by using Laplace and Fourier transforms. The Laplace transforms of the displacements are defined as

$$\bar{u} = \int_0^\infty u\exp(-st)\,dt, \tag{11.13}$$

$$\bar{w} = \int_0^\infty w\exp(-st)\,dt. \tag{11.14}$$

If it is assumed that the displacements and the velocities are zero at the time of loading $t = 0$, the transformed basic equations are

$$(\lambda + 2\mu)\frac{\partial^2\bar{u}}{\partial x^2} + (\lambda + \mu)\frac{\partial^2\bar{w}}{\partial z\partial x} + \mu\frac{\partial^2\bar{u}}{\partial z^2} = \rho s^2\bar{u}, \tag{11.15}$$

$$(\lambda + 2\mu)\frac{\partial^2\bar{w}}{\partial z^2} + (\lambda + \mu)\frac{\partial^2\bar{u}}{\partial z\partial x} + \mu\frac{\partial^2\bar{w}}{\partial x^2} = \rho s^2\bar{w}. \tag{11.16}$$

Fourier transforms are defined as

$$U = \int_{-\infty}^{\infty} \bar{u}\exp(is\alpha x)\,dx, \tag{11.17}$$

$$W = \int_{-\infty}^{\infty} \bar{w}\exp(is\alpha x)\,dx, \tag{11.18}$$

with the inverse transforms

$$\overline{u} = \frac{s}{2\pi} \int_{-\infty}^{\infty} U \exp(-is\alpha x)\, d\alpha, \tag{11.19}$$

$$\overline{w} = \frac{s}{2\pi} \int_{-\infty}^{\infty} W \exp(-is\alpha x)\, d\alpha. \tag{11.20}$$

It may be noted that the usual Fourier transform variable α has been replaced by $s\alpha$, for future convenience.

If it is assumed that the displacements and their first derivative with respect to x vanish at infinity, it can be shown, using partial integration, that

$$\int_{-\infty}^{\infty} \frac{\partial \overline{u}}{\partial x} \exp(is\alpha x)\, dx = -i\alpha s \overline{U}, \tag{11.21}$$

$$\int_{-\infty}^{\infty} \frac{\partial^2 \overline{u}}{\partial x^2} \exp(is\alpha x)\, dx = -\alpha^2 s^2 \overline{U}. \tag{11.22}$$

Similar results apply to the Fourier transform of the vertical displacement.

The transformed form of the basic equations (11.15) and (11.16) is

$$c_s^2 \frac{d^2 \overline{U}}{dz^2} - is\alpha(c_p^2 - c_s^2)\frac{d\overline{W}}{dz} = s^2(1 + c_p^2\alpha^2)\overline{U}, \tag{11.23}$$

$$c_p^2 \frac{d^2 \overline{W}}{dz^2} - is\alpha(c_p^2 - c_s^2)\frac{d\overline{U}}{dz} = s^2(1 + c_s^2\alpha^2)\overline{W}, \tag{11.24}$$

where c_p and c_s are the velocities of compression waves and shear waves, respectively,

$$c_p^2 = (\lambda + 2\mu)/\rho, \tag{11.25}$$

$$c_s^2 = \mu/\rho. \tag{11.26}$$

It is assumed that the solution of the two equations (11.23) and (11.24) can be expressed as

$$\overline{U} = iA \exp(-\gamma sz), \tag{11.27}$$

$$\overline{W} = B \exp(-\gamma sz), \tag{11.28}$$

where γ is an unknown parameter at this stage, and A and B are unknown integration constants.

Substitution of (11.27) and (11.28) into (11.23) and (11.24) gives

$$(1 + c_p^2\alpha^2 - c_s^2\gamma^2)A - (c_p^2 - c_s^2)\alpha\gamma B = 0, \tag{11.29}$$

$$(c_p^2 - c_s^2)\alpha\gamma A + (1 + c_s^2\alpha^2 - c_p^2\gamma^2)B = 0. \tag{11.30}$$

This homogeneous system of linear equations has solutions only for the two values of γ^2 for which the determinant of the system is zero. These values can be written as $\gamma = \pm \gamma_p$ and $\gamma = \pm \gamma_s$, where

$$\gamma_p = \sqrt{\alpha^2 + 1/c_p^2}, \tag{11.31}$$

$$\gamma_s = \sqrt{\alpha^2 + 1/c_s^2}. \tag{11.32}$$

It is understood that in these equations the positive root is taken. The solutions with the negative sign should be omitted to ensure that the solutions remain bounded for $z \to \infty$.

The solution of the transformed problem now is found to be

$$\overline{U} = i\alpha C_p \exp(-\gamma_p s z) + i\gamma_s C_s \exp(-\gamma_s s z), \tag{11.33}$$

$$\overline{W} = \gamma_p C_p \exp(-\gamma_p s z) + \alpha C_s \exp(-\gamma_s s z). \tag{11.34}$$

Inverse Fourier transformation of these expressions gives

$$\overline{u} = \frac{is}{2\pi} \int_{-\infty}^{\infty} \{\alpha C_p \exp(-s\gamma_p z) + \gamma_s C_s \exp(-s\gamma_s z)\} \exp(-is\alpha x)\, d\alpha, \tag{11.35}$$

$$\overline{w} = \frac{-s}{2\pi} \int_{-\infty}^{\infty} \{\gamma_p C_p \exp(-s\gamma_p z) + \alpha C_s \exp(-s\gamma_s z)\} \exp(-is\alpha x)\, d\alpha. \tag{11.36}$$

The notations C_p and C_s are used to indicate the strength of the compression wave and the shear wave, respectively, as suggested by the parameters γ_p and γ_s in the two parts of the solution.

In order to determine the coefficients C_p and C_s the boundary conditions must be used. As these are expressed in terms of the stresses it is convenient at this stage to obtain expressions for the Laplace transforms of the stresses, using the definitions (11.3), (11.4) and (11.5). This gives

$$\overline{\sigma}_{xx} = \frac{s^2}{2\pi} \int_{-\infty}^{\infty} \{(2\mu\alpha^2 - \lambda/c_p^2)C_p \exp(-\gamma_p s z) + 2\mu\alpha\gamma_s C_s \exp(-\gamma_s s z)\}$$
$$\times \exp(-is\alpha x)\, d\alpha, \tag{11.37}$$

$$\overline{\sigma}_{zz} = -\frac{s^2\mu}{2\pi} \int_{-\infty}^{\infty} \{(2\alpha^2 + 1/c_s^2)C_p \exp(-\gamma_p s z) + 2\alpha\gamma_s C_s \exp(-\gamma_s s z)\}$$
$$\times \exp(-is\alpha x)\, d\alpha, \tag{11.38}$$

$$\overline{\sigma}_{zx} = -\frac{i\mu s^2}{2\pi} \int_{-\infty}^{\infty} \{2\alpha\gamma_p C_p \exp(-\gamma_p s z) + (2\alpha^2 + 1/c_s^2)C_s \exp(-\gamma_s s z)\}$$
$$\times \exp(-is\alpha x)\, d\alpha. \tag{11.39}$$

For future reference the isotropic stress σ is given as well. This quantity is defined as

$$\sigma = \frac{1}{2}(\sigma_{xx} + \sigma_{zz}). \tag{11.40}$$

It follows from (11.37) and (11.38) that its Laplace transform is

$$\overline{\sigma} = -\frac{(\lambda + \mu)s^2}{2\pi c_p^2} \int_{-\infty}^{\infty} C_p \exp(-\gamma_p sz - is\alpha x) \, d\alpha. \tag{11.41}$$

The boundary conditions (11.10) and (11.11) can be expressed as Laplace transforms as

$$z = 0 : \overline{\sigma}_{zx} = 0, \tag{11.42}$$

$$z = 0 : \overline{\sigma}_{zz} = -\frac{Qs}{2\pi} \int_{-\infty}^{\infty} \exp(-is\alpha x) d\alpha. \tag{11.43}$$

Using these boundary conditions the coefficients C_p and C_s can be determined. The result is

$$C_p = \frac{Q}{\mu s} \frac{2\alpha^2 + 1/c_s^2}{(2\alpha^2 + 1/c_s^2)^2 - 4\alpha^2 \gamma_p \gamma_s}, \tag{11.44}$$

$$C_s = -\frac{Q}{\mu s} \frac{2\alpha \gamma_p}{(2\alpha^2 + 1/c_s^2)^2 - 4\alpha^2 \gamma_p \gamma_s}. \tag{11.45}$$

11.1.3 The Vertical Displacement

The vertical displacement is of particular interest. It is found from (11.36) that its Laplace transform is

$$\overline{w} = \overline{w}_1 + \overline{w}_2 + \overline{w}_3, \tag{11.46}$$

where

$$\overline{w}_1 = \frac{Q}{2\pi\mu} \int_{-\infty}^{\infty} \frac{\gamma_p(2\alpha^2 + 1/c_s^2)}{(2\alpha^2 + 1/c_s^2)^2 - 4\alpha^2 \gamma_p \gamma_s} \exp[-s(\gamma_p z + i\alpha x)] \, d\alpha, \tag{11.47}$$

$$\overline{w}_2 + \overline{w}_2 = -\frac{Q}{2\pi\mu} \int_{-\infty}^{\infty} \frac{2\alpha^2 \gamma_p}{(2\alpha^2 + 1/c_s^2)^2 - 4\alpha^2 \gamma_p \gamma_s} \exp[-s(\gamma_s z + i\alpha x)] \, d\alpha. \tag{11.48}$$

The two integrals (11.47) and (11.48) will be evaluated separately, using De Hoop's method (see Appendix A). The second integral will be separated into two parts, w_2 and w_3. For this first problem of the chapter, the analysis will be given in full detail.

The First Integral

Using the substitution $p = i\alpha$, the first integral, (11.47), can be written as

$$\overline{w}_1 = \frac{Q}{2\pi i \mu} \int_{-i\infty}^{i\infty} \frac{\gamma_p(1/c_s^2 - 2p^2)}{(1/c_s^2 - 2p^2)^2 + 4p^2\gamma_p\gamma_s} \exp[-s(\gamma_p z + px)]\,dp, \quad (11.49)$$

where now

$$\gamma_p = \sqrt{1/c_p^2 - p^2}, \quad (11.50)$$

$$\gamma_s = \sqrt{1/c_s^2 - p^2}. \quad (11.51)$$

The appearance of the factor γ_p in the exponential function in (11.49) suggests that it represents the contribution of the compression waves.

In the method of De Hoop (1960) it is attempted to transform the integration path in the complex p-plane in such a way that the integral obtains the form of a Laplace transform integral. For this purpose a parameter t is introduced (later to be identified with the time), defined as

$$t = \gamma_p z + px, \quad (11.52)$$

with t being real and positive, by assumption. The shape of the transformed integration path remains undetermined in this stage.

The integrand of the integral in (11.49) has singularities in the form of branch points in the points $p = \pm 1/c_p$ and $p = \pm 1/c_s$ and simple poles in the points $p = \pm 1/c_r$, where c_r is the Rayleigh wave velocity, which is slightly smaller than the shear wave velocity. It may be noted that $c_p > c_s > c_r$, so that $1/c_p < 1/c_s < 1/c_r$. The integration path from $p = -i\infty$ to $p = \infty$ is now transformed to the two paths p_1 and p_2 shown in Fig. 11.2, with the parameter t varying along these two curves from some initial value to infinity.

It may also be noted that the branch cut is necessary because the factors γ_p and γ_s are multiple valued. In the denominator of (11.49) the product $\gamma_p\gamma_s$ could be made single valued by a branch cut between $p = 1/c_p$ and $p = 1/c_s$ only, but the appearance of a factor γ_p in the numerator requires that the branch cut should extend towards infinity.

It follows from (11.50) and (11.52) that

$$r^2p^2 - 2tpx + t^2 - z^2/c_p^2 = 0, \quad (11.53)$$

where

$$r^2 = x^2 + z^2. \quad (11.54)$$

Equation (11.53) is a quadratic expression in p, with the two solutions

$$p_1 = \frac{tx}{r^2} + \frac{iz}{r^2}\sqrt{t^2 - t_p^2}, \quad (11.55)$$

Fig. 11.2 Original and
transformed integration path
for the first integral

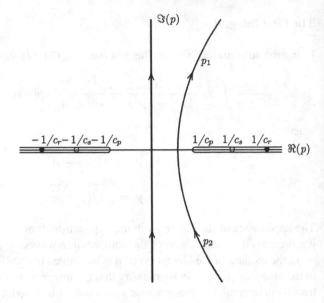

$$p_2 = \frac{tx}{r^2} - \frac{iz}{r^2}\sqrt{t^2 - t_p^2}, \tag{11.56}$$

where

$$t_p = r/c_p. \tag{11.57}$$

If it is assumed that the parameter t varies in the interval $t_p < t < \infty$, it follows that
the two paths in the complex p-plane are continuous, intersecting at the real axis in
the point $p = t_p x/r^2 = (x/r)(1/c_p)$ (which is to the left of the first branch point
because $x \le r$), and approaching infinity at the positive and negative sides of the
real axis, respectively. The precise shape of the curves p_1 and p_2 depends upon the
values of x and z, i.e. the location of the point considered in the physical plane.

Actually, the two branches, p_1 and p_2, of the transformed integration path are
hyperbolas, with the slope of the asymptote of p_1 at infinity being z/x.

The Upper Part of the Integration Path

It follows from (11.55) that on the part p_1 of the integration path

$$p = p_1 : \frac{dp}{dt} = \frac{x}{r^2} + \frac{iz}{r^2}\frac{t}{\sqrt{t^2 - t_p^2}}. \tag{11.58}$$

Furthermore, it follows from (11.52) that

$$p = p_1 : \gamma_p = \frac{tz}{r^2} - \frac{ix}{r^2}\sqrt{t^2 - t_p^2} = -i\sqrt{t^2 - t_p^2}\left\{ \frac{x}{r^2} + \frac{iz}{r^2}\frac{t}{\sqrt{t^2 - t_p^2}} \right\}. \tag{11.59}$$

It may be noted that on this part of the transformed integration path and for $x > 0$ (which will later appear to be the main branch considered), $\Re(\gamma_p) > 0$ and $\Im(\gamma_p) < 0$, so that $\arg(\gamma_p) < 0$. This is in agreement with the definition in (11.50) and its analytic continuation into the upper right quarter of the complex p-plane.

It follows from (11.58) and (11.59) that

$$p = p_1 \; : \; \frac{dp}{dt} = \frac{i\gamma_p}{\sqrt{t^2 - t_p^2}}. \tag{11.60}$$

Because $\Re(\gamma_p) > 0$ and $\Im(\gamma_p) < 0$ it follows that $\Re(dp/dt) > 0$ and $\Im(dp/dt) > 0$, which is in agreement with the shape of the part p_1 of the transformed integration path in Fig. 11.2.

The upper part of the integral (11.49) can now be written as

$$\overline{w}_{11} = \frac{Q}{2\pi\mu} \int_{t_p}^{\infty} \frac{\gamma_p^2(1/c_s^2 - 2p^2)}{(1/c_s^2 - 2p^2)^2 + 4p^2\gamma_p\gamma_s} \frac{\exp(-st)}{\sqrt{t^2 - t_p^2}} dt, \quad p = p_1. \tag{11.61}$$

The Lower Part of the Integration Path

It follows from (11.56) that on the part p_2 of the integration path

$$p = p_2 \; : \; \frac{dp}{dt} = \frac{x}{r^2} - \frac{iz}{r^2} \frac{t}{\sqrt{t^2 - t_p^2}}. \tag{11.62}$$

Furthermore, it follows from (11.52) that

$$p = p_2 \; : \; \gamma_p = \frac{tz}{r^2} + \frac{ix}{r^2}\sqrt{t^2 - t_p^2} = i\sqrt{t^2 - t_p^2} \left\{ \frac{x}{r^2} - \frac{iz}{r^2} \frac{t}{\sqrt{t^2 - t_p^2}} \right\}. \tag{11.63}$$

It now follows from (11.62) and (11.63) that

$$p = p_2 \; : \; \frac{dp}{dt} = -\frac{i\gamma_p}{\sqrt{t^2 - t_p^2}}. \tag{11.64}$$

The lower part of the integral (11.49) can now be written as

$$\overline{w}_{12} = \frac{Q}{2\pi\mu} \int_{t_p}^{\infty} \frac{\gamma_p^2(1/c_s^2 - 2p^2)}{(1/c_s^2 - 2p^2)^2 + 4p^2\gamma_p\gamma_s} \frac{\exp(-st)}{\sqrt{t^2 - t_p^2}} dt, \quad p = p_2, \tag{11.65}$$

where a minus sign has been omitted because the integration path has been reversed.

The Total Integration Path

On the two parts p_1 and p_2 of the integration path the values of p, γ_p and γ_s are complex conjugates. This means that one may write

$$\overline{w}_1 = \frac{Q}{\pi\mu} \Re \int_0^\infty \frac{\gamma_p^2(1/c_s^2 - 2p^2)}{(1/c_s^2 - 2p^2)^2 + 4p^2\gamma_p\gamma_s} \frac{H(t - t_p)}{\sqrt{t^2 - t_p^2}} \exp(-st)\,dt, \quad p = p_1,$$

(11.66)

where $H(t - t_p)$ is the Heaviside unit step function, defined as

$$H(t - t_p) = \begin{cases} 0, & \text{if } t < t_p, \\ 1, & \text{if } t > t_p. \end{cases}$$

(11.67)

The unit step function $H(t - t_p)$ has been introduced to ensure that the integration is actually from $t = t_p$ to $t = \infty$.

The integral (11.66) happens to be in the form of a Laplace transform, which was precisely the aim of the transformation of the integration path. It may be noted that the first term in the integral may be a (complex) function of the parameter t, but the Laplace transform parameter s occurs only in the factor $\exp(-st)$. It can be concluded that the inverse Laplace transform is

$$w_1 = \frac{Q}{\pi\mu} \Re \left\{ \frac{\gamma_p^2(1/c_s^2 - 2p^2)}{(1/c_s^2 - 2p^2)^2 + 4p^2\gamma_p\gamma_s} \right\} \frac{H(t - t_p)}{\sqrt{t^2 - t_p^2}}, \quad p = p_1.$$

(11.68)

The Second Integral

Using the substitution $p = i\alpha$, the second integral, (11.48), can be written as

$$\overline{w}_2 + \overline{w}_3 = \frac{Q}{2\pi i\mu} \int_{-i\infty}^{i\infty} \frac{2p^2\gamma_p}{(1/c_s^2 - 2p^2)^2 + 4p^2\gamma_p\gamma_s} \exp[-s(\gamma_s z + px)]\,dp,$$

(11.69)

where, as before,

$$\gamma_p = \sqrt{1/c_p^2 - p^2},$$

(11.70)

$$\gamma_s = \sqrt{1/c_s^2 - p^2}.$$

(11.71)

The appearance of the factor γ_s in the exponential function in (11.69) indicates that this represents the contribution of the shear waves.

Again it will be attempted to transform the integration path in the complex p-plane in such a way that the integral obtains the form of a Laplace transform integral. In this case a parameter t is introduced (later to be identified with time), defined as

$$t = \gamma_s z + px,$$

(11.72)

with t being real and positive, by assumption.

It follows from (11.71) and (11.72) that

$$r^2 p^2 - 2tpx + t^2 - z^2/c_s^2 = 0, \tag{11.73}$$

where, as before,

$$r^2 = x^2 + z^2. \tag{11.74}$$

Equation (11.73) is a quadratic expression in p, with the two solutions

$$p_3 = \frac{tx}{r^2} + \frac{iz}{r^2}\sqrt{t^2 - t_s^2}, \tag{11.75}$$

$$p_4 = \frac{tx}{r^2} - \frac{iz}{r^2}\sqrt{t^2 - t_s^2}, \tag{11.76}$$

where

$$t_s = r/c_s. \tag{11.77}$$

If it is assumed that the parameter t varies in the interval $t_s < t < \infty$, it follows that the two paths in the complex p-plane are continuous, intersecting at the real axis in the point $p = t_s x/r^2 = (x/r)(1/c_s)$ (which is to the left of the second branch point because $x \leq r$), and approaching infinity at the positive and negative sides of the real axis, respectively. The precise shape of the curves p_3 and p_4 depends upon the values of x and z, i.e. the location of the point considered in the physical plane. Because $t_s > t_p$ the two integration paths may approach the real axis at opposite points of the branch cut between $1/c_p$ and $1/c_s$. The integration path must then be extended with a loop around the first branch point, see Fig. 11.3.

The integral is separated into four parts : along the branches p_3, p_4, p_5 and p_6, where the last two are the two possible branches of the loop around the branch point $1/c_p$. The contributions of p_3 and p_4 together form the part w_2 of the integral, and the contributions of p_5 and p_6 together form the part w_3.

The Upper Part of the Integration Path

It follows from (11.75) that on the part p_3 of the integration path

$$p = p_3 : \frac{dp}{dt} = \frac{x}{r^2} + \frac{iz}{r^2}\frac{t}{\sqrt{t^2 - t_s^2}}. \tag{11.78}$$

Furthermore, it follows from (11.72) that

$$p = p_3 : \gamma_s = \frac{tz}{r^2} - \frac{ix}{r^2}\sqrt{t^2 - t_s^2} = -i\sqrt{t^2 - t_s^2}\left\{\frac{x}{r^2} + \frac{iz}{r^2}\frac{t}{\sqrt{t^2 - t_s^2}}\right\}. \tag{11.79}$$

It may be noted that on this part of the transformed integration path and for $x > 0$ (which will later appear to be the main branch considered), $\Re(\gamma_s) > 0$ and

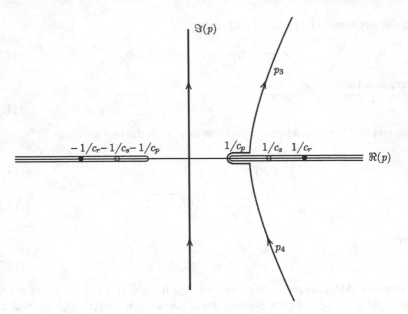

Fig. 11.3 Original and transformed integration path for the second integral

$\Im(\gamma_s) < 0$, so that $\arg(\gamma_s) < 0$. This is in agreement with the definition in (11.51) and its analytic continuation into the upper right quarter of the complex p-plane.

It follows from (11.78) and (11.79) that

$$p = p_3 \; : \; \frac{dp}{dt} = \frac{i\gamma_s}{\sqrt{t^2 - t_s^2}}. \tag{11.80}$$

The upper part of the integral (11.69) can now be written as

$$\overline{w}_{21} = \frac{Q}{2\pi\mu} \int_{t_s}^{\infty} \frac{2p^2\gamma_p\gamma_s}{(1/c_s^2 - 2p^2)^2 + 4p^2\gamma_p\gamma_s} \frac{\exp(-st)}{\sqrt{t^2 - t_s^2}} \, dt, \quad p = p_3. \tag{11.81}$$

The Lower Part of the Integration Path

It follows from (11.76) that on the part p_4 of the integration path

$$p = p_4 \; : \; \frac{dp}{dt} = \frac{x}{r^2} - \frac{iz}{r^2} \frac{t}{\sqrt{t^2 - t_s^2}}. \tag{11.82}$$

Furthermore, it follows from (11.72) that

$$p = p_4 \; : \; \gamma_s = \frac{tz}{r^2} + \frac{ix}{r^2}\sqrt{t^2 - t_s^2} = i\sqrt{t^2 - t_s^2}\left\{\frac{x}{r^2} - \frac{iz}{r^2}\frac{t}{\sqrt{t^2 - t_s^2}}\right\}. \tag{11.83}$$

It now follows from (11.82) and (11.83) that

$$p = p_4 : \quad \frac{dp}{dt} = -\frac{i\gamma_s}{\sqrt{t^2 - t_s^2}}. \tag{11.84}$$

The lower part of the integral (11.69) can now be written as

$$\overline{w}_{22} = \frac{Q}{2\pi\mu} \int_{t_s}^{\infty} \frac{2p^2\gamma_p\gamma_s}{(1/c_s^2 - 2p^2)^2 + 4p^2\gamma_p\gamma_s} \frac{\exp(-st)}{\sqrt{t^2 - t_s^2}} dt, \quad p = p_4, \tag{11.85}$$

where a minus sign has been omitted because the integration path has been reversed.

The Sum of the Upper and Lower Paths

On the two parts p_3 and p_4 the values of p, γ_p and γ_s are complex conjugates. This means that one may write for the sum of the integrals along these two parts of the integration path

$$\overline{w}_2 = \frac{Q}{\pi\mu} \Re \int_0^{\infty} \frac{2p^2\gamma_p\gamma_s}{(1/c_s^2 - 2p^2)^2 + 4p^2\gamma_p\gamma_s} \frac{H(t - t_s)}{\sqrt{t^2 - t_s^2}} \exp(-st) dt, \quad p = p_3. \tag{11.86}$$

Again, the integral happens to be in the form of a Laplace transform, and it can be concluded that the inverse Laplace transform is

$$w_2 = \frac{Q}{\pi\mu} \Re \left\{ \frac{2p^2\gamma_p\gamma_s}{(1/c_s^2 - 2p^2)^2 + 4p^2\gamma_p\gamma_s} \right\} \frac{H(t - t_s)}{\sqrt{t^2 - t_s^2}}, \quad p = p_3. \tag{11.87}$$

The Contribution of the Loop

The intersection of the branches p_3 and p_4 with the real axis $\Im(p) = 0$ can be found by determining the value of p_3 or p_4 for $t = t_s$. With (11.75) or (11.76) this gives

$$t = t_s : \quad p = p_s = \frac{t_s x}{r^2} = \frac{x}{r} \frac{1}{c_s}. \tag{11.88}$$

This point is always located to the left of the branch point $1/c_s$, whatever the values of x and z are. The point p_s may be located to the left or right of the branch point $1/c_p$, however. It is located to the left of that branch point if

$$\frac{x}{r} \frac{1}{c_s} < \frac{1}{c_p}, \tag{11.89}$$

or

$$\frac{x}{r} = \frac{x}{\sqrt{x^2 + z^2}} < \eta, \tag{11.90}$$

where

$$\eta = \frac{c_s}{c_p} = \sqrt{\frac{\mu}{\lambda + 2\mu}} = \sqrt{\frac{1 - 2v}{2(1 - v)}}. \tag{11.91}$$

If the depth z is sufficiently large for the condition (11.90) to be satisfied, the loop around the branch point $1/c_p$ is not needed, and there is no further contribution to the integral w_2.

On the other hand, if

$$\frac{x}{r} = \frac{x}{\sqrt{x^2 + z^2}} > \eta, \tag{11.92}$$

the loop around the branch point $1/c_p$ is necessary to ensure the applicability of the transformation of the integration path without passing any singularities, and there are two more contributions to the integral w_2. The additional contribution will be denoted by w_3.

The Upper Branch of the Loop

Along the upper branch of the loop $t < t_s$ and $p < 1/c_s$. This means that γ_s, as defined by (11.51),

$$\gamma_s = \sqrt{1/c_s^2 - p^2}, \tag{11.93}$$

now is real. Because $t = px + \gamma_s z$, see (11.72), it now follows that γ_s can also be expressed as

$$\gamma_s = \frac{t}{z} - \frac{px}{z}. \tag{11.94}$$

It follows from (11.93) and (11.94) that the value of p can be determined from

$$r^2 p^2 - 2tpx + t^2 - z^2/c_s^2 = 0, \tag{11.95}$$

which now gives

$$p_5 = \frac{xt}{r^2} - \frac{z}{r^2}\sqrt{t_s^2 - t^2}, \tag{11.96}$$

where the minus-sign has been taken to ensure that $p < p_s$. Actually, it follows from (11.96) that

$$p = p_5 : \frac{dp}{dt} = \frac{x}{r^2} + \frac{z}{r^2}\frac{t}{\sqrt{t_s^2 - t^2}}, \tag{11.97}$$

which shows that $dp/dt > 0$ if $0 < t < t_s$.

The smallest value of t along the upper part of the loop occurs in the point $p = 1/c_p$. If this value is denoted by t_q, it follows from (11.96) that

$$\frac{1}{c_p} = \frac{t_q x}{r^2} - \frac{z}{r^2}\sqrt{t_s^2 - t_q^2}, \tag{11.98}$$

or, with $t_p = r/c_p$,

$$t_p = (x/r) t_q - (z/r)\sqrt{t_s^2 - t_q^2}. \qquad (11.99)$$

It follows from this equation that

$$t_q = (x/r) t_p + (z/r)\sqrt{t_s^2 - t_p^2}, \qquad (11.100)$$

where the plus-sign has been chosen to ensure that $t_q \geq t_p$. Equation (11.100) can also be written as

$$t_q/t_s = (x\eta + z\sqrt{1 - \eta^2})/r, \quad t_q/t_s < 1. \qquad (11.101)$$

It can be shown that this is always smaller than 1, in the region where the condition (11.92) is satisfied.

It follows from (11.93) and (11.96) that on the upper part of the loop

$$\gamma_s = \sqrt{t_s^2 - t^2}\left\{\frac{x}{r^2} + \frac{z}{r^2}\frac{t}{\sqrt{t_s^2 - t^2}}\right\}. \qquad (11.102)$$

Comparison with (11.97) shows that

$$p = p_5 : \quad \frac{dp}{dt} = \frac{\gamma_s}{\sqrt{t_s^2 - t^2}}, \qquad (11.103)$$

which enables to write the integral along this part of the loop as an integral over the variable t. Actually, with (11.69) one obtains

$$\overline{w}_{31} = \frac{Q}{2\pi i \mu} \int_{t_q}^{t_s} \frac{2p^2 \gamma_p \gamma_s}{(1/c_s^2 - 2p^2)^2 + 4p^2 \gamma_p \gamma_s} \frac{\exp(-st)}{\sqrt{t_s^2 - t^2}} dt, \quad p = p_5. \qquad (11.104)$$

In this case all the parameters in the integrand are real, except for γ_p. Because on this part of the integration path $1/c_p < p < 1/c_s$ it follows that $\arg(1/c_p^2 - p^2) = -\pi$, so that γ_p is purely imaginary, and its argument is $-\pi/2$.

The Lower Branch of the Loop

Along the lower branch of the loop all quantities are the same as on the upper part of the loop, except for γ_p, which now is the complex conjugate of the previous value. Furthermore the integration path is from $p = 1/c_s$ to $p = 1/c_p$, i.e. from the right to the left. By reversing the integration path the result will be

$$\overline{w}_{32} = -\frac{Q}{2\pi i \mu} \int_{t_q}^{t_s} \frac{2p^2 \gamma_p \gamma_s}{(1/c_s^2 - 2p^2)^2 + 4p^2 \gamma_p \gamma_s} \frac{\exp(-st)}{\sqrt{t_s^2 - t^2}} dt, \quad p = p_6 \qquad (11.105)$$

where it should be noted that the value of γ_p is the complex conjugate of the value in the integral (11.104).

The Sum of the Upper and Lower Branches

If the integral (11.104) is written as $\overline{w}_{31} = (a+ib)/i$, the integral (11.105) will be of the form $\overline{w}_{32} = -(a-ib)/i$. The sum of these two integrals then is $\overline{w}_{31} + \overline{w}_{32} = 2b$. This means that

$$\overline{w}_3 = \frac{Q}{\pi\mu}\,\Im\int_{t_q}^{t_s} \frac{2p^2\gamma_p\gamma_s}{(1/c_s^2 - 2p^2)^2 + 4p^2\gamma_p\gamma_s}\,\frac{\exp(-st)}{\sqrt{t_s^2 - t^2}}\,dt, \qquad p = p_5. \qquad (11.106)$$

The integral is again in the form of a Laplace transform. The Laplace transform parameter s appears only in the factor $\exp(-st)$, but the time t may appear in various forms in the integrand. It can be concluded that the original function is

$$w_3 = \frac{Q}{\pi\mu}\,\Im\left\{\frac{2p^2\gamma_p\gamma_s}{(1/c_s^2 - 2p^2)^2 + 4p^2\gamma_p\gamma_s}\right\}\frac{H(t - t_q)H(t_s - t)}{\sqrt{t_s^2 - t^2}}, \qquad p = p_5.$$

$$(11.107)$$

It may be noted that the function $H(t - t_q)H(t_s - t)$ is equal to 1 only in the interval $t_q < t < t_s$, elsewhere it is zero.

Calculation of Numerical Data

The three components of the vertical displacement are defined by (11.68), (11.87) and (11.107). In order to calculate numerical data it is most convenient to express these equations in dimensionless form.

The first component of the vertical displacement is, from (11.68),

$$w_1 = \frac{Q}{\pi\mu}\,\Re\left\{\frac{\gamma_p^2(1/c_s^2 - 2p^2)}{(1/c_s^2 - 2p^2)^2 + 4p^2\gamma_p\gamma_s}\right\}\frac{H(t - t_p)}{\sqrt{t^2 - t_p^2}}, \qquad p = p_1, \qquad (11.108)$$

where $t_p = r/c_p$, $r = \sqrt{x^2 + z^2}$, and p_1 is defined by (11.55),

$$p_1 = \frac{tx}{r^2} + \frac{iz}{r^2}\sqrt{t^2 - t_p^2}. \qquad (11.109)$$

The quantities γ_p and γ_s are related to the variable p and the velocities of compression waves and shear waves by (11.50) and (11.51),

$$\gamma_p = \sqrt{1/c_p^2 - p^2}, \qquad \gamma_s = \sqrt{1/c_s^2 - p^2}. \qquad (11.110)$$

In general the quantities p, γ_p and γ_s are complex.

A suitable choice of basic dimensionless parameters seems to be

$$\xi = x/z, \qquad \tau = c_s t/z, \qquad \tau_p = c_s t_p/z, \qquad \tau_s = c_s t_s/z,$$

$$\tau_q = c_s t_q/z, \qquad a = p_1 c_s. \qquad (11.111)$$

Some derived parameters are

$$r = z\sqrt{1+\xi^2}, \qquad g_s = c_s\gamma_s = \sqrt{1-a^2}, \qquad g_p = c_s\gamma_p = \sqrt{\eta^2 - a^2}. \quad (11.112)$$

It now follows from equation (11.108) that

$$\frac{w_1\pi\mu z}{Qc_s} = \Re\left\{\frac{(1-2a^2)g_p^2}{(1-2a^2)^2 + 4a^2 g_p g_s}\right\}\frac{H(\tau-\tau_p)}{\sqrt{\tau^2 - \tau_p^2}}, \quad (11.113)$$

where the parameter a is defined by

$$a = (\xi\tau + i\sqrt{\tau^2 - \tau_p^2})/(1+\xi^2). \quad (11.114)$$

The second component of the vertical displacement is, from (11.87),

$$w_2 = \frac{Q}{\pi\mu}\Re\left\{\frac{2p^2\gamma_p\gamma_s}{(1/c_s^2 - 2p^2)^2 + 4p^2\gamma_p\gamma_s}\right\}\frac{H(t-t_s)}{\sqrt{t^2 - t_s^2}}, \quad p = p_3, \quad (11.115)$$

where $t_s = r/c_s$, and p_3 is defined by (11.75),

$$p_3 = \frac{tx}{r^2} + \frac{iz}{r^2}\sqrt{t^2 - t_s^2}. \quad (11.116)$$

The dimensionless form of p_3 is defined as

$$b = p_3 c_s. \quad (11.117)$$

The dimensionless form of (11.115) is

$$\frac{w_2\pi\mu z}{Qc_s} = \Re\left\{\frac{2b^2 g_p g_s}{(1-2b^2)^2 + 4b^2 g_p g_s}\right\}\frac{H(\tau-\tau_s)}{\sqrt{\tau^2 - \tau_s^2}}, \quad (11.118)$$

where the parameter b is defined by

$$b = (\xi\tau + i\sqrt{\tau^2 - \tau_s^2})/(1+\xi^2). \quad (11.119)$$

The third component of the displacement is, from (11.107),

$$w_3 = \frac{Q}{\pi\mu}\Im\left\{\frac{2p^2\gamma_p\gamma_s}{(1/c_s^2 - 2p^2)^2 + 4p^2\gamma_p\gamma_s}\right\}\frac{H(t-t_q)H(t_s-t)}{\sqrt{t_s^2 - t^2}}, \quad p = p_5, \quad (11.120)$$

where $t_q/t_s = (x\eta + z\sqrt{1-\eta^2})/r$ and p_5 is defined by (11.96),

$$p_5 = \frac{xt}{r^2} - \frac{z}{r^2}\sqrt{t_s^2 - t^2}. \quad (11.121)$$

The dimensionless form of p_5 is defined as

$$c = p_5 c_s. \tag{11.122}$$

The dimensionless form of (11.120) is

$$\frac{w_3 \pi \mu z}{Q c_s} = \Im \left\{ \frac{2c^2 g_p g_s}{(1 - 2c^2)^2 + 4c^2 g_p g_s} \right\} \frac{H(\tau - \tau_q) H(\tau_s - \tau)}{\sqrt{\tau_s^2 - \tau_2}}, \tag{11.123}$$

where the parameter c is defined by

$$c = (\xi \tau - \sqrt{\tau_s^2 - \tau^2})/(1 + \xi^2). \tag{11.124}$$

Computer Program

A function (in C, using complex calculus) to calculate the value of the dimensionless parameter $w \pi \mu z / Q c_s$ as a function of the parameters $\xi = x/z$ (with $\xi \geq 0$), $\tau = c_p t / z$ and Poisson's ratio ν, is shown below. The function consists of three parts, as given by (11.113), (11.118) and (11.123). Great care must be taken to verify that the arguments of the square roots are calculated correctly, in agreement with the range determined by the analytic continuation of the original definitions of γ_p and γ_s. This may require some preliminary verification of intermediate computations.

```
double LinePulseW(double xi,double tau,double nu)
{
  double w,w1,w2,w3,n,nn,xi1,taup2,taus2,taus,tauq,tau2;
  complex b,bb,b1,gp,gs,d,e,c,cc,c1;
  nn=(1-2*nu)/(2*(1-nu));n=sqrt(nn);xi1=1+xi*xi;
  taus2=xi1;taup2=nn*taus2;taus=sqrt(taus2);tauq=n*xi+sqrt(1-nn);tau2=tau*tau;
  if (tau2<=taup2) w1=0;else
    {
      b=complex(xi*tau/xi1,(sqrt(tau2-taup2))/xi1);bb=b*b;b1=1-2*bb;
      gp=sqrt(nn-bb);gs=sqrt(1-bb);d=b1*gp*gp;e=b1*b1+4*bb*gp*gs;
      w1=real(d/e)/sqrt(tau2-taup2);
    }
  if (tau2<=taus2) w2=0;else
    {
      b=complex(xi*tau/xi1,(sqrt(tau2-taus2))/xi1);bb=b*b;b1=1-2*bb;
      gp=sqrt(nn-bb);gs=sqrt(1-bb);d=2*bb*gp*gs;e=b1*b1+4*bb*gp*gs;
      w2=real(d/e)/sqrt(tau2-taus2);
    }
  if ((tau<=tauq)||(tau>=taus)||(xi*sqrt(1-nn)<nn)) w3=0;else
    {
      c=complex((xi*tau-sqrt(taus2-tau2))/taus2,0);cc=c*c;c1=1-2*cc;
      gp=sqrt(nn-cc);gs=sqrt(1-cc);d=2*cc*gp*gs;e=c1*c1+4*cc*gp*gs;
      w3=imag(d/e)/sqrt(taus2-tau2);
    }
  w=w1+w2+w3;
  return(w);
}
```

Some examples are shown in Figs. 11.4 and 11.5, for $\nu = 0$, and $c_s t/z = 5$ and $c_s t/z = 40$, respectively. In Fig. 11.4 the first wave, the compression wave, has

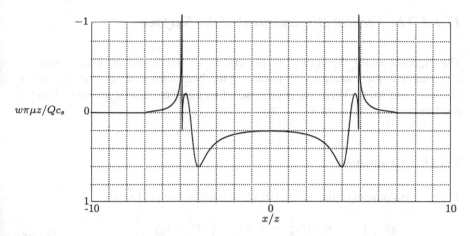

Fig. 11.4 Line pulse—vertical displacement, $v = 0$, $c_s t/z = 5$

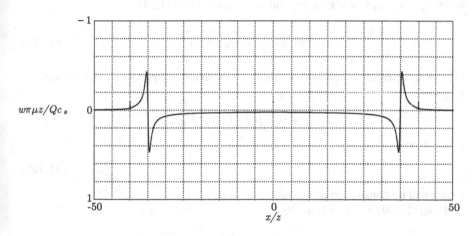

Fig. 11.5 Line pulse—vertical displacement, $v = 0$, $c_s t/z = 40$

reached the distance $x/z = 7.01$, and the second wave, the shear wave, has reached a distance $x/z = 4.91$. It can easily be verified that these values agree well with the theoretical values $c_p t/r = 1$ and $c_s t/r = 1$, respectively. The Rayleigh wave can be seen to follow some time after the shear wave in Fig. 11.5. By considering other values of time it can be seen that the shape and amplitude of the Rayleigh wave disturbance are practically independent of the horizontal distance. This is in agreement with theoretical analysis of Rayleigh waves in the two-dimensional case, see for instance Achenbach (1975), or the analysis of Rayleigh waves in Chap. 9 of this book.

11.1.4 The Vertical Displacement of the Surface

The expressions for the vertical displacement derived in the previous section are not suitable at the surface $z = 0$, because z has been used as a factor in the dimensionless parameters. Therefore another set of parameters must be introduced for the displacements of the surface.

The First Component

The first component of the vertical displacement is, from (11.68),

$$w_1 = \frac{Q}{\pi \mu} \Re \left\{ \frac{\gamma_p^2 (1/c_s^2 - 2p^2)}{(1/c_s^2 - 2p^2)^2 + 4p^2 \gamma_p \gamma_s} \right\} \frac{H(t - t_p)}{\sqrt{t^2 - t_p^2}}, \quad p = p_1, \quad (11.125)$$

where $t_p = r/c_p$, $r = \sqrt{x^2 + z^2}$, and p_1 is defined by (11.55),

$$p_1 = \frac{tx}{r^2} + \frac{iz}{r^2} \sqrt{t^2 - t_p^2}. \quad (11.126)$$

For $z = 0$ the radial coordinate is simply $r = x$, and the expression (11.126) reduces to

$$z = 0 : \quad p = t/x = \tau/c_s, \quad (11.127)$$

where now

$$\tau = c_s t/x, \quad (11.128)$$

which will be considered as the basic variable.

From (11.59) it now follows that

$$z = 0 : \quad \gamma_p = -\frac{ix}{r^2} \sqrt{t^2 - t_p^2} = -i \sqrt{\tau^2 - \eta^2}/c_s, \quad (11.129)$$

where, as before,

$$\eta^2 = \frac{c_s^2}{c_p^2} = \frac{\mu}{\lambda + 2\mu} = \frac{1 - 2v}{2(1 - v)}. \quad (11.130)$$

Using the parameter η the value of t_p, as defined by (11.57), can be written as

$$z = 0 : \quad t_p = r/c_p = \eta x/c_s, \quad (11.131)$$

so that

$$z = 0 : \quad \sqrt{t^2 - t_p^2} = (x/c_s) \sqrt{\tau^2 - \eta^2}. \quad (11.132)$$

Also, it follows from (11.129) that

$$z = 0 : \gamma_p^2 = -(\tau^2 - \eta^2)/c_s^2. \tag{11.133}$$

Furthermore, it follows from (11.51) that

$$z = 0 : \gamma_s^2 = 1/c_s^2 - p^2 = (1 - \tau^2)/c_s^2. \tag{11.134}$$

It has been seen before that the first integral is unequal to zero only if $t > t_p$, or $\tau > \eta$, where η is a constant smaller than 1. The value of γ_s now appears to depend upon the value of τ with respect to 1,

$$z = 0, \ \tau < 1 : \gamma_s = \sqrt{1 - \tau^2}/c_s, \tag{11.135}$$

$$z = 0, \ \tau > 1 : \gamma_s = -i\sqrt{\tau^2 - 1}/c_s. \tag{11.136}$$

The minus sign in the last expression has been taken because $\arg(1/c_s^2 - p^2) = -\pi$ if $p > 1/c_s^2$ along the path p_1 just above the real axis.

Finally, it follows that the expression (11.125) can be calculated as

$$z = 0, \ \tau < \eta : w_1 = 0, \tag{11.137}$$

$$z = 0, \ \eta < \tau < 1 : w_1 = -\frac{Qc_s}{\pi\mu x}\Re\left\{\frac{(1 - 2\tau^2)\sqrt{\tau^2 - \eta^2}}{(1 - 2\tau^2)^2 - 4i\tau^2\sqrt{\tau^2 - \eta^2}\sqrt{1 - \tau^2}}\right\}, \tag{11.138}$$

$$z = 0, \ \tau > 1 : w_1 = -\frac{Qc_s}{\pi\mu x}\left\{\frac{(1 - 2\tau^2)\sqrt{\tau^2 - \eta^2}}{(1 - 2\tau^2)^2 - 4\tau^2\sqrt{\tau^2 - \eta^2}\sqrt{\tau^2 - 1}}\right\}. \tag{11.139}$$

This defines the value of the first contribution to the surface displacements, as a function of the dimensionless variable $\tau = c_s t/x$.

The Second Component

The second component of the vertical displacement is, from (11.87),

$$w_2 = \frac{Q}{\pi\mu}\Re\left\{\frac{2p^2\gamma_p\gamma_s}{(1/c_s^2 - 2p^2)^2 + 4p^2\gamma_p\gamma_s}\right\}\frac{H(t - t_s)}{\sqrt{t^2 - t_s^2}}, \quad p = p_3, \tag{11.140}$$

where $t_s = r/c_s$, and p_3 is defined by (11.75),

$$p_3 = \frac{tx}{r^2} + \frac{iz}{r^2}\sqrt{t^2 - t_s^2}. \tag{11.141}$$

At the surface $z = 0$ the radial coordinate is $r = x$, and the expression (11.141) reduces to

$$z = 0 : p = t/x = \tau/c_s, \tag{11.142}$$

where, as before, see (11.128),

$$\tau = c_s t/x, \tag{11.143}$$

which is the basic variable. It may be noted that in the part of the solution considered here, see (11.140), the variable $t > t_s$, where now $t_s = x/c_s$, see (11.77). This means that $\tau > 1$.

The values of γ_p and γ_s, as defined in general by (11.70) and (11.71), now are, because the argument of the expressions $1/c_p^2 - p^2$ and $1/c_s^2 - p^2$ is $-\pi$ for points p following a path just above the real axis,

$$\gamma_p = -i\sqrt{\tau^2 - \eta^2}/c_s, \tag{11.144}$$

$$\gamma_s = -i\sqrt{\tau^2 - 1}/c_s. \tag{11.145}$$

Furthermore,

$$\sqrt{t^2 - t_s^2} = (x/c_s)\sqrt{\tau^2 - 1}. \tag{11.146}$$

Using these results it follows that the expression (11.140) can be calculated as

$$z = 0, \ \tau < 1 : \ w_2 = 0, \tag{11.147}$$

$$z = 0, \ \tau > 1 : \ w_2 = -\frac{Qc_s}{\pi\mu x}\left\{\frac{2\tau^2\sqrt{\tau^2 - \eta^2}}{(1 - 2\tau^2)^2 - 4\tau^2\sqrt{\tau^2 - \eta^2}\sqrt{\tau^2 - 1}}\right\}. \tag{11.148}$$

This defines the value of the second contribution to the surface displacements, as a function of the dimensionless variable $\tau = c_s t/x$.

The Third Component

The third component of the displacement is, from (11.107),

$$w_3 = \frac{Q}{\pi\mu}\Im\left\{\frac{2p^2\gamma_p\gamma_s}{(1/c_s^2 - 2p^2)^2 + 4p^2\gamma_p\gamma_s}\right\}\frac{H(t - t_q)H(t_s - t)}{\sqrt{t_s^2 - t^2}}, \quad p = p_5, \tag{11.149}$$

where $t_q/t_s = (x\eta + z\sqrt{1 - \eta^2})/r$ and p_5 is defined by (11.96),

$$p_5 = \frac{xt}{r^2} - \frac{z}{r^2}\sqrt{t_s^2 - t^2}. \tag{11.150}$$

At the surface $z = 0$ the radial coordinate is $r = x$, and the expression (11.150) reduces to

$$z = 0 : \ p = t/x = \tau/c_s, \tag{11.151}$$

where, as before, see (11.128) and (11.143),

$$\tau = c_s t/x, \tag{11.152}$$

which is the basic variable. It may be noted that in the part of the solution considered here, see (11.149), the variable t varies in the range $t_q < t < t_s$. Because for $z = 0$ it follows from (11.100) that $t_q = t_p$, the range of t is $t_p < t < t_s$. This means that $\eta < \tau < 1$. It may also be noted that for $z = 0$ the condition (11.92), which is necessary for this contribution to be applicable, is always satisfied.

In this case

$$\gamma_p = -i\sqrt{\tau^2 - \eta^2}/c_s, \tag{11.153}$$

$$\gamma_s = \sqrt{1 - \tau^2}/c_s, \tag{11.154}$$

$$\sqrt{t_s^2 - t^2} = (x/c_s)\sqrt{1 - \tau^2}. \tag{11.155}$$

Using these results it follows that the expression (11.149) can be calculated as

$$z = 0, \ \tau < \eta \ : \ w_3 = 0, \tag{11.156}$$

$$z = 0, \ \eta < \tau < 1 \ : \ w_3 = -\frac{Qc_s}{\pi\mu x}\Im\left\{\frac{2i\tau^2\sqrt{\tau^2 - \eta^2}}{(1 - 2\tau^2)^2 - 4i\tau^2\sqrt{\tau^2 - \eta^2}\sqrt{1 - \tau^2}}\right\},$$
$$\tag{11.157}$$

$$z = 0, \ \tau > 1 \ : \ w_3 = 0. \tag{11.158}$$

This defines the value of the third contribution to the surface displacements, as a function of the dimensionless variable $\tau = c_s t/x$. This part of the solution is often denoted as the *head wave*.

Total Surface Displacements

Adding the three contributions to the surface displacements, the final expressions for the displacements of the surface $z = 0$ are, with $\tau = c_s t/x$,

$$z = 0, \ \tau < \eta \ : \ w = 0, \tag{11.159}$$

$$z = 0, \ \eta < \tau < 1 \ : \ w = -\frac{Qc_s}{\pi\mu x}\left\{\frac{(1 - 2\tau^2)^2\sqrt{\tau^2 - \eta^2}}{(1 - 2\tau^2)^4 + 16\tau^4(\tau^2 - \eta^2)(1 - \tau^2)}\right\}, \tag{11.160}$$

$$z = 0, \ \tau > 1 \ : \ w = -\frac{Qc_s}{\pi\mu x}\left\{\frac{\sqrt{\tau^2 - \eta^2}}{(1 - 2\tau^2)^2 - 4\tau^2\sqrt{\tau^2 - \eta^2}\sqrt{\tau^2 - 1}}\right\}. \tag{11.161}$$

This completely defines the surface displacements, as a function of the dimensionless variable $\tau = c_s t/x$. The displacement is a continuous function of this variable, but there is a singularity for the value $\tau = \beta$, where β denotes the arrival time of the Rayleigh wave. Mathematically, this singularity is caused by a zero of the denominator of the functions (11.160) and (11.161). This zero occurs in the range $\tau > 1$, indicating that the Rayleigh wave arrives (shortly) after the shear wave.

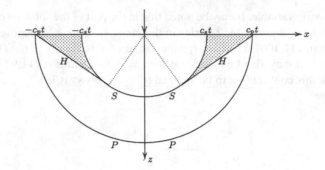

Fig. 11.6 The three wave fronts

The solution derived here, as given in (11.159)–(11.161), is in agreement with the solution given by Eringen and Suhubi (1975), and with Lamb's original solution (Lamb, 1904). The present solution method has the advantage that it gives the solution for every point x, z in the half-plane, see the previous section.

It may also be interesting to note that the first two parts of the solution represent a compression wave (or P-wave), propagating at velocity c_p, and a shear wave (or S-wave), propagating at velocity c_s. The P-wave part of the solution is defined by (11.68), and the S-wave part of the solution is defined by (11.87). It has appeared, however, that between the arrival of the compression wave and the shear wave an additional solution is needed, near the surface (often denoted as the *head wave*), in order to satisfy the zero stress boundary condition at the surface. This part of the solution is defined by (11.107).

The three wave fronts are indicated in Fig. 11.6, see also Achenbach (1975). The area affected by the head wave is indicated in the figure by shading. This area is defined by the condition (11.92), or

$$\frac{z^2}{x^2} < \frac{c_p^2}{c_s^2} - 1 = \frac{1}{1 - 2v}. \tag{11.162}$$

The figure has been drawn for $v = 0.25$. The distances $c_s t$ and $c_p t$ in the figure indicate that the figure expands linearly with time.

Computer Program

A function (in C) that calculates the vertical displacement of the surface for given values of v and $c_s t / x$ is reproduced below. It is assumed that $x \geq 0$. For values of $x < 0$ the displacements can be obtained using the symmetry of the solution.

The parameters v and $c_s t / x$ are denoted by nu and tau in the program. The quantity tr denotes the parameter $c_s t_r / x$, where t_r is the Rayleigh wave velocity.

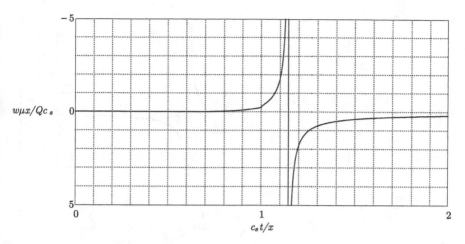

Fig. 11.7 Vertical displacement of the surface, $v = 0.00$

```
double LinePulseWS(double nu, double tau)
{
  double pi,fac,n,nn,e,f,a,b,b1,b2,w,t,tt,tr,eps;
  pi=4*atan(1.0);fac=1/pi;eps=0.0001;t=tau;
  nn=(1-2*nu)/(2*(1-nu));n=sqrt(nn);e=0.000001;e*=e;f=1;b=(1-nu)/8;
  if (nu>0.1) {while(f>e) {a=b;b=(1-nu)/(8*(1+a)*(nu+a));f=fabs(b-a);}}
  else {while (f>e) {a=b;b=sqrt((1-nu)/(8*(1+a)*(1+nu/a)));f=fabs(b-a);}}
  tr=sqrt(1+b);if (t<=n) w=0;else if (t<=1)
    {
      tt=t*t;a=sqrt(tt-nn);
      b1=(1-2*tt)*(1-2*tt);b2=4*tt*sqrt((tt-nn)*(1-tt));b=b1*b1+b2*b2;
      w=-fac*a*b1/b;
    }
  else
    {
      if (fabs(t-tr)<eps) {if (t<tr) t=tr-eps;else t=tr+eps;}
      tt=t*t;a=sqrt(tt-nn);b=(1-2*tt)*(1-2*tt)-4*tt*sqrt((tt-nn)*(tt-1));
      w=-fac*a/b;
    }
  return(w);
}
```

The surface displacements are shown in graphical form in the Figs. 11.7, 11.8 and 11.9, for three values of v. In each case the displacements remain zero until the arrival of the compression wave, and there is a singularity at the passage of the Rayleigh wave. At the time of arrival of the shear wave, $c_s t/x = 1$, there is a discontinuity in the slope of the curves.

11.1.5 The Horizontal Displacement

The general solution for the horizontal displacement u has been given in (11.35), in the form of the Laplace transform of a Fourier integral,

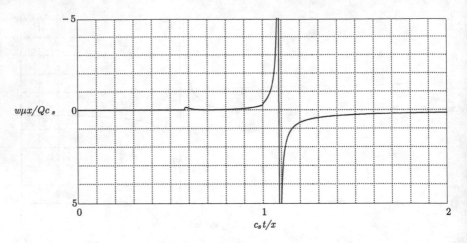

Fig. 11.8 Vertical displacement of the surface, $\nu = 0.25$

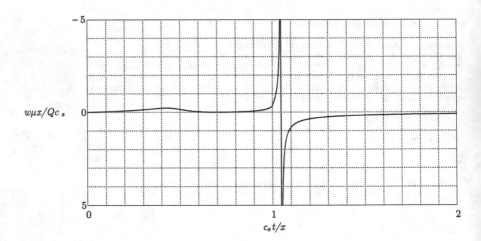

Fig. 11.9 Vertical displacement of the surface, $\nu = 0.50$

$$\bar{u} = \frac{is}{2\pi} \int_{-\infty}^{\infty} \{\alpha C_p \exp(-s\gamma_p z) + \gamma_s C_s \exp(-s\gamma_s z)\} \exp(-is\alpha x)\, d\alpha, \quad (11.163)$$

where the constants C_p and C_s have been given in (11.44) and (11.45),

$$C_p = \frac{Q}{\mu s} \frac{2\alpha^2 + 1/c_s^2}{(2\alpha^2 + 1/c_s^2)^2 - 4\alpha^2 \gamma_p \gamma_s}, \quad (11.164)$$

$$C_s = -\frac{Q}{\mu s} \frac{2\alpha \gamma_p}{(2\alpha^2 + 1/c_s^2)^2 - 4\alpha^2 \gamma_p \gamma_s}. \quad (11.165)$$

Substitution of these results in (11.163) gives

$$\bar{u} = \bar{u}_1 + \bar{u}_2 + \bar{u}_3, \tag{11.166}$$

where

$$\bar{u}_1 = \frac{iQ}{2\pi\mu} \int_{-\infty}^{\infty} \frac{\alpha(2\alpha^2 + 1/c_s^2)}{(2\alpha^2 + 1/c_s^2)^2 - 4\alpha^2\gamma_p\gamma_s} \exp[-s(\gamma_p z + i\alpha x)] \, d\alpha, \tag{11.167}$$

$$\bar{u}_2 + \bar{u}_3 = -\frac{iQ}{2\pi\mu} \int_{-\infty}^{\infty} \frac{2\alpha\gamma_p\gamma_s}{(2\alpha^2 + 1/c_s^2)^2 - 4\alpha^2\gamma_p\gamma_s} \exp[-s(\gamma_s z + i\alpha x)] \, d\alpha. \tag{11.168}$$

As in the analysis of the vertical displacement, the variable α is replaced by $p = i\alpha$, so that

$$\bar{u}_1 = \frac{iQ}{2\pi\mu} \int_{-i\infty}^{i\infty} \frac{p(1/c_s^2 - 2p^2)}{(1/c_s^2 - 2p^2)^2 + 4p^2\gamma_p\gamma_s} \exp[-s(\gamma_p z + px)] \, dp, \tag{11.169}$$

$$\bar{u}_2 + \bar{u}_3 = -\frac{iQ}{2\pi\mu} \int_{-i\infty}^{i\infty} \frac{2p\gamma_p\gamma_s}{(1/c_s^2 - 2p^2)^2 + 4p^2\gamma_p\gamma_s} \exp[-s(\gamma_s z + px)] \, dp, \tag{11.170}$$

where, as before,

$$\gamma_p = \sqrt{1/c_p^2 - p^2}, \qquad \gamma_s = \sqrt{1/c_s^2 - p^2}. \tag{11.171}$$

These integrals can be evaluated in the same way as the integrals for the vertical displacement, using a transformation of the integration path so that the Fourier integral is modified into a Laplace transform integral, and the inverse transform can immediately be found.

In this case it would be sufficient to introduce branch cuts in the p-plane between the points $p = 1/c_p$ and $p = 1/c_s$, and between the points $p = -1/c_s$ and $p = -1/c_p$, see Fig. 11.10, because this would make the integrands of (11.169) and (11.170) single valued. Care should be taken to avoid passing the poles at $1/c_r$ or $-1/c_r$, as was already pointed out by Lamb (1904).

The first component of the solution, obtained from (11.169) by transforming the integration path into two parabolic curves, is found to be

$$u_1 = \frac{Q}{\pi\mu} \Re\left\{ \frac{p\gamma_p(1/c_s^2 - 2p^2)}{(1/c_s^2 - 2p^2)^2 + 4p^2\gamma_p\gamma_s} \right\} \frac{H(t - t_p)}{\sqrt{t^2 - t_p^2}}, \quad p = p_1, \tag{11.172}$$

where p_1 is defined by (11.109),

$$p_1 = \frac{tx}{r^2} + \frac{iz}{r^2}\sqrt{t^2 - t_p^2}. \tag{11.173}$$

Fig. 11.10 Branch cuts in the
p-plane

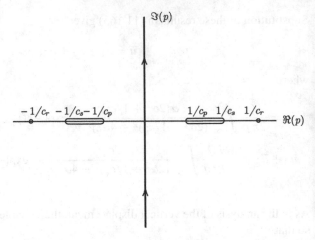

Fig. 11.10 Branch cuts in the p-plane

The second component of the solution can be obtained from (11.170) by transforming the integration into two parabolic curves and a loop around the first branch point. The contribution of the two parabolic parts is

$$u_2 = -\frac{Q}{\pi\mu}\Re\left\{\frac{2p\gamma_p\gamma_s^2}{(1/c_s^2 - 2p^2)^2 + 4p^2\gamma_p\gamma_s}\right\}\frac{H(t-t_s)}{\sqrt{t^2 - t_s^2}}, \qquad p = p_3, \qquad (11.174)$$

where p_3 is defined by (11.116),

$$p_3 = \frac{tx}{r^2} + \frac{iz}{r^2}\sqrt{t^2 - t_s^2}. \tag{11.175}$$

The third component of the solution is the contribution of the loop around the first branch point,

$$u_3 = -\frac{Q}{\pi\mu}\Im\left\{\frac{2p\gamma_p\gamma_s^2}{(1/c_s^2 - 2p^2)^2 + 4p^2\gamma_p\gamma_s}\right\}\frac{H(t-t_q)H(t_s-t)}{\sqrt{t_s^2 - t^2}}, \qquad p = p_5, \tag{11.176}$$

where p_5 is defined by (11.121),

$$p_5 = \frac{xt}{r^2} - \frac{z}{r^2}\sqrt{t_s^2 - t^2}. \tag{11.177}$$

The dimensionless form of (11.172) is

$$\frac{u_1\pi\mu z}{Qc_s} = \Re\left\{\frac{ag_p(1 - 2a^2)}{(1 - 2a^2)^2 + 4a^2 g_p g_s}\right\}\frac{H(\tau - \tau_p)}{\sqrt{\tau^2 - \tau_p^2}}, \tag{11.178}$$

where the parameter a is defined by (11.114), i.e.

$$a = (\xi\tau + i\sqrt{\tau^2 - \tau_p^2})/(1 + \xi^2). \tag{11.179}$$

The dimensionless form of (11.174) is

$$\frac{u_2\pi\mu z}{Qc_s} = -\Re\left\{\frac{bg_pg_s^2}{(1-2b^2)^2+4b^2g_pg_s}\right\}\frac{H(\tau-\tau_s)}{\sqrt{\tau^2-\tau_s^2}}, \tag{11.180}$$

where the parameter b is defined by (11.119), i.e.

$$b = (\xi\tau + i\sqrt{\tau^2-\tau_s^2})/(1+\xi^2). \tag{11.181}$$

The dimensionless form of (11.176) is

$$\frac{u_3\pi\mu z}{Qc_s} = -\Im\left\{\frac{2cg_pg_s^2}{(1-2c^2)^2+4c^2g_pg_s}\right\}\frac{H(\tau-\tau_q)H(\tau_s-\tau)}{\sqrt{\tau_s^2-\tau^2}}, \tag{11.182}$$

where the parameter c is defined by (11.124), i.e.

$$c = (\xi\tau - \sqrt{\tau_s^2-\tau^2})/(1+\xi^2). \tag{11.183}$$

Calculation of Numerical Values

A function (in C, using complex calculus) to calculate the value of the dimensionless parameter $u\pi\mu z/Qc_s$ as a function of the parameters $\xi = x/z$ (with $\xi \geq 0$), $\tau = c_pt/z$ and Poisson's ratio v, is shown below. The function consists of three parts, as given by (11.178), (11.180) and (11.182).

```
double LinePulseU(double xi,double tau,double nu)
{
 double u,u1,u2,u3,n,nn,tau2,taup2,taus2,taus,tauq;
 complex a,aa,a1,b,bb,b1,gp,gs,d,e,c,cc,c1;
 nn=(1-2*nu)/(2*(1-nu));n=sqrt(nn);tau2=tau*tau;
 taus2=1+xi*xi;taup2=nn*taus2;taus=sqrt(taus2);tauq=n*xi+sqrt(1-nn);
 if (tau2<=taup2) u1=0;else
   {
    a=complex(xi*tau/taus2,sqrt(tau2-taup2)/taus2);aa=a*a;a1=1-2*aa;
    gp=sqrt(nn-aa);gs=sqrt(1-aa);d=a*gp*a1;
    e=a1*a1+4*aa*gp*gs;u1=real(d/e)/(sqrt(tau2-taup2));
   }
 if (tau2<=taus2) u2=0;else
   {
    b=complex(xi*tau/taus2,sqrt(tau2-taus2)/taus2);bb=b*b;b1=1-2*bb;
    gp=sqrt(nn-bb);gs=sqrt(1-bb);d=2*b*gp*gs*gs;
    e=b1*b1+4*bb*gp*gs;u2=-real(d/e)/(sqrt(tau2-taus2));
   }
 if ((tau<=tauq)||(tau>=taus)||(xi*sqrt(1-nn)<nn)) u3=0;else
   {
    c=complex((xi*tau-sqrt(taus2-tau2))/taus2,0);cc=c*c;c1=1-2*cc;
    gp=sqrt(nn-cc);gs=sqrt(1-cc);d=2*c*gp*gs*gs;e=c1*c1+4*cc*gp*gs;
    u3=-imag(d/e)/sqrt(taus2-tau2);
   }
 u=u1+u2+u3;
 return(u);
}
```

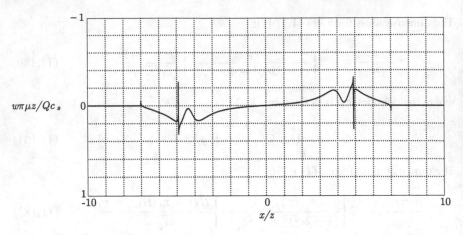

Fig. 11.11 Line pulse—horizontal displacement, $v = 0$, $c_s t/z = 5$

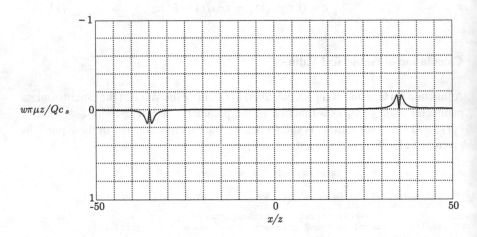

Fig. 11.12 Line pulse—horizontal displacement, $v = 0$, $c_s t/z = 40$

Some examples are shown in Figs. 11.11 and 11.12, for $v = 0$, and $c_s t/z = 5$ and $c_s t/z = 40$, respectively. In Fig. 11.11 the first wave, the compression wave, has reached the distance $x/z = 7.01$, and the second wave, the shear wave, has reached a distance $x/z = 4.91$. It can easily be verified that these values agree well with the theoretical values $c_p t/r = 1$ and $c_s t/r = 1$, respectively. For large values of time, practically the only effect remaining is the Rayleigh wave, see Fig. 11.12, arriving shortly after the passage of the shear wave.

11.1.6 The Horizontal Displacement of the Surface

The expressions for the horizontal displacement derived in the previous section can not be used at the surface $z = 0$, because z has been used as the scaling factor in the dimensionless parameters. Therefore another set of dimensionless parameters must be introduced for the displacements of the surface.

The First Component

The first component of the horizontal displacement is, from (11.172),

$$u_1 = \frac{Q}{\pi\mu}\Re\left\{\frac{p\gamma_p(1/c_s^2 - 2p^2)}{(1/c_s^2 - 2p^2)^2 + 4p^2\gamma_p\gamma_s}\right\}\frac{H(t - t_p)}{\sqrt{t^2 - t_p^2}}, \qquad p = p_1, \qquad (11.184)$$

where

$$p_1 = \frac{tx}{r^2} + \frac{iz}{r^2}\sqrt{t^2 - t_p^2}. \qquad (11.185)$$

At the surface $z = 0$ the radial coordinate r coincides with x, $r = x$, so that $p_1 = t/x$. Introducing a dimensionless time variable τ, defined as

$$\tau = c_s t/x, \qquad (11.186)$$

the parameter p_1 can be written as $p_1 = \tau/c_s$.

Because the integral (11.184) contains a factor $H(t - t_p)$, and $t_p = x/c_p = x\eta/c_s = \eta t/\tau$, non-zero results will be obtained only for $\tau > \eta$.

The parameters γ_p and γ_s now are

$$z = 0, \ \tau > \eta \ : \ \gamma_p = -i\sqrt{\tau^2 - \eta^2}/c_s, \qquad (11.187)$$

$$z = 0, \ \tau < 1 \ : \ \gamma_s = \sqrt{1 - \tau^2}/c_s, \quad \tau > 1 \ : \ \gamma_s = -i\sqrt{\tau^2 - 1}/c_s. \quad (11.188)$$

And the factor $\sqrt{t^2 - t_p^2}$ is

$$z = 0, \ \tau > \eta \ : \ \sqrt{t^2 - t_p^2} = (x/c_s)\sqrt{\tau^2 - \eta^2}, \qquad (11.189)$$

which shows that for all values $\tau > \eta$ the factor $\gamma_p/\sqrt{t^2 - t_p^2} = -i/x$.

As stated above, the result will be zero if $\tau < \eta$. If $\eta < \tau < 1$ the factor $\gamma_p\gamma_s$ will be imaginary, so that the denominator in the expression between brackets in (11.184) will be complex. This leads to the result

$$z = 0, \ \eta < \tau < 1 \ : \ u_1 = \frac{Qc_s}{\pi\mu x}\Im\left\{\frac{\tau(1 - 2\tau^2)}{(1 - 2\tau^2)^2 - 4i\tau^2\sqrt{\tau^2 - \eta^2}\sqrt{1 - \tau^2}}\right\}. \qquad (11.190)$$

Fig. 11.13 Modified
integration path for $z = 0$

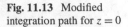

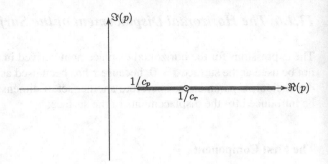

Fig. 11.13 Modified
integration path for $z = 0$

If $\tau > 1$ the factor $\gamma_p \gamma_s$ will be real, and the factor between brackets in (11.184) will be imaginary, so that the real part is zero. This suggests that the expression $u_1 = 0$ for $\tau > 1$, but this may not be correct, because the denominator of (11.184) passes through zero, at the value of p corresponding to the Rayleigh wave singularity. The simplest way to investigate this is to return to the original integral (11.169), with $z = 0$,

$$\bar{u}_1 = \frac{iQ}{2\pi\mu} \int_{-i\infty}^{i\infty} \frac{p(1/c_s^2 - 2p^2)}{(1/c_s^2 - 2p^2)^2 + 4p^2\gamma_p\gamma_s} \exp(-spx)\,dp, \qquad (11.191)$$

and to analyze the behaviour of the integral along the modified integration path when it passes the singularity at $p = 1/c_r$, for $z = 0$, see Fig. 11.13. The integrand of (11.169) is real along the real axis for $p > 1/c_s$ (this corresponds to $\tau > 1$), and therefore the contributions of the path just below the real axis (from right to left) and the path just above the real axis (from left to right) cancel, as was also obtained above. This requires, however, that the integral be considered as a Cauchy principal value, because the integrand of (11.169) has a singularity on the positive part of the real axis at $p = 1/c_r$. The possible contribution of integrating around this singularity can be determined by calculating the contribution to the integral of a small circle surrounding the pole at $p = 1/c_r$.

Therefore, let the denominator of the integral, the Rayleigh function, be denoted by $R(p^2)$,

$$R(p^2) = (2p^2 - 1/c_s^2)^2 - 4p^2\sqrt{p^2 - 1/c_p^2}\sqrt{p^2 - 1/c_s^2}, \qquad (11.192)$$

where γ_p and γ_s have been given their appropriate values for $z = 0$ and large real values of p. Introducing a dimensionless parameter $q = pc_s$ one may write, with $\eta = c_s/c_p$,

$$R(q^2) = c_s^4 R(p^2) = (2q^2 - 1)^2 - 4q^2\sqrt{q^2 - \eta^2}\sqrt{q^2 - 1}. \qquad (11.193)$$

For real values of q, such that $q > 1$, the function $R(q^2)$ is real, and it has a zero for $q = \beta = c_s/c_r$, where c_r is the Rayleigh wave velocity. It follows that

$$\sqrt{\beta^2 - \eta^2}\sqrt{\beta^2 - 1} = (2\beta^2 - 1)^2/4\beta^2. \qquad (11.194)$$

In the vicinity of this zero, the function $R(q^2)$ can be written as

$$R(q^2) = (q^2 - \beta^2)R'(q^2)|_{q^2=\beta^2},$$ (11.195)

where $R'(q^2) = dR(q^2)/dq^2$, or

$$R'(q^2) = 4(2q^2 - 1) - 4\sqrt{q^2 - \eta^2}\sqrt{q^2 - 1} + \frac{2q^2(2q^2 - 1 - \eta^2)}{\sqrt{q^2 - \eta^2}\sqrt{q^2 - 1}}.$$ (11.196)

Using (11.194) it follows, after some simple algebraic operations, that in the vicinity of β

$$R(q^2) = -2(q - \beta)\frac{1 - 4\beta^2 + 8(1 - \eta^2)\beta^6}{\beta(2\beta^2 - 1)^2}.$$ (11.197)

It now follows that integration along a small circle surrounding the pole $p = 1/c_r$, in clockwise direction, gives a contribution

$$\bar{u}_1 = \frac{Q}{2\mu}\frac{\beta^2(2\beta^2 - 1)^3}{1 - 4\beta^2 + 8(1 - \eta^2)\beta^6}\exp(-sx/c_r).$$ (11.198)

Inverse Laplace transformation gives

$$z = 0, \ \tau > 1 \ : \ u_1 = \frac{Q}{2\mu}\frac{\beta^2(2\beta^2 - 1)^3}{1 - 4\beta^2 + 8(1 - \eta^2)\beta^6}\delta(t - x/c_r),$$ (11.199)

where $\delta(t - x/c_r)$ is Dirac's delta function. It appears that there is indeed a non-zero contribution due to the pole at $p = 1/c_r$.

The Second Component

The second component of the horizontal displacement is, from (11.174),

$$u_2 = -\frac{Q}{\pi\mu}\Re\left\{\frac{2p\gamma_p\gamma_s^2}{(1/c_s^2 - 2p^2)^2 + 4p^2\gamma_p\gamma_s}\right\}\frac{H(t - t_s)}{\sqrt{t^2 - t_s^2}}, \quad p = p_3,$$ (11.200)

where

$$p_3 = \frac{tx}{r^2} + \frac{iz}{r^2}\sqrt{t^2 - t_s^2}.$$ (11.201)

Again using the dimensionless time parameter $\tau = c_s t/x$ for points on the surface $z = 0$, where $p_3 = t/x$, and noting that contributions can only be expected for $t > t_s$, or $\tau > 1$, it can be seen that the denominator of the term between brackets is real, and that the denominator is imaginary, so that the result would be zero, if it were not for a possible contribution from the pole at $p = 1/c_r$. This contribution can be determined in the same way as in the case of the first component. The expression for $\bar{u}_2 + \bar{u}_3$ in the integral (11.170) is

$$\bar{u}_2 + \bar{u}_3 = -\frac{iQ}{2\pi\mu} \int_{-i\infty}^{i\infty} \frac{2p\gamma_p\gamma_s}{(1/c_s^2 - 2p^2)^2 + 4p^2\gamma_p\gamma_s} \exp[-s(\gamma_s z + px)] \, dp, \tag{11.202}$$

with

$$\gamma_p = \sqrt{1/c_p^2 - p^2}, \qquad \gamma_s = \sqrt{1/c_s^2 - p^2}. \tag{11.203}$$

The parts of the integral along the real axis $z = 0$ are, if $p > 1/c_s$,

$$\bar{u}_2 = \frac{iQ}{2\pi\mu} \int \frac{2p\sqrt{p^2 - 1/c_p}\sqrt{p^2 - 1/c_s}}{(2p^2 - 1/c_s^2)^2 - 4p^2\sqrt{p^2 - 1/c_p}\sqrt{p^2 - 1/c_s}} \exp(-spx) \, dp. \tag{11.204}$$

Along the real axis, for $p > 1/c_s$, the integrand appears to be real, so that the two parts of the integral just above and just below the real axis cancel, provided that the two parts of the integral are considered as Cauchy principal values, because of the singularity at $p = 1/c_r$. Again, the possible contribution of integrating around this singularity can be determined by calculating the contribution to the integral of a small circle surrounding the pole at $p = 1/c_r$. In this case this gives, using the same type of analysis as previously,

$$\bar{u}_2 = -\frac{Q}{4\mu} \frac{(2\beta^2 - 1)^4}{1 - 4\beta^2 + 8(1 - \eta^2)\beta^6} \exp(-sx/c_r). \tag{11.205}$$

Inverse Laplace transformation gives

$$z = 0, \ \tau > 1 : u_2 = -\frac{Q}{4\mu} \frac{(2\beta^2 - 1)^4}{1 - 4\beta^2 + 8(1 - \eta^2)\beta^6} \delta(t - x/c_r). \tag{11.206}$$

Again there appears to be a non-zero contribution due to the pole at $p = 1/c_r$.

The Third Component

The third component of the horizontal displacement is, from (11.176),

$$u_3 = -\frac{Q}{\pi\mu} \Im \left\{ \frac{2p\gamma_p\gamma_s^2}{(1/c_s^2 - 2p^2)^2 + 4p^2\gamma_p\gamma_s} \right\} \frac{H(t - t_q)H(t_s - t)}{\sqrt{t_s^2 - t^2}}, \qquad p = p_5, \tag{11.207}$$

where

$$p_5 = \frac{xt}{r^2} - \frac{z}{r^2}\sqrt{t_s^2 - t^2}. \tag{11.208}$$

Using the dimensionless time parameter $\tau = c_s t / x$ for points on the surface $z = 0$, this reduces to

$$z = 0,\ \eta < \tau < 1 : u_3 = \frac{Qc_s}{\pi \mu x} \Im \left\{ \frac{2i\tau \sqrt{\tau^2 - \eta^2}\sqrt{1 - \tau_2}}{(1 - 2\tau^2)^2 - 4i\tau^2\sqrt{\tau^2 - \eta^2}\sqrt{1 - \tau^2}} \right\}.$$

(11.209)

For other values of τ, notably $\tau < \eta$ or $\tau > 1$, there is no contribution of this component, $u_3 = 0$.

Total Surface Displacements

Adding the contributions to the surface displacements, the final expressions for the displacements of the surface $z = 0$ are, with $\tau = c_s t / x$,

$$z = 0,\ \tau < \eta : u = 0,$$ (11.210)

$$z = 0,\ \eta < \tau < 1 : u = \frac{Qc_s}{\pi \mu x} \frac{2\tau(1 - 2\tau^2)\sqrt{\tau^2 - \eta^2}\sqrt{1 - \tau_2}}{(1 - 2\tau^2)^4 + 16\tau^4(\tau^2 - \eta^2)(1 - \tau^2)},$$ (11.211)

$$z = 0,\ \tau > 1 : u = \frac{Qc_s}{4\mu x} \frac{(2\tau^3 - 1)^3}{1 - 4\beta^2 + 8(1 - \eta^2)\beta^6} \delta(\tau - \beta).$$ (11.212)

In the last equation it has been used that

$$\delta(\tau - \beta) = (x/c_s)\delta(t - x/c_r),$$ (11.213)

which can be derived from the definition of Dirac's delta function, (11.12).

The surface displacements are shown in graphical form in the Figs. 11.14, 11.15 and 11.16, for three values of v. In each case the displacements are zero before the

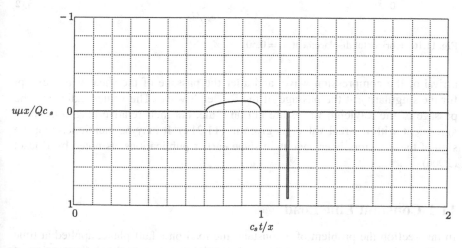

Fig. 11.14 Horizontal displacement, $v = 0.00$

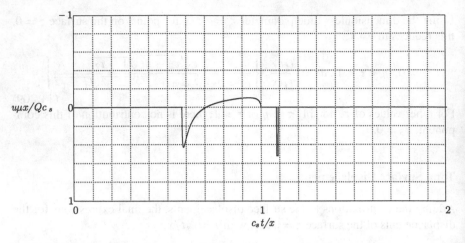

Fig. 11.15 Horizontal displacement, $\nu = 0.25$

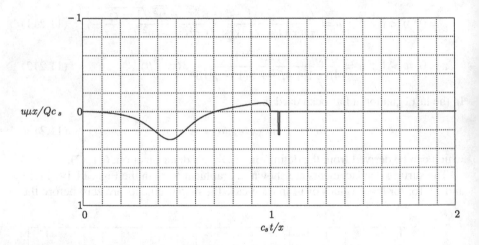

Fig. 11.16 Horizontal displacement, $\nu = 0.50$

arrival of the compression wave, and after the passage of the shear wave, except for the singularity at the passage of the Rayleigh wave. The values indicating the passage of the Rayleigh waves are not on scale, but their relative magnitude is in agreement with the real value, as given in (11.212). It may be mentioned that the sign of these factors is erroneous in some earlier publications, as noted by Kausel (2006).

11.2 Constant Line Load

In this section the problem of a constant line load on a half plane, applied at time $t = 0$ is considered, see Fig. 11.17, with special attention to the determination of

Fig. 11.17 Half plane with line load

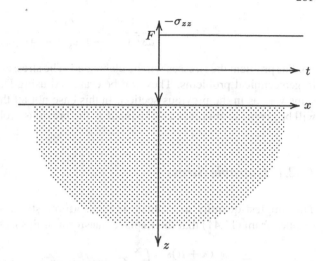

the stress components (Verruijt, 2008a). This is the dynamic equivalent of the classical Flamant problem of elastostatics (Timoshenko and Goodier, 1970). It can be expected that in the dynamic case compression waves and shear waves will be generated, and probably Rayleigh waves near the surface $z = 0$. It can also be expected that for very large values of time the elastostatic solution will be recovered.

In this case the boundary condition for the normal stress on the boundary is

$$z = 0 : \sigma_{zz} = -FH(t)\delta(x), \tag{11.214}$$

where F is the magnitude of the load (per unit length), $\delta(x)$ is a Dirac delta function, and $H(t)$ is Heaviside's unit step function. This boundary condition expresses that a line load of magnitude F is applied at time $t = 0$, and that this load then remains constant.

The Laplace transform of the boundary condition (11.214) is

$$z = 0 : \overline{\sigma}_{zz} = -\frac{F}{s}\delta(x), \tag{11.215}$$

or, when the delta-function is expressed as a Fourier integral,

$$z = 0 : \overline{\sigma}_{zz} = -\frac{F}{2\pi}\int_{-\infty}^{\infty} \exp(-is\alpha x)d\alpha. \tag{11.216}$$

The difference with the impulse problem considered in the previous section is that the quantity Qs is now replaced by F. This is in agreement with the difference in the loading function. Dirac's delta function is the derivative of Heaviside's unit step function, and in the Laplace transforms this results in multiplication by s.

It can be concluded that in this case the two constants in the general solution of the problem are, compare (11.44) and (11.45),

$$C_p = \frac{F}{\mu s^2} \frac{2\alpha^2 + 1/c_s^2}{(2\alpha^2 + 1/c_s^2)^2 - 4\alpha^2\gamma_p\gamma_s}, \tag{11.217}$$

$$C_s = -\frac{F}{\mu s^2} \frac{2\alpha \gamma_p}{(2\alpha^2 + 1/c_s^2)^2 - 4\alpha^2 \gamma_p \gamma_s}. \tag{11.218}$$

For this problem the stresses will be elaborated. The stresses are of particular interest in geotechnical problems. They can be evaluated using De Hoop's method, in the same way as in the previous section. In this case not all the details of the analysis will be given, as reference can be made to the previous problem.

11.2.1 Isotropic Stress

The simplest quantity to evaluate is the isotropic stress $\sigma = (\sigma_{xx} + \sigma_{zz})/2$. It is recalled from (11.41) that the Laplace transform of this isotropic stress is

$$\bar{\sigma} = -\frac{(\lambda + \mu)s^2}{2\pi c_p^2} \int_{-\infty}^{\infty} C_p \exp(-\gamma_p sz - is\alpha x)\, d\alpha. \tag{11.219}$$

Substitution of (11.217) into this expression gives

$$\bar{\sigma} = -\frac{(1 - \eta^2)F}{2\pi c_s^2} \int_{-\infty}^{\infty} \frac{2\alpha^2 + 1/c_s^2}{(2\alpha^2 + 1/c_s^2)^2 - 4\alpha^2 \gamma_p \gamma_s} \exp[-s(\gamma_p z + i\alpha x)]\, d\alpha, \tag{11.220}$$

where

$$\eta^2 = \frac{c_s^2}{c_p^2} = \frac{\mu}{\lambda + 2\mu} = \frac{1 - 2v}{2(1 - v)}. \tag{11.221}$$

Using the same methods as in the previous section this integral can be transformed into the form

$$\bar{\sigma} = -\frac{(1 - \eta^2)F}{\pi c_s^2} \int_0^{\infty} \Re\left\{ \frac{(1/c_s^2 - 2p^2)\gamma_p}{(1/c_s^2 - 2p^2)^2 + 4p^2 \gamma_p \gamma_s} \right\} \frac{H(t - t_p)}{\sqrt{t^2 - t_p^2}} \exp(-st)\, dt, \tag{11.222}$$

where $p = p_1$, which is defined by the equation

$$p_1 = \frac{tx}{r^2} + \frac{iz}{r^2}\sqrt{t^2 - t_p^2}, \tag{11.223}$$

and, as before, $t_p = r/c_p$, $\gamma_p = \sqrt{1/c_p^2 - p^2}$ and $\gamma_s = \sqrt{1/c_s^2 - p^2}$.

The integral (11.222) is of the form of a Laplace transform, which means that the inverse Laplace transform is

$$\sigma = -\frac{(1 - \eta^2)F}{\pi c_s^2}\Re\left\{ \frac{(1/c_s^2 - 2p^2)\gamma_p}{(1/c_s^2 - 2p^2)^2 + 4p^2 \gamma_p \gamma_s} \right\} \frac{H(t - t_p)}{\sqrt{t^2 - t_p^2}}, \quad p = p_1. \tag{11.224}$$

This is the final expression for the isotropic stress.

Calculation of Numerical Values

Using the dimensionless variables

$$\xi = x/z, \qquad \tau = c_s t/z, \qquad \tau_p = c_s t_p/z = \eta\sqrt{1 + \xi^2}, \qquad b = p c_s, \qquad (11.225)$$

it follows that suitable dimensionless forms of the quantities γ_p and γ_s are

$$g_p = \gamma_p c_s = \sqrt{\eta^2 - b^2}, \qquad g_s = \gamma_s c_s = \sqrt{1 - b^2}. \qquad (11.226)$$

The dimensionless form of (11.224) is

$$\frac{\sigma \pi z}{F} = -(1 - \eta^2)\Re\left\{\frac{(1 - 2b^2)g_p}{(1 - 2b^2)^2 + 4b^2 g_p g_s}\right\}\frac{H(\tau - \tau_p)}{\sqrt{\tau^2 - \tau_p^2}}, \qquad (11.227)$$

where

$$b = \frac{\xi \tau + i\sqrt{\tau^2 - \tau_p^2}}{1 + \xi^2}. \qquad (11.228)$$

It may be noted that $\tau_p = \eta\sqrt{1 + \xi^2}$, the dimensionless arrival time of the compression wave.

Equation (11.227) can be used to produce numerical results, by a simple computer program. A function (in C, using complex calculus) to determine the value of the dimensionless isotropic stress $\sigma \pi z/F$ as a function of the parameters $\xi = x/z$, $\tau = c_p t/z$ and Poisson's ratio ν, is shown below.

```
double LineLoadS(double xi,double tau,double nu)
{
  double s,n,nn,x1,tp2,t2;complex b,bb,b1,gp,gs,d,e;
  nn=(1-2*nu)/(2*(1-nu));n=sqrt(nn);x1=1+x*x;tp2=nn*x1;t2=t*t;
  if (t2<=tp2) s=0;else
    {
    b=complex(x*t/x1,(sqrt(t2-tp2))/x1);bb=b*b;b1=1-2*bb;
    gp=sqrt(nn-bb);gs=sqrt(1-bb);d=b1*gp;e=b1*b1+4*bb*gp*gs;
    s=-(1-nn)*real(d/e)/sqrt(t2-tp2);
    }
  return(s);
}
```

It may be noted that the argument of the parameters g_p and g_s should be taken in the range $(-\pi, 0)$, and that the square roots should be calculated separately, to ensure that the arguments are determined correctly.

Some examples are shown in Figs. 11.18, 11.19 and 11.20. The figures show the isotropic stress as a function of x/z, for $\nu = 0$ and for three increasing values of time, $c_s t/z = 2$, $c_s t/z = 10$ and $c_s t/z = 20$. Although in these figures the maximum

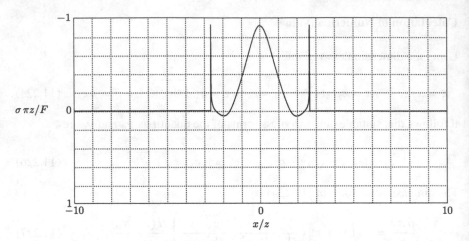

Fig. 11.18 Line load—isotropic stress, $\nu = 0$, $c_s t/z = 2$

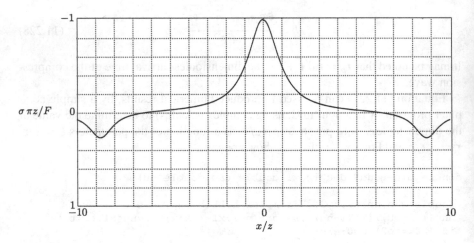

Fig. 11.19 Line load—isotropic stress, $\nu = 0$, $c_s t/z = 10$

value of the dimensionless stress seems to be 1, larger values may occur in the immediate vicinity of the wave front. In this case of the isotropic stress the wave front travels with the velocity of compression waves. For $\nu = 0$ the ratio of compression waves and shear waves is $c_p/c_s = \sqrt{2} = 1.4142$. This can be seen in Fig. 11.18, where the wave front has reached a distance of about $x/z = 2.66$ at time $c_s t/z = 2$, which corresponds to $c_p t/r = 1$, approximately. In Fig. 11.19, at time $c_s t/z = 10$, the disturbance at $c_s t/z \approx 8.8$ indicates the Rayleigh wave. In Fig. 11.20 the value of time is so large that the static results are approached. Actually, the static solution

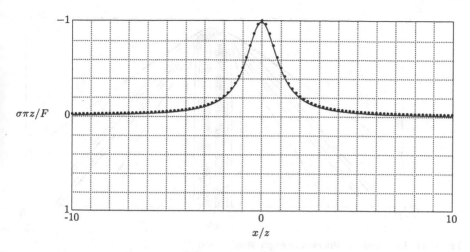

Fig. 11.20 Line load—isotropic stress, $\nu = 0$, $c_s t/z = 20$

of this problem is (Timoshenko and Goodier, 1970)

$$t \rightarrow \infty : \frac{\sigma \pi z}{F} = -\frac{1}{1 + x^2/z^2}. \tag{11.229}$$

This static solution is shown in Fig. 11.20 by small asterisks. The agreement appears to be excellent.

In general the numerical values depend upon the value of Poisson's ratio, but for very large values of time no such dependence is found, in agreement with the static solution.

A comprehensive view of the results is shown in Fig. 11.21. This figure shows contours of the isotropic stress as a function of x/z and $c_s t/z$. The interval between successive contours is $\Delta \sigma \pi z/F = 0.1$. It is interesting to note that the Rayleigh wave can clearly be distinguished in the results, as the tensile zone, progressing towards infinity along the lines $c_s t/x \approx \pm 1.1$. It appears that at very large distances from the point of application of the load the shape and the magnitude of the Rayleigh wave are preserved, even when near the center the static results have long been reached. This is in agreement with the classical analysis of Rayleigh waves in the two-dimensional case, see for instance Achenbach (1975), or the analysis of Rayleigh waves in Chap. 9 of this book. The results shown in Fig. 11.21 suggest that, for sufficiently large values of time, the solution consists of the elastostatic stresses plus a moving stress distribution representing the Rayleigh wave,

$$c_s t/z \gg 1 : \frac{\sigma \pi z}{F} = -\frac{1}{1 + x^2/z^2} + \frac{m}{1 + (x - c_r t)^2/(wz)^2}, \tag{11.230}$$

where c_r is the velocity of the Rayleigh wave, and the coefficients m and w can be determined by fitting the curve with the exact results. For $\nu = 0$ it follows that $m \approx 0.28$ and $w \approx 0.7$.

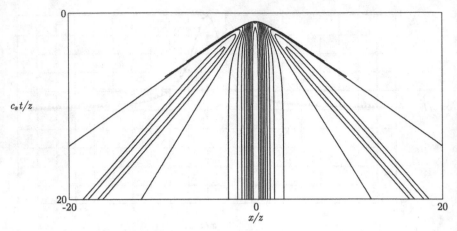

Fig. 11.21 Line load—contours of isotropic stress, $\nu = 0$

Verification of the Elastostatic Limit

Although the numerical results have been found to be in agreement with the elastostatic limit for very large values of time, it may be illustrative to also verify this agreement analytically, by considering the limiting behaviour of the solution (11.227) of the elastodynamic problem,

$$\frac{\sigma \pi z}{F} = -(1 - \eta^2)\Re\left\{\frac{(1 - 2b^2)g_p}{(1 - 2b^2)^2 + 4b^2 g_p g_s}\right\}\frac{H(\tau - \tau_p)}{\sqrt{\tau^2 - \tau_p^2}},\qquad(11.231)$$

where

$$b = \frac{\xi \tau + i\sqrt{\tau^2 - \tau_p^2}}{1 + \xi^2},\qquad(11.232)$$

$$g_p = \gamma_p c_s = \sqrt{\eta^2 - b^2},\qquad g_s = \gamma_s c_s = \sqrt{1 - b^2},\qquad \tau_p = \eta\sqrt{1 + \xi^2}.\quad(11.233)$$

If $\tau \gg \tau_p$ it follows from (11.232) that $b = \frac{(\xi + i)\tau}{1 + \xi^2}$, indicating that b tends to infinity, if the dimensionless time parameter τ tends to infinity, assuming that the dimensionless distance ξ remains finite, and does not also tend to infinity.

Taking into account that the real parts of g_p and g_s (the dimensionless forms of the parameters γ_p and γ_s) should be positive, and their imaginary parts should be negative, as indicated in Sect. 11.1, it now follows from (11.233) that

$$g_p = -ib(1 - \eta^2/2b^2),\qquad g_s = -ib(1 - 1/2b^2).\qquad(11.234)$$

In order to evaluate the limiting form of the denominator in (11.231) it may be noted that

$$(1 - 2b^2)^2 = 4b^4 - 4b^2 + 1,\qquad 4b^2 g_p g_s = -4b^4 + 2b^2(1 + \eta^2),\quad(11.235)$$

so that

$$(1 - 2b^2)^2 + 4b^2 g_p g_s = -2b^2(1 - \eta^2). \tag{11.236}$$

On the other hand, the numerator in (11.231) is

$$(1 - 2b^2)g_p = 2ib^3 = 2b^2 \frac{i(\xi + i)\tau}{1 + \xi^2}. \tag{11.237}$$

It now follows that

$$\Re\left\{\frac{(1 - 2b^2)g_p}{(1 - 2b^2)^2 + 4b^2 g_p g_s}\right\} = \frac{\tau}{(1 - \eta^2)(1 + \xi^2)}. \tag{11.238}$$

Finally it follows, substituting this result into (11.231), that

$$\tau \to \infty : \quad \frac{\sigma \pi z}{F} = -\frac{1}{1 + \xi^2}. \tag{11.239}$$

This is indeed the correct elastostatic solution, see (11.229).

Approximate Analysis of the Rayleigh Wave

It has already been observed from the complete solution that the line load produces Rayleigh waves, see Fig. 11.21. These waves travel at a constant speed c_r and with constant shape to infinity, at both ends. This part of the solution may be further analyzed by investigating the behaviour of the general solution (11.227) in the vicinity of the Rayleigh wave (Verruijt, 2008a).

For this purpose the general solution (11.227) is written in the form

$$\frac{\sigma \pi z}{F} = \Im\left\{\frac{Q(b^2)}{R(b^2)}\right\} H(\tau - \tau_p), \tag{11.240}$$

where $Q(b^2)$ is defined by

$$Q(b^2) = (1 - \eta^2)(2b^2 - 1)h_p / \sqrt{\tau^2 - \tau_p^2}, \tag{11.241}$$

and $R(b^2)$ is the Rayleigh function,

$$R(b^2) = (2b^2 - 1)^2 - 4b^2 h_p h_s. \tag{11.242}$$

In these functions the following parameters have been used

$$b = \frac{\xi \tau + i\sqrt{\tau^2 - \tau_p^2}}{1 + \xi^2}, \quad \xi = x/z, \quad \tau = c_s t/z, \quad \tau_p = \eta\sqrt{1 + \xi^2}, \tag{11.243}$$

and

$$h_p = ig_p = \sqrt{b^2 - \eta^2}, \qquad h_s = ig_s = \sqrt{b^2 - 1}. \tag{11.244}$$

The function $R(b^2)$ assumes a zero value for the real value $b = \beta$, where

$$\beta = c_s/c_r, \tag{11.245}$$

the ratio of the shear wave velocity c_s to the Rayleigh wave velocity c_r, see Chap. 9. The value of β depends upon the value of Poisson's ratio, but is always somewhat larger than 1, indicating that the Rayleigh wave is always slightly slower than the shear wave.

In the solution considered here the value of b is complex, see (11.243), but it can be expected that for values of b close to β the absolute value of $R(b^2)$ will be small, perhaps very small, so that the isotropic stress will be large. Using Taylor's expansion formula the function $R(b^2)$ for values of b close to β may be written as

$$R(b^2) = -M(b^2 - \beta^2) \approx -2M\beta(b - \beta), \tag{11.246}$$

where

$$M = -\frac{dR}{db^2}\Big|_{b=\beta}. \tag{11.247}$$

The minus sign has been included in this definition to ensure that $M > 0$.

It follows from (11.242) and (11.244) that

$$\frac{dR}{db^2} = 4(2b^2 - 1) - 4h_p h_s - \frac{2b^2 h_s}{h_p} - \frac{2b^2 h_p}{h_s}. \tag{11.248}$$

The parameter M can be calculated by taking $b = \beta$. With (11.247) this gives, making use of the knowledge that $R(\beta^2) = 0$,

$$M = \frac{1 - 4\beta^2 + 8(1 - \eta^2)\beta^6}{\beta^2(2\beta^2 - 1)^2}. \tag{11.249}$$

It may be noted that this is a real (and positive) value. Table 11.1 shows the values of $\eta = c_s/c_p$, $\beta = c_s/c_r$ and M, as functions of Poisson's ratio ν, together with some other parameters, to be introduced later.

The Rayleigh wave is especially prominent for large values of ξ and τ, with ξ in the vicinity of τ/β. If it is assumed that $\tau \gg 1$ and that $\xi \approx \tau/\beta$ it follows that $\xi \gg 1$, so that $1 + \xi^2 \approx \xi^2$. It then follows, with the last of (11.243), that $\tau_p \approx \eta\tau/\beta$, so that $\tau^2 - \tau_p^2 \approx \tau^2(1 - \eta^2/\beta^2)$. Furthermore, the value of h_p, defined in the first of (11.244), can be approximated by $h_p \approx \sqrt{\beta^2 - \eta^2}$, so that $h_p/\sqrt{\tau^2 - \tau_p^2} \approx \beta/\tau$. It follows that the expression for $Q(b^2)$, see (11.241), can be approximated by

$$\xi \gg 1, \ \tau \gg 1 : Q(b^2) \approx Q(\beta^2) \approx (1 - \eta^2)(2\beta^2 - 1)\beta/\tau. \tag{11.250}$$

Table 11.1 Rayleigh wave velocities, and some derived parameters

v	$\eta = c_s/c_p$	$\beta = c_s/c_r$	M	m	w_p	w_s
0.00	0.707107	1.144123	1.381966	0.284432	0.786151	0.485868
0.05	0.688247	1.131612	1.512235	0.267257	0.793783	0.468064
0.10	0.666667	1.119688	1.664979	0.249677	0.803426	0.449846
0.15	0.641689	1.108377	1.844774	0.231905	0.815367	0.431277
0.20	0.612372	1.097700	2.057248	0.214161	0.829929	0.412415
0.25	0.577350	1.087664	2.309401	0.196660	0.847487	0.393320
0.30	0.534522	1.078269	2.610083	0.179596	0.868481	0.374040
0.35	0.480384	1.069504	2.970690	0.163133	0.893448	0.354613
0.40	0.408248	1.061351	3.406234	0.147398	0.923063	0.335064
0.45	0.301511	1.053786	3.936997	0.132478	0.958193	0.315397
0.50	0.000000	1.046778	4.591195	0.118420	1.000000	0.295598

In (11.246) the variable is b. With the first of (11.243), and assuming that $\xi \gg 1$ and $\tau \gg 1$, this quantity can be expressed as

$$\xi \gg 1, \ \tau \gg 1 : \ b \approx \frac{\tau}{\xi} + \frac{i\sqrt{\tau^2 - \eta^2 \xi^2}}{\xi^2}. \tag{11.251}$$

Substitution into (11.246) gives

$$\xi \gg 1, \ \tau \gg 1 : \ R(b^2) \approx -2M\beta \left\{ \frac{\tau - \xi\beta}{\xi} + \frac{i\sqrt{\tau^2 - \eta^2 \xi^2}}{\xi^2} \right\}. \tag{11.252}$$

Taking $\xi = \tau/\beta$ everywhere, except in the factor $\tau - \xi\beta$, gives

$$\xi \gg 1, \ \tau \gg 1 : \ R(b^2) \approx \frac{2M\beta^3}{\tau} \left[(\xi - \tau/\beta) - i\sqrt{1 - \eta^2/\beta^2} \right]. \tag{11.253}$$

Substitution of (11.250) and (11.253) into (11.240) gives

$$\xi \gg 1, \ \tau \gg 1 : \ \frac{\sigma \pi z}{F} \approx \frac{(1 - \eta^2)(2\beta^2 - 1)}{2a\beta^2 \sqrt{1 - \eta^2/\beta^2}} \frac{1}{1 + (\xi - \tau/\beta)^2/(1 - \eta^2/\beta^2)}. \tag{11.254}$$

Using the original variables this can be written as

$$x/z \gg 1, \ c_s t/z \gg 1 : \ \frac{\sigma \pi z}{F} \approx \frac{m}{1 + (x - c_r t)^2/(w_p z)^2}, \tag{11.255}$$

where m is the maximum value of the Rayleigh wave disturbance,

$$m = \frac{(1 - \eta^2)(2\beta^2 - 1)}{2M\beta^2 w_p}, \tag{11.256}$$

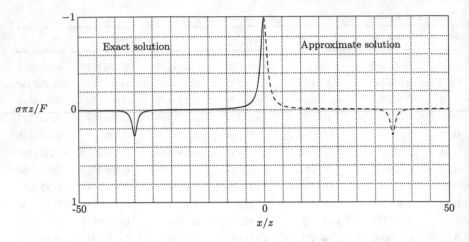

Fig. 11.22 Line load—isotropic stress, $\nu = 0$, $c_s t/z = 40$

and w_p is a measure for the width of the Rayleigh wave disturbance,

$$w_p = \sqrt{1 - \eta^2/\beta^2}. \tag{11.257}$$

It may be noted that this form of the approximation is in agreement with the suggestion of (11.230), but now the equation has been derived analytically.

It can be seen from (11.255) that for $x = c_r t$, the stress parameter $\sigma \pi z/F = m$, and that for $x = c_r t \pm w_p z$ this value is $\sigma \pi z/F = m/2$, indicating that the distance $2w_p z$ is the total width of the disturbance at its medium height. The parameters m and w_p depend upon the value of Poisson's ratio, see Table 11.1, but they do not depend upon x/z or $c_s t/z$, thus confirming the well known property of Rayleigh waves in the two dimensional case, that they are independent of the distance travelled.

To demonstrate the accuracy of the approximation two examples are shown in Figs. 11.22 and 11.23. These figures show the values of the isotropic stress as a function of x/z, for $c_s t/z = 40$, and for two values of Poisson's ratio, $\nu = 0$ and $\nu = 0.5$. The fully drawn lines show the exact results, calculated by (11.227), and the dashed lines indicate the approximate results, calculated by a superposition of the steady state solution (11.229) and the approximate formula for the Rayleigh wave disturbance (11.255). This final approximate formula, valid for any value of $\xi = x/z$, is

$$c_s t/z \gg 1 : \quad \frac{\sigma \pi z}{F} \approx -\frac{1}{1 + x^2/z^2} + \frac{m}{1 + (x - c_r t)^2/(w_p z)^2}. \tag{11.258}$$

It appears that the exact results and the approximate results can hardly be distinguished from each other, indicating that the approximate results are very accurate. Actually, the maximum difference between the exact solution and the approximate solution is less than 0.01, in both cases.

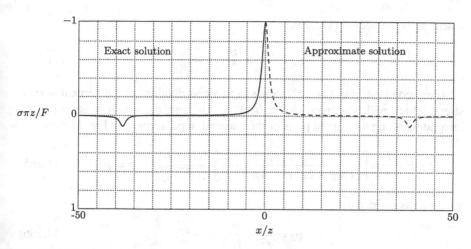

Fig. 11.23 Line load—isotropic stress, $\nu = 0.5$, $c_s t/z = 40$

11.2.2 The Vertical Normal Stress

Probably the most interesting quantity for soil mechanics practice is the vertical normal stress σ_{zz}. It is recalled from (11.38) that its Laplace transform is

$$\overline{\sigma}_{zz} = -\frac{s^2 \mu}{2\pi} \int_{-\infty}^{\infty} \{(2\alpha^2 + 1/c_s^2)C_p \exp(-\gamma_p s z) + 2\alpha \gamma_s C_s \exp(-\gamma_s s z)\}$$
$$\times \exp(-is\alpha x)\, d\alpha. \tag{11.259}$$

Substitution of the expressions (11.217) and (11.218) for the two constants C_p and C_s gives

$$\overline{\sigma}_{zz} = \overline{\sigma}_1 + \overline{\sigma}_2 + \overline{\sigma}_3, \tag{11.260}$$

where

$$\overline{\sigma}_1 = -\frac{F}{2\pi} \int_{-\infty}^{\infty} \frac{(2\alpha^2 + 1/c_s^2)^2}{(2\alpha^2 + 1/c_s^2)^2 - 4\alpha^2 \gamma_p \gamma_s} \exp[-s(\gamma_p z + i\alpha x)]\, d\alpha, \tag{11.261}$$

$$\overline{\sigma}_2 + \overline{\sigma}_3 = \frac{F}{2\pi} \int_{-\infty}^{\infty} \frac{4\alpha^2 \gamma_p \gamma_s}{(2\alpha^2 + 1/c_s^2)^2 - 4\alpha^2 \gamma_p \gamma_s} \exp[-s(\gamma_s z + i\alpha x)]\, d\alpha. \tag{11.262}$$

The evaluation of the two integrals (11.261) and (11.262) can be performed using the same procedures as for the first problem of this chapter, the vertical displacement due to a pulse load.

The first integral, (11.261), leads to the expression

$$\sigma_1 = -\frac{F}{\pi} \Re \left\{ \frac{(1/c_s^2 - 2p^2)^2 \gamma_p}{(1/c_s^2 - 2p^2)^2 + 4p^2 \gamma_p \gamma_s} \right\} \frac{H(t - t_p)}{\sqrt{t^2 - t_p^2}}, \qquad p = p_1, \tag{11.263}$$

where p_1 is defined by (11.223),

$$p_1 = \frac{tx}{r^2} + \frac{iz}{r^2}\sqrt{t^2 - t_p^2}. \tag{11.264}$$

The second integral, (11.262), leads to two contributions to the vertical normal stress, which will be denoted by σ_2 and σ_3. The part σ_2, which is generated by integrating along the curved parts of the integration path shown in Fig. 11.3, is

$$\sigma_2 = -\frac{F}{\pi}\Re\left\{\frac{4p^2\gamma_p\gamma_s^2}{(1/c_s^2 - 2p^2)^2 + 4p^2\gamma_p\gamma_s}\right\}\frac{H(t - t_s)}{\sqrt{t^2 - t_s^2}}, \quad p = p_2, \tag{11.265}$$

where p_2 is defined by

$$p_2 = \frac{tx}{r^2} + \frac{iz}{r^2}\sqrt{t^2 - t_s^2}. \tag{11.266}$$

If

$$\frac{x^2}{z^2} > \frac{\eta^2}{1 - \eta^2}, \tag{11.267}$$

there is an additional contribution, denoted by σ_3, to the second integral, produced by integrating along the loop in the integration path shown in Fig. 11.3. This contribution is

$$\sigma_3 = -\frac{F}{\pi}\Im\left\{\frac{4p^2\gamma_p\gamma_s^2}{(1/c_s^2 - 2p^2)^2 + 4p^2\gamma_p\gamma_s}\right\}\frac{H(t - t_q)H(t_s - t)}{\sqrt{t_s^2 - t^2}}, \quad p = p_3, \tag{11.268}$$

where p_3 is defined by

$$p_3 = \frac{xt}{r^2} - \frac{z}{r^2}\sqrt{t_s^2 - t^2}, \tag{11.269}$$

and t_q is defined by

$$t_q/t_s = (\eta x + z\sqrt{1 - \eta^2})/r, \quad t_q/t_s < 1. \tag{11.270}$$

It may be noted that the function $H(t - t_q)H(t_s - t)$ is equal to 1 in the interval $t_q < t < t_s$ only, elsewhere it is zero. A factor $H(x\sqrt{1 - \eta^2} - \eta z)$ might be added to take into account that a contribution of the loop should be included only if the condition (11.267) is satisfied, but this is not really necessary as it is ensured by the condition $t_q < t_s$, which is ensured by the factor $H(t - t_q)H(t_s - t)$.

Calculation of Numerical Values

The vertical normal stress σ_{zz} can be written as

$$\sigma_{zz} = \sigma_1 + \sigma_2 + \sigma_3, \tag{11.271}$$

where σ_1 is given by (11.263), σ_2 by (11.265) and σ_3 by (11.268). For the computation of numerical results these formulas can be made dimensionless by introducing a reference stress F/z and a reference time z/c_p, and using the dimensionless parameters

$$\xi = x/z, \qquad \tau = c_s t/z, \qquad \tau_p = \eta\sqrt{1+\xi^2}, \qquad \tau_s = \sqrt{1+\xi^2},$$

$$\tau_q = \eta\xi + \sqrt{1-\eta^2}, \qquad a = c_s p, \qquad g_p = c_s\gamma_p = \sqrt{\eta^2 - a^2}, \quad (11.272)$$

$$g_s = c_s\gamma_s = \sqrt{1-a^2}.$$

The dimensionless form of the first term is

$$\frac{\sigma_1\pi z}{F} = -\Re\left\{\frac{(1-2a^2)^2\sqrt{\eta^2-a^2}}{(1-2a^2)^2+4a^2\sqrt{1-a^2}\sqrt{\eta^2-a^2}}\right\}\frac{H(\tau-\tau_p)}{\sqrt{\tau^2-\tau_p^2}}, \qquad (11.273)$$

where

$$a = \frac{\tau\xi + i\sqrt{\tau^2-\tau_p^2}}{1+\xi^2}. \qquad (11.274)$$

The dimensionless form of the second term is

$$\frac{\sigma_2\pi z}{F} = -\Re\left\{\frac{4b^2(1-b^2)\sqrt{\eta^2-b^2}}{(1-2b^2)^2+4b^2\sqrt{1-b^2}\sqrt{\eta^2-b^2}}\right\}\frac{H(\tau-\tau_s)}{\sqrt{\tau^2-\tau_s^2}}, \qquad (11.275)$$

where

$$b = \frac{\tau\xi + i\sqrt{\tau^2-\tau_s^2}}{1+\xi^2}. \qquad (11.276)$$

The dimensionless form of the third term is

$$\frac{\sigma_3\pi z}{F} = \frac{4\eta c^2(1-c^2)(1-2c^2)^2\sqrt{c^2-\eta^2}}{(1-2c^2)^4+16c^4(c^2-\eta^2)(1-c^2)}\frac{H(\tau-\tau_q)H(\tau_s-\tau)}{\sqrt{\tau_s^2-\tau^2}}, \qquad (11.277)$$

where

$$c = \frac{\xi\tau - \sqrt{\tau_s^2-\tau^2}}{1+\xi^2}. \qquad (11.278)$$

All factors in the expression (11.277) are real.

It should be noted that in the derivation of these expressions it has been assumed that $x > 0$, so that $\xi > 0$. Values for $\xi < 0$ can be obtained by using the symmetry of the problem.

A function to calculate the value of the dimensionless parameter $\sigma_{zz}\pi z/2F$ as a function of the parameters $\xi = x/z$ (with $\xi \geq 0$), $\tau = c_p t/z$ and Poisson's ratio v, is shown below. In this function the dimensionless parameters ξ, τ and v are denoted

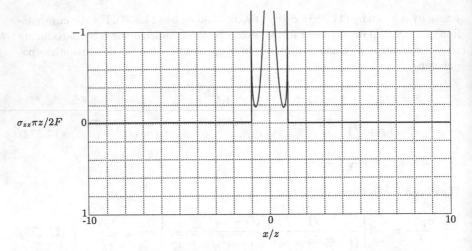

Fig. 11.24 Line load—vertical normal stress, $\nu = 0$, $c_s t/z = 1$

by x, t and nu. The function consists of three parts, as given by (11.273), (11.275) and (11.277).

```
double LineLoadSzz(double x,double t,double nu)
{
  double s,s1,s2,s3,n,nn,nn1,tp2,ts2,ts,tq,t2,c,cc,f,g,h;
  complex a,aa,a1,b,bb,b1,gp,gs,d,e;
  nn=(1-2*nu)/(2*(1-nu));n=sqrt(nn);nn1=sqrt(1-nn);
  ts2=1+x*x;tp2=nn*ts2;ts=sqrt(ts2);t2=t*t;tq=n*x+nn1;
  if (t2<=tp2) s1=0;else
    {
      a=complex(x*t/ts2,sqrt(t2-tp2)/ts2);aa=a*a;a1=1-2*aa;
      gp=sqrt(nn-aa);gs=sqrt(1-aa);d=a1*a1*gp;e=a1*a1+4*aa*gp*gs;
      s1=-real(d/e)/(2*sqrt(t2-tp2));
    }
  if (t2<=ts2) s2=0;else
    {
      b=complex(x*t/ts2,sqrt(t2-ts2)/ts2);bb=b*b;b1=1-2*bb;
      gp=sqrt(nn-bb);gs=sqrt(1-bb);d=4*bb*(1-bb)*gp;e=b1*b1+4*bb*gp*gs;
      s2=-real(d/e)/(2*sqrt(t2-ts2));
    }
  if ((t<=tq)||(t>=ts)||(tq>ts)) s3=0;else
    {
      c=(x*t-sqrt(ts2-t2))/ts2;cc=c*c;f=(1-2*cc)*(1-2*cc);
      g=4*n*cc*(1-cc)*f*sqrt(cc-nn);h=f*f+16*cc*cc*(1-cc)*(cc-nn);
      s3=g/(2*h*sqrt(ts2-t2));
    }
  s=s1+s2+s3;
  return(s);
}
```

Some results are shown in Figs. 11.24, 11.25 and 11.26. These figures show the vertical normal stress σ_{zz} as a function of $\xi = x/z$, for $\nu = 0$, and for the values $c_s t/z = 1$, 8 and 40, respectively. The results for $c_s t/z = 1$ indicate that for relatively small values of time non-zero values are obtained only if $t > t_p$. For larger values, for instance $c_s t/z = 8$, a discontinuity can be observed when the shear wave

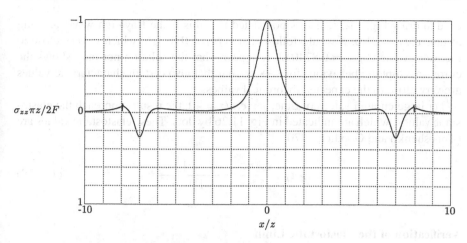

Fig. 11.25 Line load—vertical normal stress, $\nu = 0$, $c_s t/z = 8$

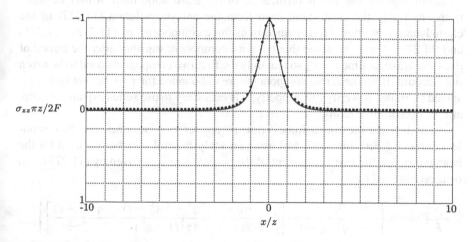

Fig. 11.26 Line load—vertical normal stress, $\nu = 0$, $c_s t/z = 40$

passes, and relatively large values are observed somewhat later, probably at the passing of the Rayleigh wave.

It may be noted from Fig. 11.24 that the first wave arrives at $x/z = 1$. This means that then $r/x = 1.414214$. If it is assumed that this is the compression wave, arriving at time $c_p t/r = 1.0$, it would follow that $c_p/c_s = 1.414214$, which is precisely the known value of this ratio for $\nu = 0$.

Furthermore, from the numerical data used to produce Fig. 11.25 it follows that for $c_s t/z = 8$ the shear wave discontinuity appears for $x/z = 7.935$, which can be verified, approximately, by inspection of the figure. This means that $r/z = 7.998$, so that this discontinuity appears if $c_s t/r = 8/7.998 = 1.00025$, which is very close to the expected value of $c_s t/r = 1$. Secondly, it can be seen that a local maximum of the stress appears if $x/z = 6.99$, again determined from the actual numerical data, used to draw Fig. 11.25. If it is postulated that this is the Rayleigh wave, arriving at time $c_r t/x = 1.0$, it would follow that $c_r/c_s = 0.87375$, which is very close to the exact value $c_r/c_s = 0.874032$. It appears that the solution not only shows the correct asymptotic behaviour for $t \to \infty$, but that also certain characteristic values are very close to the expected theoretical values.

For large enough values of time, say $c_s t/z = 40$, the results approach the elastostatic values, which are indicated in Fig. 11.26 by dots. These elastostatic values are (Timoshenko and Goodier, 1970)

$$\tau \gg \tau_p : \frac{\sigma_{zz}\pi z}{2F} = -\frac{1}{(1+\xi^2)^2}. \tag{11.279}$$

Verification of the Elastostatic Limit

In addition to the numerical verification of the elastostatic limit, it may be illustrative to verify the elastostatic limit from the analytic solution (11.227) of the elastodynamic problem, as described in its three components by (11.273), (11.275) and (11.277). It may be noted that the third component vanishes after the arrival of the shear wave, at time $t = t_s$, so that only the first two components need to be taken into account. Both these two components are unbounded for $\tau \to \infty$, but their sum should be finite. To verify this property, it is necessary to use series expansions with at least two or three terms.

Assuming that the dimensionless time parameter $\tau \to \infty$, while the dimensionless distance ξ remains finite, and using an analysis similar to the one used for the isotropic stress, the first component of the vertical stress, as given by (11.273), can be approximated by

$$\frac{\sigma_1 \pi z}{F} = -\Re\left\{\frac{2i\tau^2}{(1-\eta^2)(\xi-i)^2}\left[1 - \frac{3i\eta^2}{2\tau^2} + \frac{(3\eta^4-1)(\xi-i)}{4\tau^2(1-\eta^2)} + \frac{\eta^2(\xi+i)}{2\tau^2}\right]\right\}. \tag{11.280}$$

The limiting behaviour of the second component, (11.275) for large values of τ and finite values of ξ is found to be

$$\frac{\sigma_2 \pi z}{F} = \Re\left\{\frac{2i\tau^2}{(1-\eta^2)(\xi-i)^2}\left[1 - \frac{3i}{2\tau^2} + \frac{(3\eta^4-1)(\xi-i)}{4\tau^2(1-\eta^2)} + \frac{\xi+i}{2\tau^2}\right]\right\}. \tag{11.281}$$

It follows from (11.280) and (11.281) that for very large values of time the vertical normal stress, which is the sum of the components σ_1 and σ_2, is

$$\tau \to \infty : \frac{\sigma_{zz}\pi z}{F} = \Re\left\{\frac{-2+3i\xi+i\xi^3}{(1+\xi^2)^2}\right\} = -\frac{2}{(1+\xi^2)^2}. \tag{11.282}$$

This is indeed the correct elastostatic solution, see (11.279).

Approximate Analysis of the Rayleigh Wave

For large values of time and the lateral coordinate x the solution for the vertical normal stress appears to have all the characteristics of a Rayleigh wave, as is suggested by Fig. 11.25. This property can be further analyzed by considering the behaviour of the analytical solution near the points where the denominator of the expressions can be expected to become very small.

Using the same procedures as used for the analysis of the behaviour of the solution for the isotropic stress, earlier in this chapter, leading to (11.255), it can be derived that an approximation for the vertical normal stress for large values of the time parameter $\tau = c_s t/z$ and the lateral coordinate $\xi = x/z$ is, expressed in the original variables,

$$x/z \gg 1, \ c_s t/z \gg 1 \ : \ \frac{\sigma_{zz} \pi z}{2F} \approx -\frac{m_1}{1+(x-c_r t)^2/(w_p z)^2}$$
$$+ \frac{m_2}{1+(x-c_r t)^2/(w_s z)^2}, \quad (11.283)$$

where

$$m_1 = \frac{(2\beta^2-1)^2}{4M\beta^2 w_p}, \quad w_p = \sqrt{1-\eta^2/\beta^2}, \quad (11.284)$$

$$m_2 = \frac{\beta^2 w_p}{M}, \quad w_s = \sqrt{1-1/\beta^2}, \quad (11.285)$$

where M is given by (11.249),

$$M = \frac{1-\beta^2+8(1-\eta^2)\beta^6}{\beta^2(2\beta^2-1)^2}. \quad (11.286)$$

The two expressions in (11.283) are the approximations of (11.273) and (11.275) for large values of $c_s t/z$, and values of x/z in the neighbourhood of $c_r t$, where the Rayleigh wave has its peak value. The third part of the solution, (11.277), does not play a role, because this applies only for values of time before the passage of the shear wave.

To demonstrate the accuracy of the approximation two examples are shown in Figs. 11.27 and 11.28. These figures show the values of the vertical normal stress as a function of x/z, for $c_s t/z = 40$, and for two values of Poisson's ratio, $\nu = 0$ and $\nu = 0.5$. The fully drawn lines show the exact results, and the dashed lines show the approximate results, calculated by a superposition of the steady state solution (11.279) and the approximate formula for the Rayleigh wave disturbance (11.283). This final approximate formula, valid for any value of x/z, is

$$c_s t/z \gg 1 \ : \ \frac{\sigma_{zz}\pi z}{2F} \approx -\frac{1}{(1+x^2/z^2)^2} - \frac{m_1}{1+(x-c_r t)^2/(w_p z)^2}$$
$$+ \frac{m_2}{1+(x-c_r t)^2/(w_s z)^2}. \quad (11.287)$$

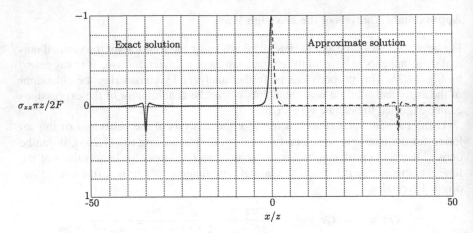

Fig. 11.27 Line load—vertical normal stress, $\nu = 0$, $c_s t/z = 40$

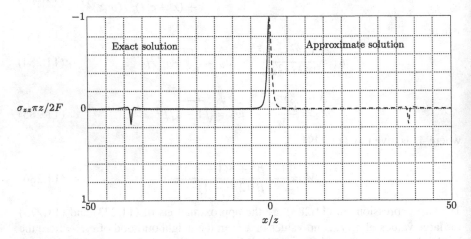

Fig. 11.28 Line load—vertical normal stress, $\nu = 0.5$, $c_s t/z = 40$

As in the case of the isotropic stress the approximation appears to be very good. The maximum difference between the exact solution and the approximate solution is smaller than 0.01.

11.2.3 The Horizontal Normal Stress

It is recalled from (11.37) that the general expression for the Laplace transform of the horizontal normal stress σ_{xx} in an elastic half plane is

$$\overline{\sigma}_{xx} = \frac{s^2 \mu}{2\pi} \int_{-\infty}^{\infty} \{(2\alpha^2 - \lambda/\mu c_p^2)C_p \exp(-\gamma_p s z) + 2\alpha \gamma_s C_s \exp(-\gamma_s s z)\}$$
$$\times \exp(-i s \alpha x)\, d\alpha. \tag{11.288}$$

Substitution of the expressions (11.217) and (11.218) for the two constants C_p and C_s gives

$$\overline{\sigma}_{xx} = \overline{\sigma}_1 + \overline{\sigma}_2 + \overline{\sigma}_3, \tag{11.289}$$

where

$$\overline{\sigma}_1 = \frac{F}{2\pi} \int_{-\infty}^{\infty} \frac{(2\alpha^2 + 1/c_s^2)(2\alpha^2 - \lambda/\mu c_p^2)}{(2\alpha^2 + 1/c_s^2)^2 - 4\alpha^2 \gamma_p \gamma_s} \exp[-s(\gamma_p z + i\alpha x)] \, d\alpha, \tag{11.290}$$

$$\overline{\sigma}_2 + \overline{\sigma}_3 = -\frac{F}{2\pi} \int_{-\infty}^{\infty} \frac{4\alpha^2 \gamma_p \gamma_s}{(2\alpha^2 + 1/c_s^2)^2 - 4\alpha^2 \gamma_p \gamma_s} \exp[-s(\gamma_s z + i\alpha x)] \, d\alpha. \tag{11.291}$$

The two integrals can be evaluated using the techniques of De Hoop's method, in the same way as in the previous problems of this chapter.

The result is that the horizontal normal stress σ_{xx} can be written as the sum of three contributions,

$$\sigma_{xx} = \sigma_1 + \sigma_2 + \sigma_3. \tag{11.292}$$

The first term is, in dimensionless form,

$$\frac{\sigma_1 \pi z}{F} = -\Re \left\{ \frac{(1 - 2a^2)(1 - 2\eta^2 + 2a^2)\sqrt{\eta^2 - a^2}}{(1 - 2a^2)^2 + 4a^2\sqrt{1 - a^2}\sqrt{\eta^2 - a^2}} \right\} \frac{H(\tau - \tau_p)}{\sqrt{\tau^2 - \tau_p^2}}, \tag{11.293}$$

where

$$a = (\tau \xi + i\sqrt{\tau^2 - \tau_p^2})/(1 + \xi^2). \tag{11.294}$$

The second term is

$$\frac{\sigma_2 \pi z}{F} = \Re \left\{ \frac{4b^2(1 - b^2)\sqrt{\eta^2 - b^2}}{(1 - 2b^2)^2 + 4b^2\sqrt{1 - b^2}\sqrt{\eta^2 - b^2}} \right\} \frac{H(\tau - \tau_s)}{\sqrt{\tau^2 - \tau_s^2}}, \tag{11.295}$$

where

$$b = (\tau \xi + i\sqrt{\tau^2 - \tau_s^2})/(1 + \xi^2). \tag{11.296}$$

And the third term is

$$\frac{\sigma_3 \pi z}{F} = -\frac{4\eta c^2(1 - c^2)(1 - 2c^2)^2\sqrt{c^2 - \eta^2}}{(1 - 2c^2)^4 + 16c^4(c^2 - \eta^2)(1 - c^2)} \frac{H(\tau - \tau_q)H(\tau_s - \tau)}{\sqrt{\tau_s^2 - \tau^2}}, \tag{11.297}$$

where

$$c = (\xi \tau - \sqrt{\tau_s^2 - \tau^2})/(1 + \xi^2). \tag{11.298}$$

Apart from the usual parameter $\eta = c_s/c_p$, the following dimensionless parameters have been used,

$$\xi = x/z, \qquad \tau = c_s t/z, \qquad \tau_p = \eta\sqrt{1+\xi^2}, \qquad \tau_s = \sqrt{1+\xi^2}, \tag{11.299}$$
$$\tau_q = \xi\eta + \sqrt{1-\eta^2}.$$

In the derivation of the expressions it has been assumed that $x > 0$, so that $\xi > 0$. Values for $\xi < 0$ can be obtained using the symmetry of the problem.

Calculation of Numerical Values

A function to calculate the value of the dimensionless parameter $\sigma_{xx}\pi z/2F$ as a function of the parameters $\xi = x/z$ (with $\xi \geq 0$), $\tau = c_s t/z$ and Poisson's ratio v, is shown below. In this function the dimensionless parameters ξ, τ and v are denoted by x, t and nu. The function consists of three parts, as given by (11.293), (11.295) and (11.297).

```
double LineLoadSxx(double x,double t,double nu)
{
  double s,s1,s2,s3,n,nn,nn1,tp2,ts2,ts,tq,t2,c,cc,f,g,h;
  complex a,aa,a1,b,bb,b1,gp,gs,d,e;
  nn=(1-2*nu)/(2*(1-nu));nn1=sqrt(1-nn);n=sqrt(nn);
  ts2=1+x*x;tp2=nn*ts2;ts=sqrt(ts2);t2=t*t;tq=n*x+nn1;
  if (t2<=tp2) s1=0;else
    {
    a=complex(x*t/ts2,sqrt(t2-tp2)/ts2);aa=a*a;a1=1-2*aa;
    gp=sqrt(nn-aa);gs=sqrt(1-aa);d=a1*(1+2*aa-2*nn)*gp;e=a1*a1+4*aa*gp*gs;
    s1=-real(d/e)/(2*sqrt(t2-tp2));
    }
  if (t2<=ts2) s2=0;else
    {
    b=complex(x*t/ts2,sqrt(t2-ts2)/ts2);bb=b*b;b1=1-2*bb;
    gp=sqrt(nn-bb);gs=sqrt(1-bb);d=4*bb*(1-bb)*gp;e=b1*b1+4*bb*gp*gs;
    s2=real(d/e)/(2*sqrt(t2-ts2));
    }
  if ((t<=tq)||(t>=ts)||(tq>ts)) s3=0;else
    {
    c=(x*t-sqrt(ts2-t2))/ts2;cc=c*c;f=(1-2*cc)*(1-2*cc);
    g=4*n*cc*(1-cc)*f*sqrt(cc-nn);h=f*f+16*cc*cc*(cc-nn)*(1-cc);
    s3=-g/(2*h*sqrt(ts2-t2));
    }
  s=s1+s2+s3;
  return(s);
}
```

Some examples are shown in Figs. 11.29, 11.30 and 11.31. These figures show the horizontal normal stress σ_{xx} as a function of x/z for $v = 0$, and for three values of time: $c_s t/z = 1$, $c_s t/z = 8$ and $c_s t/z = 40$. As in the case of the vertical normal stress, the ratio of the various waves, the compression wave, the shear wave, and the Rayleigh wave, may be verified from the results shown in Figs. 11.29 and 11.30, for the cases $c_s t/z = 1$ and $c_s t/z = 8$.

The results for large values of $c_s t/z$ are approaching the elastostatic values. In general the numerical values depend upon the value of Poisson's ratio, but for very

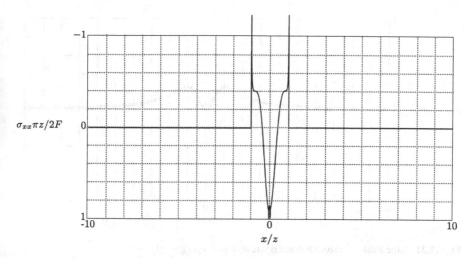

Fig. 11.29 Line load—horizontal normal stress, $v = 0$, $c_s t/z = 1$

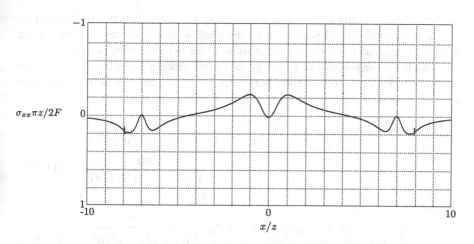

Fig. 11.30 Line load—horizontal normal stress, $v = 0$, $c_s t/z = 8$

large values of time no such dependence is found, in agreement with the elastostatic solution,

$$t \to \infty : \quad \frac{\sigma_{xx} \pi z}{2F} = -\frac{x^2/z^2}{(1 + x^2/z^2)^2}. \qquad (11.300)$$

The elastostatic solution is also shown in Fig. 11.31, by dots. The agreement with the numerical results for $c_s t/z = 40$ appears to be excellent.

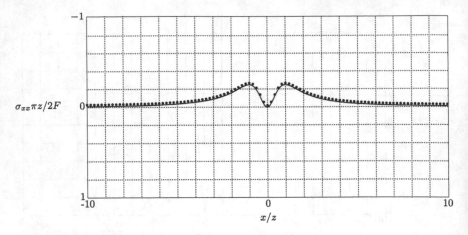

Fig. 11.31 Line load—horizontal normal stress, $\nu = 0$, $c_s t/z = 40$

Approximate Analysis of the Rayleigh Wave

For large values of time and the lateral coordinate x an approximation of the solution for the horizontal normal stress can be obtained in the same way as used for the isotropic stress and the vertical normal stress in the two preceding sections.

Using the same procedures as before, an approximation of the horizontal normal stress for large values of the time parameter $\tau = c_s t/z$ and the lateral coordinate $\xi = x/z$ is found to be

$$x/z \gg 1, \ c_s t/z \gg 1 \ : \ \frac{\sigma_{xx} \pi z}{2F} \approx \frac{m_3}{1 + (x - c_r t)^2/(w_p z)^2}$$
$$- \frac{m_4}{1 + (x - c_r t)^2/(w_s z)^2}, \quad (11.301)$$

where now

$$m_3 = \frac{(2\beta^2 - 1)(2\beta^2 + 1 - 2\eta^2)}{4M\beta^2 w_p}, \quad w_p = \sqrt{1 - \eta^2/\beta^2}, \quad (11.302)$$

$$m_4 = \frac{\beta^2 w_p}{M}, \quad w_s = \sqrt{1 - 1/\beta^2}, \quad (11.303)$$

and M is defined in (11.249).

The two expressions in (11.301) are the approximations of (11.293) and (11.295) for large values of $c_s t/z$, and values of x/z in the neighbourhood of $c_r t$, where the Rayleigh wave has its peak. The third part of the solution, (11.297), can be disregarded, because this part vanishes after the passage of the shear wave.

To demonstrate the accuracy of the approximation two examples are shown in Figs. 11.32 and 11.33. These figures show the values of the horizontal normal stress as a function of x/z, for $c_s t/z = 40$, and for two values of Poisson's ratio, $\nu = 0$ and

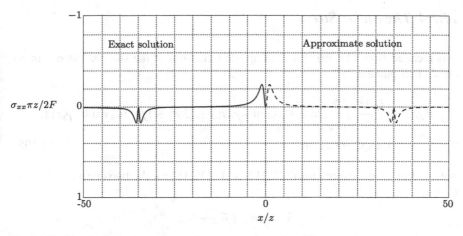

Fig. 11.32 Line load—horizontal normal stress, $\nu = 0$, $c_s t/z = 40$

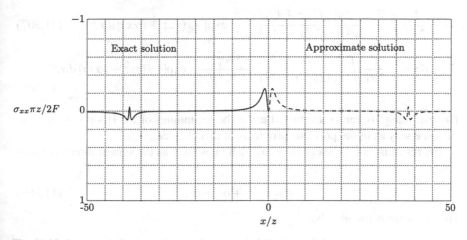

Fig. 11.33 Line load—horizontal normal stress, $\nu = 0.5$, $c_s t/z = 40$

$\nu = 0.5$. The fully drawn lines show the exact results, and the dashed lines show the approximate results, calculated by a superposition of the steady state solution (11.300) and the approximate formula for the Rayleigh wave disturbance (11.301). This final approximate formula, valid for any value of x/z, is

$$c_s t/z \gg 1 : \quad \frac{\sigma_{xx}\pi z}{2F} \approx -\frac{x^2/z^2}{(1+x^2/z^2)^2} + \frac{m_3}{1+(x-c_r t)^2/(w_p z)^2}$$
$$-\frac{m_4}{1+(x-c_r t)^2/(w_s z)^2}. \quad (11.304)$$

As in the case of the isotropic stress and the vertical normal stress the approximation appears to be very good. The maximum difference between the exact solution and the approximate solution in this case is smaller than 0.02.

11.2.4 The Shear Stress

It is recalled from (11.39) that the Laplace transform of the shear stress σ_{xz} in an elastic half plane is

$$\overline{\sigma}_{zx} = -\frac{i\mu s^2}{2\pi} \int_{-\infty}^{\infty} \{2\alpha\gamma_p C_p \exp(-\gamma_p sz) + (2\alpha^2 + 1/c_s^2)C_s \exp(-\gamma_s sz)\}$$

$$\times \exp(-is\alpha x)\, d\alpha. \tag{11.305}$$

Substitution of the expressions (11.217) and (11.218) for the two constants C_p and C_s gives

$$\overline{\sigma}_{xz} = \overline{\sigma}_1 + \overline{\sigma}_2 + \overline{\sigma}_3, \tag{11.306}$$

where

$$\overline{\sigma}_1 = \frac{F}{2\pi i} \int_{-\infty}^{\infty} \frac{2\alpha\gamma_p(2\alpha^2 + 1/c_s^2)}{(2\alpha^2 + 1/c_s^2)^2 - 4\alpha^2\gamma_p\gamma_s} \exp[-s(\gamma_p z + i\alpha x)]\, d\alpha, \tag{11.307}$$

$$\overline{\sigma}_2 + \overline{\sigma}_3 = -\frac{F}{2\pi i} \int_{-\infty}^{\infty} \frac{2\alpha\gamma_p(2\alpha^2 + 1/c_s^2)}{(2\alpha^2 + 1/c_s^2)^2 - 4\alpha^2\gamma_p\gamma_s} \exp[-s(\gamma_s z + i\alpha x)]\, d\alpha. \tag{11.308}$$

The two integrals can be evaluated using the techniques of De Hoop's method, in the same way as in the previous problems of this chapter.

The result is that the shear stress σ_{xz} can be written as the sum of three contributions,

$$\sigma_{xz} = \sigma_1 + \sigma_2 + \sigma_3. \tag{11.309}$$

The first term is found to be

$$\sigma_1 = -\frac{2F}{\pi}\Re\left\{\frac{p\gamma_p^2(1/c_s^2 - 2p^2)}{(1/c_s^2 - 2p^2)^2 + 4p^2\gamma_p\gamma_s}\right\}\frac{H(t - t_p)}{\sqrt{t^2 - t_p^2}}, \qquad p = p_1, \tag{11.310}$$

where

$$p_1 = \left(tx + iz\sqrt{t^2 - t_p^2}\right)/r^2, \tag{11.311}$$

and

$$\gamma_p^2 = 1/c_p^2 - p^2, \qquad \gamma_s^2 = 1/c_s^2 - p^2. \tag{11.312}$$

In dimensionless form, (11.310) can be written as

$$\frac{\sigma_1 \pi z}{2F} = -\Re\left\{\frac{a(\eta^2 - a^2)(1 - 2a^2)}{(1 - 2a^2)^2 + 4a^2\sqrt{1 - a^2}\sqrt{\eta^2 - a^2}}\right\}\frac{H(\tau - \tau_p)}{\sqrt{\tau^2 - \tau_p^2}}, \tag{11.313}$$

where

$$a = \left(\xi\tau + i\sqrt{\tau^2 - \tau_p^2}\right)/(1 + \xi^2). \qquad (11.314)$$

The second term is found to be

$$\sigma_2 = \frac{2F}{\pi}\Re\left\{\frac{p\gamma_p\gamma_s(1/c_s^2 - 2p^2)}{(1/c_s^2 - 2p^2)^2 + 4p^2\gamma_p\gamma_s}\right\}\frac{H(t - t_s)}{\sqrt{t^2 - t_s^2}}, \quad p = p_2, \qquad (11.315)$$

where

$$p_2 = \left(tx + iz\sqrt{t^2 - t_s^2}\right)/r^2, \qquad (11.316)$$

and γ_p and γ_s can be expressed into the variable p by the same relations as before, see (11.312).

The dimensionless form of (11.315) is

$$\frac{\sigma_2\pi z}{2F} = \Re\left\{\frac{b\sqrt{1 - b^2}\sqrt{\eta^2 - b^2}(1 - 2b^2)}{(1 - 2b^2)^2 + 4b^2\sqrt{1 - b^2}\sqrt{\eta^2 - b^2}}\right\}\frac{H(\tau - \tau_s)}{\sqrt{\tau^2 - \tau_s^2}}, \qquad (11.317)$$

where

$$b = (\xi\tau + i\sqrt{\tau^2 - \tau_s^2})/(1 + \xi^2). \qquad (11.318)$$

Finally, the third term is found to be

$$\sigma_3 = \frac{2F}{\pi}\Im\left\{\frac{p\gamma_p\gamma_s(1/c_s^2 - 2p^2)}{(1/c_s^2 - 2p^2)^2 + 4p^2\gamma_p\gamma_s}\right\}\frac{H(t - t_q)H(t_s - t)}{\sqrt{t_s^2 - t^2}}, \quad p = p_3,$$

$$(11.319)$$

where

$$p_3 = \left(xt - z\sqrt{t_s^2 - t^2}\right)/r^2, \qquad (11.320)$$

and the time t_q is given by

$$t_q/t_s = \left(\eta x + z\sqrt{1 - \eta^2}\right)/r, \quad t_q/t_s < 1. \qquad (11.321)$$

The dimensionless form of (11.319) is

$$\frac{\sigma_3\pi z}{2F} = -\frac{c(1 - 2c^2)^3\sqrt{1 - c^2}\sqrt{c^2 - \eta^2}}{(1 - 2c^2)^4 + 16c^4(1 - c^2)(c^2 - \eta^2)}\frac{H(\tau - \tau_q)H(\tau_s - \tau)}{\sqrt{t_s^2 - \tau^2}}, \qquad (11.322)$$

where c is defined by

$$c = \left(\xi\tau - \sqrt{\tau_s^2 - \tau^2}\right)/(1 + \xi^2). \qquad (11.323)$$

In the equations given above the following dimensionless parameters have been used,

$$\xi = x/z, \qquad \tau = c_s t/z, \qquad \tau_s = \sqrt{1+\xi^2}, \qquad \tau_p = \eta\sqrt{1+\xi^2},$$

$$\tau_q = \xi\eta + \sqrt{1-\eta^2}. \tag{11.324}$$

This means that the depth z is used as the length scale, and that the value of z/c_s is used as the time scale.

In the derivation of the expressions it has been assumed that $x > 0$, so that $\xi > 0$. Values for $\xi < 0$ can be obtained using the antisymmetry of the shear stress.

Calculation of Numerical Values

A function to calculate the value of the dimensionless parameter $\sigma_{xz}\pi z/2F$ as a function of the parameters $\xi = x/z$ (with $\xi \geq 0$), $\tau = c_p t/z$ and Poisson's ratio ν, is shown below. In this function the dimensionless parameters ξ, τ and ν are denoted by x, t and nu. The function consists of three parts, as given by (11.313), (11.317) and (11.322).

```
double LineLoadSxz(double x,double t,double nu)
{
 double s,s1,s2,s3,n,nn,nn1,tp2,ts2,ts,tq,t2,c,cc,f,g,h;
 complex a,aa,a1,b,bb,b1,gp,gs,d,e;
 nn=(1-2*nu)/(2*(1-nu));n=sqrt(nn);nn1=sqrt(1-nn);
 ts2=1+x*x;tp2=nn*ts2;ts=sqrt(ts2);t2=t*t;tq=n*x+nn1;
 if (t2<=tp2) s1=0;else
   {
    a=complex(x*t/ts2,sqrt(t2-tp2)/ts2);aa=a*a;a1=1-2*aa;
    gp=sqrt(nn-aa);gs=sqrt(1-aa);d=a*(nn-aa)*a1;e=a1*a1+4*aa*gp*gs;
    s1=-real(d/e)/sqrt(t2-tp2);
   }
 if (t2<=ts2) s2=0;else
   {
    b=complex(x*t/ts2,sqrt(t2-ts2)/ts2);bb=b*b;b1=1-2*bb;
    gp=sqrt(nn-bb);gs=sqrt(1-bb);d=b*b1*gp*gs;e=b1*b1+4*bb*gp*gs;
    s2=real(d/e)/sqrt(t2-ts2);
   }
 if ((t<=tq)||(t>=ts)||(tq>ts)) s3=0;else
   {
    c=(x*t-sqrt(ts2-t2))/ts2;cc=c*c;f=(1-2*cc)*(1-2*cc);
    g=n*c*(1-2*cc)*f*sqrt(1-cc)*sqrt(cc-nn);h=f*f+16*cc*cc*(1-cc)*(cc-nn);
    s3=-g/(h*sqrt(ts2-t2));
   }
 s=s1+s2+s3;
 return(s);
}
```

Some examples are shown in Figs. 11.34 and 11.35, for $\nu = 0$, and for two values of time: $c_s t/z = 5$ and $c_s t/z = 20$. The results for large values of $c_s t/z$ are approaching the elastostatic solution,

$$t \to \infty : \quad \frac{\sigma_{xz}\pi z}{2F} = -\frac{x/z}{(1+x^2/z^2)^2}. \tag{11.325}$$

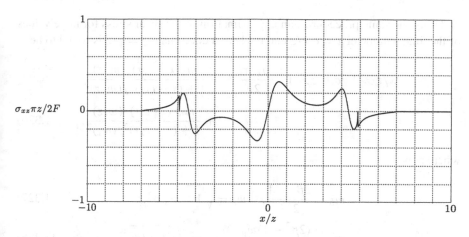

Fig. 11.34 Line load—shear stress, $\nu = 0$, $c_s t/z = 5$

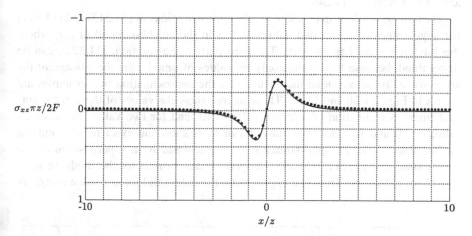

Fig. 11.35 Line load—shear stress, $\nu = 0$, $c_s t/z = 20$

The elastostatic solution is also shown in Fig. 11.35, by small asterisks. As before, the agreement is excellent, which is also the case for other values of Poisson's ratio.

Approximate Analysis of the Rayleigh Wave

For large values of time and the lateral coordinate x an approximation of the solution for the shear stress can be obtained in the same way as used for the other stress components in the preceding sections.

Using the same procedures, an approximation of the shear stress for large values of the time parameter $\tau = c_s t/z$ and the lateral coordinate $\xi = x/z$ is found to be

$$x/z \gg 1, \ c_s t/z \gg 1 \ : \ \frac{\sigma_{xz}\pi z}{2F} \approx -\frac{m_5(x - c_r t)/(w_p z)}{1 + (x - c_r t)^2/(w_p z)^2}$$

$$+ \frac{m_6(x - c_r t)/(w_s z)}{1 + (x - c_r t)^2/(w_s)^2}, \quad (11.326)$$

where now

$$m_5 = \frac{2\beta^2 - 1}{2M}, \quad w_p = \sqrt{1 - \eta^2/\beta^2}, \quad (11.327)$$

$$m_6 = \frac{(2\beta^2 - 1)w_p}{2M w_s}, \quad w_s = \sqrt{1 - 1/\beta^2}, \quad (11.328)$$

and M is defined in (11.249).

The two expressions in (11.326) are the approximations of (11.313) and (11.317) for large values of $c_s t/z$, and values of x/z in the neighbourhood of $c_r t$, where the Rayleigh wave has its peak. The third part of the solution, (11.322), can be disregarded, because this applies only for values of time before the passage of the shear wave. To demonstrate the accuracy of the approximation two examples are shown in Figs. 11.36 and 11.37. These figures show the values of the vertical normal stress as a function of x/z, for $c_s t/z = 40$, and for two values of Poisson's ratio, $\nu = 0$ and $\nu = 0.5$. The fully drawn lines show the exact results, and the dashed lines show the approximate results, calculated by a superposition of the steady state solution (11.325) and the approximate formula for the Rayleigh wave disturbance (11.326). This final approximate formula, valid for any value of x/z, is

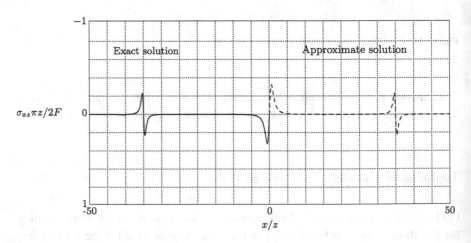

Fig. 11.36 Line load—shear stress, $\nu = 0$, $c_s t/z = 40$

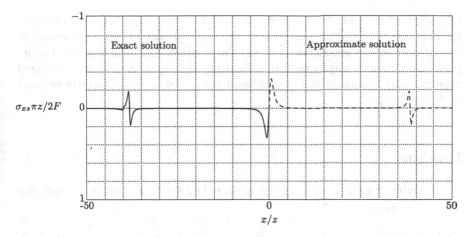

Fig. 11.37 Line load—shear stress, $\nu = 0.5$, $c_s t/z = 40$

$$c_s t/z \gg 1 \; : \; \frac{\sigma_{xz}\pi z}{2F} \approx -\frac{x/z}{(1+x^2/z^2)^2} - \frac{m_5(x-c_r t)/(w_p z)}{1+(x-c_r t)^2/(w_p z)^2}$$
$$+ \frac{m_6(x-c_r t)/(w_s z)}{1+(x-c_r t)^2/(w_s z)^2}. \tag{11.329}$$

As in the case of the isotropic stress and the vertical normal stress the approximation appears to be good. The maximum difference between the exact solution and the approximate solution in this case is smaller than 0.03. This is somewhat larger than for the other stress components. The largest error occurs for $\nu = 0.5$ just before the arrival of the Rayleigh wave. It seems that the error is caused by the magnitude of the shear wave, which has just passed.

Conclusion

In this chapter the solution of the problem of a line pulse on an elastic half space has been considered. The known solutions by De Hoop (1960) and Eringen and Suhubi (1975) for the displacements of the surface have been rederived, and expressions for the displacements in an arbitrary point of the half space have been derived.

As a second problem the elastodynamic equivalent of the Flamant problem, a constant line load on the elastic half space, has been considered. Closed form expressions for the stress components have been derived.

The solutions have been validated by considering the limiting state for $t \to \infty$. Special attention has been given to the generation of Rayleigh waves in the vicinity of the surface. In general, for large values of time the solutions appear to consist of the elastostatic stress distribution plus a constant Rayleigh wave disturbance near the surface. Although the constant value of the amplitude and the shape is a well known property of Rayleigh waves in the two dimensional case, it has been shown that simple expressions for the magnitude and the width of the Rayleigh wave disturbance can be derived, for each of the stress components. These approximations appear to be in very good agreement with the complete analytical solution. The approximate solutions may be useful for the (approximate) analysis of the solution of problems with a more general type of loading. Some examples of this will be given in the next chapter.

Problems

11.1 Verify that the solutions given in Sect. 11.2 for σ, σ_{zz} and σ_{xx} satisfy that $\sigma = (\sigma_{zz} + \sigma_{xx})/2$.

11.2 Verify the same property for the approximate solutions, given in (11.258), (11.287) and (11.304).

11.3 Verify, analytically, that the solution for the vertical normal stress caused by a line load, as given in (11.227), with its three components given in (11.273), (11.275) and (11.277), satisfies the boundary condition that $\sigma_{zz} = 0$ for $z = 0$, and $|x| > 0$.

11.4 Verify that the solution for the shear stress caused by a line load, as given in (11.309), satisfies the boundary condition that for $z = 0$ the shear stress is zero, $\sigma_{zx} = 0$.

Chapter 12
Strip Load on Elastic Half Space

In this chapter the problem of a strip load on the surface of an elastic half plane is considered, i.e. problems in which the elastic half space is loaded by a load that is constant over an area in the form of a strip of finite width, see Fig. 12.1. As a function of time the load may be an impulse load or a step load. The case of an impulse load will be considered first. The step load will be considered later, with the solution being derived from the solution for the impulse load by integration over time.

The solutions will be obtained as an application of De Hoop's method (De Hoop, 1960, 1970), using an extension due to Stam (1990) to generalize the line load to a loading over a strip of finite width, see also Verruijt (2008a).

Emphasis will be on the determination of the stress components as functions of depth and time, as these are of main interest in soil engineering. For very large values of time the results for a step load should be in agreement with the elastostatic solutions, and this condition will be used as a validation of the elastodynamic solutions. Also, the Rayleigh wave should be recovered at large distances from the load, and should conform to its required behaviour.

A computer program for the constant strip load is available as the program STRIPLOAD.

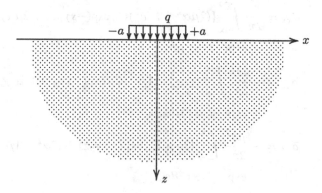

Fig. 12.1 Half plane with strip load

A. Verruijt, *An Introduction to Soil Dynamics*,
Theory and Applications of Transport in Porous Media 24,
© Springer Science+Business Media B.V. 2010

12.1 Strip Pulse

The first problem to be considered is the case of a strip pulse on an elastic half plane, see Fig. 12.1. In this case the boundary conditions are

$$z = 0 : \sigma_{zx} = 0, \tag{12.1}$$

$$z = 0 : \sigma_{zz} = \begin{cases} -q\,\delta(t), & \text{if } |x| < a, \\ 0, & \text{if } |x| > a. \end{cases} \tag{12.2}$$

The second boundary condition expresses that there is a homogeneous load on the strip between the points $x = -a$ and $x = a$ in the form of a pulse of very short duration.

The Laplace transform of this condition is

$$z = 0 : \overline{\sigma}_{zz} = \begin{cases} -q, & \text{if } |x| < a, \\ 0, & \text{if } |x| > a. \end{cases} \tag{12.3}$$

This can also be written in the form of a Fourier integral,

$$z = 0 : \overline{\sigma}_{zz} = -\frac{q}{\pi} \int_{-\infty}^{\infty} \frac{\sin(s\alpha a)\exp(-is\alpha x)}{\alpha}\,d\alpha. \tag{12.4}$$

It is recalled from the previous chapter, see (11.35) and (11.36), that the general solution of the elastodynamic problem for a half plane is, in the form of the Laplace transforms of the two displacement components,

$$\overline{u} = \frac{is}{2\pi} \int_{-\infty}^{\infty} \{\alpha C_p \exp(-s\gamma_p z) + \gamma_s C_s \exp(-s\gamma_s z)\}\exp(-is\alpha x)\,d\alpha, \tag{12.5}$$

$$\overline{w} = \frac{s}{2\pi} \int_{-\infty}^{\infty} \{\gamma_p C_p \exp(-s\gamma_p z) + \alpha C_s \exp(-s\gamma_s z)\}\exp(-is\alpha x)\,d\alpha. \tag{12.6}$$

Furthermore, the Laplace transforms of the stress components are, from (11.37), (11.38) and (11.39),

$$\overline{\sigma}_{xx} = \frac{s^2}{2\pi} \int_{-\infty}^{\infty} \{(2\mu\alpha^2 - \lambda/c_p^2)C_p \exp(-s\gamma_p z) + 2\mu\alpha\gamma_s C_s \exp(-s\gamma_s z)\}$$
$$\times \exp(-is\alpha x)\,d\alpha, \tag{12.7}$$

$$\overline{\sigma}_{zz} = -\frac{\mu s^2}{2\pi} \int_{-\infty}^{\infty} \{(2\alpha^2 + 1/c_s^2)C_p \exp(-s\gamma_p z) + 2\alpha\gamma_s C_s \exp(-s\gamma_s z)\}$$
$$\times \exp(-is\alpha x)\,d\alpha, \tag{12.8}$$

$$\overline{\sigma}_{xz} = -\frac{i\mu s^2}{2\pi} \int_{-\infty}^{\infty} \{2\alpha\gamma_p C_p \exp(-s\gamma_p z) + (2\alpha^2 + 1/c_s^2)C_s \exp(-s\gamma_s z)\}$$
$$\times \exp(-is\alpha x)\,d\alpha. \tag{12.9}$$

The Laplace transform of the isotropic stress $\sigma = \frac{1}{2}(\sigma_{xx} + \sigma_{zz})$ is given by (11.41),

$$\bar{\sigma} = -\frac{(\lambda + \mu)s^2}{2\pi c_p^2} \int_{-\infty}^{\infty} C_p \exp(-s\gamma_p z - is\alpha x)\, d\alpha. \tag{12.10}$$

Using the boundary conditions (12.1) and (12.4) the two integration constants C_p and C_s in the general solution are found to be

$$C_p = \frac{2q}{\mu s^2} \frac{\sin(s\alpha a)}{\alpha} \frac{2\alpha^2 + 1/c_s^2}{(2\alpha^2 + 1/c_s^2)^2 - 4\alpha^2 \gamma_p \gamma_s}, \tag{12.11}$$

$$C_s = -\frac{2q}{\mu s^2} \frac{\sin(s\alpha a)}{\alpha} \frac{2\alpha\gamma_p}{(2\alpha^2 + 1/c_s^2)^2 - 4\alpha^2 \gamma_p \gamma_s}. \tag{12.12}$$

The stress components will next be evaluated.

12.1.1 The Isotropic Stress

The simplest quantity to evaluate is the isotropic stress. With (12.10) and (12.11) it is found that

$$\frac{\bar{\sigma}}{q} = -\frac{1 - \eta^2}{\pi \eta^2 c_p^2} \int_{-\infty}^{\infty} \frac{\sin(s\alpha a)}{\alpha} \frac{2\alpha^2 + 1/c_s^2}{(2\alpha^2 + 1/c_s^2)^2 - 4\alpha^2 \gamma_p \gamma_s} \exp[-s(\gamma_p z + i\alpha x)]\, d\alpha, \tag{12.13}$$

where $\eta^2 = c_s^2/c_p^2 = (1 - 2\nu)/[2(1 - \nu)]$.

Following a suggestion by Stam (1990), the function $\sin(s\alpha a)$ is written as $[\exp(is\alpha a) - \exp(-is\alpha a)]/2i$. This gives

$$\frac{\bar{\sigma}}{q} = \frac{1 - \eta^2}{\eta^2} \{\bar{g}(x + a) - \bar{g}(x - a)\}, \tag{12.14}$$

where

$$\bar{g}(x) = \frac{1}{2\pi i c_p^2} \int_{-\infty}^{\infty} \frac{1}{\alpha} \frac{2\alpha^2 + 1/c_s^2}{(2\alpha^2 + 1/c_s^2)^2 - 4\alpha^2 \gamma_p \gamma_s} \exp[-s(\gamma_p z + i\alpha x)]\, d\alpha. \tag{12.15}$$

The value of this integral depends upon the sign of the variable x. The two possibilities will be considered separately.

The Case $x > 0$

Using the substitution $p = i\alpha$ the integral (12.15) can be written as

$$\bar{g}(x) = \frac{1}{2\pi i c_p^2} \int_{-i\infty}^{i\infty} \frac{1}{p} \frac{1/c_s^2 - 2p^2}{(1/c_s^2 - 2p^2)^2 + 4p^2 \gamma_p \gamma_s} \exp[-s(\gamma_p z + px)]\, dp, \tag{12.16}$$

where now γ_p and γ_s are related to p by the equations

$$\gamma_p^2 = 1/c_p^2 - p^2, \qquad \gamma_s^2 = 1/c_s^2 - p^2. \tag{12.17}$$

As in the previous chapter, the integration path in the complex p-plane is modified such that the integral obtains the form of a Laplace transform integral. For this purpose a parameter t is introduced (later to be identified with the time), defined as

$$t = \gamma_p z + px, \tag{12.18}$$

with t being real and positive, by assumption. The shape of the transformed integration path remains undetermined in this stage.

The integrand of the integral in (12.16) has singularities in the form of branch points in the points $p = \pm 1/c_p$ and $p = \pm 1/c_s$, simple poles in the points $p = \pm 1/c_r$, where c_r is the Rayleigh wave velocity, which is slightly smaller than the shear wave velocity, and a simple pole in the point $p = 0$. It may be noted that $c_p > c_s > c_r$, so that $1/c_p < 1/c_s < 1/c_r$. The original integration path from $p = -i\infty$ to $p = +i\infty$ is now modified to the sum of the two paths p_2 and p_1 shown in Fig. 12.2, with the parameter t varying along these two curves from some initial (positive) value to infinity.

It follows from (12.18) and (12.17) that

$$r^2 p^2 - 2tpx + t^2 - z^2/c_p^2 = 0, \tag{12.19}$$

where

$$r^2 = x^2 + z^2. \tag{12.20}$$

The quadratic equation (12.19) has two solutions for p,

$$p_1 = \frac{tx}{r^2} + \frac{iz}{r^2}\sqrt{t^2 - t_p^2}, \tag{12.21}$$

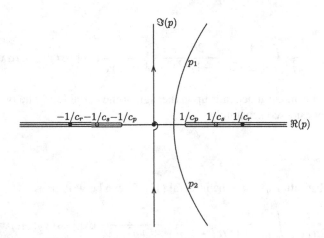

Fig. 12.2 Modified integration path, for $x > 0$

$$p_2 = \frac{tx}{r^2} - \frac{iz}{r^2}\sqrt{t^2 - t_p^2}, \tag{12.22}$$

where

$$t_p = r/c_p. \tag{12.23}$$

If it is assumed that $t_p < t < \infty$ the two branches p_1 and p_2 shown in Fig. 12.2 form a continuous path, with the two branches intersecting for $t = t_p$, where $p = p_1 = p_2 = (x/r)(1/c_p)$, which is a point on the real axis, always located between the origin and the first singularity at $p = 1/c_p$. For the two integration paths to be equivalent, the contributions of the parts of the closing contour at infinity must vanish. This will indeed be the case if $x > 0$. It has been assumed that the original integration path, along the imaginary axis, passes to the right of the pole in the origin, see Fig. 12.2. This should then also be the case for the case $x < 0$, which will have consequences for the contribution of this pole, of course.

It can be shown that along the path p_1

$$\frac{1}{p}\frac{dp}{dt} = \frac{t\sqrt{t^2 - t_p^2} + ixz/c_p^2}{(t^2 - z^2/c_p^2)\sqrt{t^2 - t_p^2}}. \tag{12.24}$$

This means that the contribution of the path p_1 to the integral (12.16) is

$$x > 0 : \bar{g}_1(x) = \frac{1}{2\pi i\, c_p^2} \int_{t_p}^{\infty} \frac{t\sqrt{t^2 - t_p^2} + ixz/c_p^2}{(t^2 - z^2/c_p^2)\sqrt{t^2 - t_p^2}}$$

$$\times \frac{1/c_s^2 - 2p^2}{(1/c_s^2 - 2p^2)^2 + 4p^2\gamma_p\gamma_s}\, \exp(-st)\, dt. \tag{12.25}$$

Along the path p_2 all quantities will be complex conjugates, but the path is in inverse direction, so that if one writes $\bar{g}_1(x) = A + iB$ then $\bar{g}_2(x) = -(A - iB)$. The sum of these two contributions is $2iB$. It now follows that the sum is

$$x > 0 : \bar{g}(x) = \frac{1}{\pi c_p^2}\Im \int_{t_p}^{\infty} \frac{tr^2\sqrt{t^2 - t_p^2} + ixzt_p^2}{r^2(t^2 - z^2/c_p^2)\sqrt{t^2 - t_p^2}}$$

$$\times \frac{1/c_s^2 - 2p^2}{(1/c_s^2 - 2p^2)^2 + 4p^2\gamma_p\gamma_s}\, \exp(-st)\, dt. \tag{12.26}$$

This expression happens to be in the form of a Laplace transform. Inverse Laplace transformation gives

$$x > 0 : g(x) = \frac{1}{\pi c_p^2}\Im \left\{ \frac{t\sqrt{t^2 - t_p^2} + ixz/c_p^2}{(t^2 - z^2/c_p^2)\sqrt{t^2 - t_p^2}} \right.$$

$$\left. \times \frac{1/c_s^2 - 2p^2}{(1/c_s^2 - 2p^2)^2 + 4p^2\gamma_p\gamma_s} \right\} H(t - t_p). \tag{12.27}$$

For the calculation of numerical values it is convenient to introduce the dimensionless parameters

$$\xi = x/a, \qquad \zeta = z/a, \qquad \tau = c_s t/a, \qquad \rho = \sqrt{\xi^2 + \zeta^2}, \qquad \tau_p = \eta\rho,$$

$$\beta_p = c_s p, \qquad \eta = c_s/c_p. \tag{12.28}$$

The parameter β_p is a dimensionless complex variable defined by

$$\beta_p = \left(\tau\xi + i\zeta\sqrt{\tau^2 - \eta^2\rho^2}\right)/\rho^2, \tag{12.29}$$

as follows immediately from (12.21). The important parameters γ_p and γ_s can now be represented by their dimensionless equivalents

$$g_p = c_s\gamma_p = \sqrt{\eta^2 - \beta_p^2}, \qquad g_s = c_s\gamma_s = \sqrt{1 - \beta_p^2}. \tag{12.30}$$

Using these parameters equation (12.27) can be written as

$$\xi > 0 : \quad g(x) = \frac{\eta^2 c_s}{a} h(\xi), \tag{12.31}$$

with

$$h(\xi) = \frac{1}{\pi}\Im\left\{\frac{\tau\sqrt{\tau^2 - \eta^2\rho^2} + i\eta^2\xi\zeta}{(\tau^2 - \eta^2\zeta^2)\sqrt{\tau^2 - \eta^2\rho^2}}\frac{(1 - 2\beta_p^2)}{(1 - 2\beta_p^2)^2 + 4\beta_p^2 g_p g_s}\right\} H(\tau - \eta\rho), \tag{12.32}$$

It may be noted that in (12.32) the parameter $\xi > 0$, and the only relevant values of τ are those for which $\tau > \eta\rho$.

The Case $x < 0$

If the parameter $x < 0$ the integration path must be transformed by moving the integration path to the left, in order that the contributions by the arcs at infinity vanish. This means that the pole at $p = 0$ will be passed, resulting in a contribution to the integral, see Fig. 12.3. In this figure the transformed integration path is indicated by the path consisting of the curves p_2 and p_1, with a loop around the pole. It can be shown that the result of the integration along p_2 and p_1 will be the same as before, see (12.27). However, to this expression the contribution by integrating around the pole must be added. Along this path the integration variable p is

$$p = \varepsilon \exp(i\theta), \tag{12.33}$$

where $\varepsilon \to 0$, and the angle θ runs from $\theta = -\pi$ to $\theta = +\pi$ along the small circle around the pole. This contribution can be determined by considering the limiting

Fig. 12.3 Modified
integration path, if $x < 0$

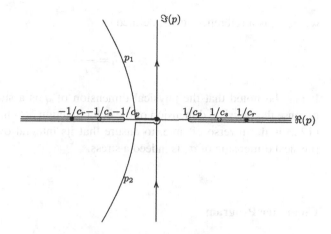

value of the Laplace transform $\overline{g}(x)$ as defined in (12.16) for $p \to 0$. This leads to
an additional contribution

$$\Delta\overline{g} = \eta^2 \exp(-sz/c_p). \tag{12.34}$$

Inverse Laplace transformation gives

$$\Delta g = \eta^2 \delta(t - z/c_p). \tag{12.35}$$

Because $\tau = c_s t/a$ this means that there is a second contribution to the function
$h(\xi)$,

$$\Delta h = \delta(\tau - \eta\zeta)\{1 - H(\xi)\}, \tag{12.36}$$

where the factor $1 - H(\xi)$ has been added to indicate that this contribution applies
only if $\xi < 0$, and where it has been assumed that $\delta(t - z/c_p) = (c_s/a)\delta(\tau - \eta\zeta)$,
because both delta functions should have an area equal to 1,

$$\int_{-\infty}^{+\infty} \delta(t - z/c_p)\,dt = \int_{-\infty}^{+\infty} \delta(\tau - \eta\zeta)\,d\tau = 1. \tag{12.37}$$

General Result

The results for $\xi > 0$ and $\xi < 0$ can be combined in the single formula

$$g(\xi, \zeta, \tau) = \frac{\eta^2 c_s}{a}\{h(\xi, \zeta, \tau) + \Delta h(\xi, \zeta, \tau)\}. \tag{12.38}$$

Using the general formula (12.38), the expression for the isotropic stress (12.14)
becomes, after inverse Laplace transformation,

$$\frac{\sigma}{\sigma_0} = (1 - \eta^2)\{h(\xi + 1, \zeta, \tau) + \Delta h(\xi + 1, \zeta, \tau) - h(\xi - 1, \zeta, \tau) - \Delta h(\xi - 1, \zeta, \tau)\}, \tag{12.39}$$

where σ_0 is a reference stress, defined by

$$\sigma_0 = \frac{q\,c_s}{a}. \tag{12.40}$$

It may be noted that the physical dimension of q is a stress multiplied by time, because the physical dimension of the delta function $\delta(t)$ in the boundary condition (12.2) is the inverse of time, to ensure that its integral over time is 1. Thus, the physical dimension of σ_0 is indeed a stress.

Computer Program

The isotropic stress can be calculated as a function of ξ, ζ, τ and ν by the function StripPulseS shown below. This function uses the functions delta and h, which are also shown. The delta function is approximated by a parabolic arc of small width and of unit area.

```
double delta(double t,double z,double e)
{
 double f;
 if ((t<z-e)||(t>z+e)) f=0;else f=3*(e*e-(t-z)*(t-z))/(4*e*e*e);
 return(f);
}
double h(double x,double z,double t,double nu)
{
 double n,nn,rr,pi,s,tt,tr,tz,xx,zz,eps;
 complex a,b,bb,b1,gp,gs,d,e;
 pi=4*atan(1.0);nn=(1-2*nu)/(2*(1-nu));n=sqrt(nn);eps=0.001;tt=t*t;
 xx=x*x;zz=z*z;rr=xx+zz;tr=tt-nn*rr;tz=tt-nn*zz;
 if (tr<=0) s=0;
 else
   {
    a=complex(t/tz,nn*x*z/(tz*sqrt(tr)));b=complex(t*x/rr,(z/rr)*sqrt(tr));
    bb=b*b;b1=1-2*bb;gp=sqrt(nn-bb);gs=sqrt(1-bb);d=a*b1;e=b1*b1+4*bb*gp*gs;
    s=imag(d/e)/pi;
   }
 if (x<0) s+=delta(t,n*z,eps);
 return(s);
}
double StripPulseS(double x,double z,double t,double nu)
{
 double s,nn;
 nn=(1-2*nu)/(2*(1-nu));s=(1-nn)*(h(x+1,z,t,nu)-h(x-1,z,t,nu));
 return(s);
}
```

Figure 12.4 shows the isotropic stress as a function of time, for $\nu = 0$, in the point $x/a = 0$, $z/a = 1$. It appears that first the compressive wave below the load arrives, at the time $t = z/c_p$, so that $c_s t/a = \eta = 1/\sqrt{2}$, and then some time later (a factor $\sqrt{2}$ later), the negative compression waves emanating from the end points of the load arrive at this point. In this case of the isotropic stress, there is no effect of shear waves.

Fig. 12.4 Strip
pulse—isotropic stress,
$v = 0$, $x/a = 0$, $z/a = 1$

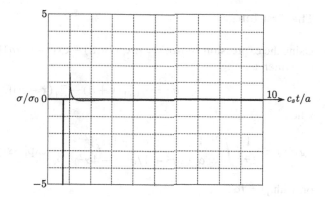

12.1.2 The Vertical Normal Stress

Another interesting quantity is the vertical normal stress σ_{zz}. It is recalled
from (12.8) that the Laplace transform of this quantity is

$$\bar{\sigma}_{zz} = -\frac{s^2\mu}{2\pi}\int_{-\infty}^{\infty}\{(2\alpha^2 + 1/c_s^2)C_p\exp(-\gamma_p sz) + 2\alpha\gamma_s C_s\exp(-\gamma_s sz)\}$$
$$\times \exp(-isax)\,d\alpha. \tag{12.41}$$

Substitution of the expressions (12.11) and (12.12) for the two constants C_p and C_s
gives

$$\bar{\sigma}_{zz} = \bar{\sigma}_1 + \bar{\sigma}_2 + \bar{\sigma}_3, \tag{12.42}$$

where

$$\bar{\sigma}_1 = -\frac{q}{\pi}\int_{-\infty}^{\infty}\frac{\sin(s\alpha a)}{\alpha}\frac{(2\alpha^2 + 1/c_s^2)^2}{(2\alpha^2 + 1/c_s^2)^2 - 4\alpha^2\gamma_p\gamma_s}\exp[-s(\gamma_p z + i\alpha x)]\,d\alpha,$$
$$\tag{12.43}$$

$$\bar{\sigma}_2 = \frac{q}{\pi}\int_{-\infty}^{\infty}\frac{\sin(s\alpha a)}{\alpha}\frac{4\alpha^2\gamma_s\gamma_p}{(2\alpha^2 + 1/c_s^2)^2 - 4\alpha^2\gamma_p\gamma_s}\exp[-s(\gamma_s z + i\alpha x)]\,d\alpha.$$
$$\tag{12.44}$$

These two integrals will be considered separately. It may be noted that the integrand
of the first integral has a singularity at $\alpha = 0$, which means that special care must be
taken when transforming the integration path. When passing the origin, the contribution of the pole must be taken into account. The integrand of the second integral
has no such singularity.

The First Integral

Using the expression $\sin(\alpha a) = [\exp(is\alpha a) - \exp(-is\alpha a)]/2i$, the first integral can be written as

$$\overline{\sigma}_1 = q\{\overline{g}_1(x+a) - \overline{g}_1(x-a)\}, \tag{12.45}$$

where

$$\overline{g}_1(x) = \frac{1}{2\pi i}\int_{-\infty}^{\infty}\frac{1}{\alpha}\frac{(2\alpha^2 + 1/c_s^2)^2}{(2\alpha^2 + 1/c_s^2)^2 - 4\alpha^2\gamma_p\gamma_s}\exp[-s(\gamma_p z + i\alpha x)]d\alpha, \tag{12.46}$$

or, with $p = i\alpha$,

$$\overline{g}_1(x) = \frac{1}{2\pi i}\int_{-i\infty}^{i\infty}\frac{1}{p}\frac{(1/c_s^2 - 2p^2)^2}{(1/c_s^2 - 2p^2)^2 + 4p^2\gamma_p\gamma_s}\exp[-s(\gamma_p z + px)]dp. \tag{12.47}$$

Comparison with the expression (12.16) for the function $g(x)$ in the case of the isotropic stress shows that the two expressions are very similar. The only differences are a constant factor c_p^2 and the square of the factor $1/c_s^2 - 2p^2$. This means that application of the same method to transform the integration path in the complex p-plane will give, in this case,

$$x > 0 \ : \ g_1(x) = \frac{1}{\pi}\Im\left\{\frac{t\sqrt{t^2 - t_p^2} + ixz/c_p^2}{(t^2 - z^2/c_p^2)\sqrt{t^2 - t_p^2}}\frac{(1/c_s^2 - 2p^2)^2}{(1/c_s^2 - 2p^2)^2 + 4p^2\gamma_p\gamma_s}\right\}$$
$$\times H(t - t_p), \tag{12.48}$$

where, as before,

$$r = \sqrt{x^2 + z^2}, \tag{12.49}$$

$$t_p = r/c_p. \tag{12.50}$$

Using the dimensionless parameters defined in (12.28) and (12.30) the function $g_1(x)$ can be expressed as

$$\xi > 0 \ : \ g_1(x) = \frac{c_s}{a}h_1(\xi), \tag{12.51}$$

with

$$h_1(\xi) = \frac{1}{\pi}\Im\left\{\frac{\tau\sqrt{\tau^2 - \eta^2\rho^2} + i\eta^2\xi\zeta}{(\tau^2 - \eta^2\zeta^2)\sqrt{\tau^2 - \eta^2\rho^2}}\frac{(1 - 2\beta_p^2)^2}{(1 - 2\beta_p^2)^2 + 4\beta_p^2 g_p g_s}\right\}H(\tau - \eta\rho). \tag{12.52}$$

Here the parameter β_p is the dimensionless complex variable defined by (12.29),

$$\beta_p = \left(\tau\xi + i\zeta\sqrt{\tau^2 - \eta^2\rho^2}\right)/\rho^2, \tag{12.53}$$

and the parameters g_p and g_s are defined by (12.30), i.e.

$$g_p = \sqrt{\eta^2 - \beta_p^2}, \qquad g_s = \sqrt{1 - \beta_p^2}. \tag{12.54}$$

For $x < 0$ (or $\xi < 0$) the modified integration path again includes the small circle around the pole at the origin $p = 0$. Using the same procedures as for the isotropic stress, the contribution of this pole is found to be

$$\Delta h_1(\xi) = \delta(\tau - \eta\zeta)\{1 - H(\xi)\}. \tag{12.55}$$

Combination of (12.52) and (12.55) finally gives

$$h_1(\xi) = \frac{1}{\pi} \Im \left\{ \frac{\tau\sqrt{\tau^2 - \eta^2\rho^2} + i\eta^2\xi\zeta}{(\tau^2 - \eta^2\zeta^2)\sqrt{\tau^2 - \eta^2\rho^2}} \frac{(1 - 2\beta_p^2)^2}{(1 - 2\beta_p^2)^2 + 4\beta_p^2 g_p g_s} \right\} H(\tau - \eta\rho)$$
$$+ \delta(\tau - \eta\zeta)\{1 - H(\xi)\}, \tag{12.56}$$

which is valid for all values of ξ.

With (12.45) and (12.51) the expression for the first term σ_1 is

$$\frac{\sigma_1}{\sigma_0} = \{h_1(\xi + 1) - h_1(\xi - 1)\}, \tag{12.57}$$

where the function $h_1(\xi)$ is given in dimensionless form in (12.56), and where σ_0 is the reference stress defined by (12.40), i.e.

$$\sigma_0 = \frac{q\, c_s}{a}. \tag{12.58}$$

The Second Integral

Using the expression $\sin(\alpha a) = [\exp(i s\alpha a) - \exp(-i s\alpha a)]/2i$, the second integral, (12.44), can be written as

$$\bar{\sigma}_2 = q\{\bar{g}_2(x + a) - \bar{g}_2(x - a)\}, \tag{12.59}$$

where

$$\bar{g}_2(x) = -\frac{1}{2\pi i} \int_{-\infty}^{\infty} \frac{4\alpha\gamma_p\gamma_s}{(2\alpha^2 + 1/c_s^2)^2 - 4\alpha^2\gamma_p\gamma_s} \exp[-s(\gamma_s z + i\alpha x)]\, d\alpha. \tag{12.60}$$

The integration parameter is renamed by the substitution $p = i\alpha$. This gives

$$\bar{g}_2(x) = \frac{1}{2\pi i} \int_{-i\infty}^{i\infty} \frac{4p\gamma_p\gamma_s}{(1/c_s^2 - 2p^2)^2 + 4p^2\gamma_p\gamma_s} \exp[-s(\gamma_s z + px)]\, dp, \tag{12.61}$$

where

$$\gamma_p = \sqrt{1/c_p^2 - p^2}, \qquad \gamma_s = \sqrt{1/c_s^2 - p^2}. \tag{12.62}$$

The integral (12.61) can be evaluated in the same way as the integral $\overline{\sigma}_2$ in (11.262) in the previous chapter, the main difference being a factor p in the numerator of the integrand. It should be noted that the integral may consist of two contributions, one from the curved path up to infinity in the complex p-plane, and one from the possible loop around the branch point at $p = 1/c_p$.

The result, which will not be presented in detail here, is that the function $g_2(x)$ can be expressed as

$$g_2(x) = \frac{c_s}{a}\{h_2(\xi) + h_3(\xi)\}, \tag{12.63}$$

where $h_2(\xi)$ is the dimensionless contribution of the integration along the curved parts p_1 and p_2 in Fig. 11.3, and $h_3(\xi)$ is the possible contribution of the integration along the loop on the real axis around the branch point $p = 1/c_p$. The expression for $h_2(\xi)$ is found to be

$$h_2(\xi) = \frac{1}{\pi}\Re\left\{\frac{4\beta_s g_p g_s^2}{(1 - 2\beta_s^2)^2 + 4\beta_s^2 g_p g_s}\right\}\frac{H(\tau - \rho)}{\sqrt{\tau^2 - \rho^2}}, \tag{12.64}$$

where the dimensionless complex parameter β_s is defined by

$$\beta_s = \left(\xi + i\zeta\sqrt{\tau^2 - \rho^2}\right)/\rho^2, \tag{12.65}$$

and the parameters g_p and g_s are defined by

$$g_p = \sqrt{\eta^2 - \beta_s^2}, \qquad g_s = \sqrt{1 - \beta_s^2}. \tag{12.66}$$

The expression for $h_3(\xi)$ is found to be

$$h_3(\xi) = -\frac{1}{\pi}\frac{4\beta_q(1 - \beta_q^2)(1 - 2\beta_q^2)^2\sqrt{\beta_q^2 - \eta^2}}{(1 - 2\beta_q^2)^4 + 16\beta_q^4(1 - \beta_q^2)(\beta_q^2 - \eta^2)}$$

$$\times \frac{H(\tau - \tau_q) - H(\tau - \tau_s)}{\sqrt{\tau_s^2 - \tau^2}} H(\xi - \eta\rho), \tag{12.67}$$

where the dimensionless real parameter β_q is defined by

$$\beta_q = \left(\xi\tau - \zeta\sqrt{\rho^2 - \tau^2}\right)/\rho^2, \tag{12.68}$$

and the dimensionless parameters τ_q and τ_s are defined by

$$\tau_q = \eta\xi + \zeta\sqrt{1 - \eta^2}, \qquad \tau_s = \rho. \tag{12.69}$$

Equations (12.64) and (12.67) apply only for $x > 0$ or $\xi > 0$. For $x < 0$ the value of $g_2(x)$ can be determined by noting that $g_2(-x) = -g_2(x)$, which can be derived from the definition (12.61) when the integration variable p is replaced by $-p$.

The vertical normal stress σ_{zz} can be obtained by substituting the results derived above into (12.42), using the further elaborations of $\overline{\sigma}_1$ and $\overline{\sigma}_2$. This gives

$$\frac{\sigma_{zz}}{\sigma_0} = h_1(\xi+1) - h_1(\xi-1) + h_2(\xi+1) - h_2(\xi-1) + h_3(\xi+1) - h_3(\xi-1),$$
(12.70)

where σ_0 is a reference stress defined as

$$\sigma_0 = \frac{qc_s}{a},$$
(12.71)

and where the functions $h_1(\xi)$, $h_2(\xi)$ and $h_3(\xi)$ are defined in (12.56), (12.64) and (12.67).

Computer Program

The vertical normal stress σ_{zz} can be calculated by the function StripPulseSzz shown below. This function uses the functions delta, h1, h2, and h3, which are also shown. The delta function is approximated by a parabolic arc of small width and of unit area.

```
double delta(double t,double z,double e)
{
 double f;
 if ((t<z-e)||(t>z+e)) f=0;else f=3*(e*e-(t-z)*(t-z))/(4*e*e*e);
 return(f);
}
double h1(double x,double z,double t,double nu)
{
double n,nn,rr,pi,s,tt,tr,tz,xx,zz,eps;complex a,b,bb,b1,gp,gs,d,e;
pi=4*atan(1.0);nn=(1-2*nu)/(2*(1-nu));n=sqrt(nn);eps=0.01;tt=t*t;
xx=x*x;zz=z*z;rr=xx+zz;tr=tt-nn*rr;tz=tt-nn*zz;
if (tr<=0) s=0;else
 {
  a=complex(t/tz,nn*x*z/(tz*sqrt(tr)));b=complex(t*x/rr,(z/rr)*sqrt(tr));
  bb=b*b;b1=1-2*bb;gp=sqrt(nn-bb);gs=sqrt(1-bb);
  d=a*b1*b1;e=b1*b1+4*bb*gp*gs;s=imag(d/e)/pi;
 }
if (x<0) s+=delta(t,n*z,eps);
return(s);
}
double h2(double x,double z,double t,double nu)
{
 double n,nn,rr,pi,s,tt,tr,xx,zz;complex b,bb,b1,gp,gs,d,e;
 pi=4*atan(1.0);nn=(1-2*nu)/(2*(1-nu));n=sqrt(nn);tt=t*t;
xx=x*x;zz=z*z;rr=xx+zz;tr=tt-rr;
 if (tr<=0) s=0;else
  {
   b=complex(t*x/rr,(z/rr)*sqrt(tr));bb=b*b;b1=1-2*bb;
   gp=sqrt(nn-bb);gs=sqrt(1-bb);d=4*b*gp*(1-bb);e=b1*b1+4*bb*gp*gs;
   s=real(d/e)/(pi*sqrt(tr));
  }
 return(s);
}
double h3(double x,double z,double t,double nu)
{
 double n,nn,rr,rt,pi,s,b,bb,tt,xx,zz,tq,b2,c,d;
 pi=4*atan(1.0);nn=(1-2*nu)/(2*(1-nu));n=sqrt(nn);tt=t*t;
```

```
xx=x*x;zz=z*z;rr=xx+zz;rt=rr-tt;tq=fabs(n*x)+z*sqrt(1-nn);
if ((rt<=0)||(t<=tq)||(xx<=nn*rr)) s=0;else
 {
  if (x>0) b=x*t/rr-(z/rr)*sqrt(rt);else b=x*t/rr+(z/rr)*sqrt(rt);
  bb=b*b;b2=(1-2*bb)*(1-2*bb);c=4*b*(1-bb)*b2*sqrt(bb-nn);
  d=b2*b2+16*bb*bb*(1-bb)*(bb-nn);
  s=-c/(pi*d*sqrt(rt));
 }
return(s);
}
double StripPulseSzz(double x,double z,double t,double nu)
{
 double s;
 s=h1(x+1,z,t,nu)-h1(x-1,z,t,nu);
 s+=h2(x+1,z,t,nu)-h2(x-1,z,t,nu);
 s+=h3(x+1,z,t,nu)-h3(x-1,z,t,nu);
 return(s);
}
```

Examples

The vertical normal stress σ_{zz} in the point $x = 0$, $z = a$, is shown, as a function of time, in the Figs. 12.5 and 12.6 for two values of Poisson's ratio: $\nu = 0$ and $\nu = 0.45$.

In these figures the first singularity indicates the arrival of the compression wave under the load, the second singularity indicates the arrival of the (negative) compression waves emanating from the end points of the load (which arrive a factor $\sqrt{2}$

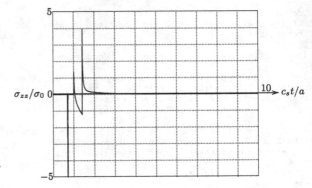

Fig. 12.5 Strip pulse—vertical normal stress, $\nu = 0$, $x/a = 0$, $z/a = 1$

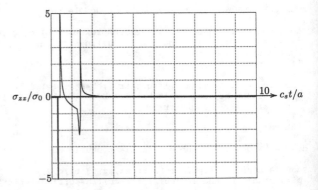

Fig. 12.6 Strip pulse—vertical normal stress, $\nu = 0.45$, $x/a = 0$, $z/a = 1$

later), and the third singularity indicates the arrival of the shear waves from these points (which arrive at time $t = a\sqrt{2}/c_s$).

12.1.3 The Horizontal Normal Stress

The next quantity to be evaluated is the horizontal normal stress σ_{xx}. It is recalled from (12.7) that the Laplace transform of this quantity is

$$\overline{\sigma}_{xx} = \frac{s^2}{2\pi} \int_{-\infty}^{\infty} \{(2\mu\alpha^2 - \lambda/c_p^2)C_p \exp(-s\gamma_p z) + 2\mu\alpha\gamma_s C_s \exp(-s\gamma_s z)\}$$
$$\times \exp(-is\alpha x)\, d\alpha. \tag{12.72}$$

Substitution of the expressions (12.11) and (12.12) for the two constants C_p and C_s gives

$$\overline{\sigma}_{xx} = \overline{\sigma}_1 + \overline{\sigma}_2, \tag{12.73}$$

where

$$\overline{\sigma}_1 = \frac{q}{\pi} \int_{-\infty}^{\infty} \frac{\sin(s\alpha a)}{\alpha} \frac{(2\alpha^2 + 1/c_s^2)[2\alpha^2 - \lambda/(\mu c_p^2)]}{(2\alpha^2 + 1/c_s^2)^2 - 4\alpha^2\gamma_p\gamma_s} \exp[-s(\gamma_p z + i\alpha x)]\, d\alpha, \tag{12.74}$$

$$\overline{\sigma}_2 = -\frac{q}{\pi} \int_{-\infty}^{\infty} \frac{\sin(s\alpha a)}{\alpha} \frac{4\alpha^2\gamma_s\gamma_p}{(2\alpha^2 + 1/c_s^2)^2 - 4\alpha^2\gamma_p\gamma_s} \exp[-s(\gamma_s z + i\alpha x)]\, d\alpha. \tag{12.75}$$

The first integral, (12.74), is very similar to the expression (12.43) obtained when considering the vertical normal stress. Using the same procedures, with a change of variable and a modification of the integration path, gives, by analogy with (12.57) and (12.56),

$$\frac{\sigma_1}{q} = \frac{c_p}{a}\{h_1(\xi + 1) - h_1(\xi - 1)\}, \tag{12.76}$$

where the function $h_1(\xi)$ is defined by the dimensionless form

$$h_1(\xi) = \frac{1}{\pi}\Im\left\{\frac{\tau\sqrt{\tau^2 - \eta^2\rho^2} + i\eta^2\xi\zeta}{(\tau^2 - \eta^2\zeta^2)\sqrt{\tau^2 - \eta^2\rho^2}} \frac{(1 - 2\beta_p^2)(1 - 2\eta^2 + 2\beta_p^2)}{(1 - 2\beta_p^2)^2 + 4\beta_p^2 g_p g_s}\right\}H(\tau - \eta\rho)$$
$$+ (1 - 2\eta^2)\delta(\tau - \eta\zeta)\{1 - H(\xi)\}. \tag{12.77}$$

The parameter β_p is a dimensionless complex variable defined by equation (12.53),

$$\beta_p = \left(\tau\xi + i\zeta\sqrt{\tau^2 - \eta^2\rho^2}\right)/\rho^2. \tag{12.78}$$

The coefficient of the last term in (12.77) is a consequence of the limiting behaviour of the integrand of (12.74) for $\alpha \to 0$, which determines the contribution of the pole for $x < 0$.

The second integral, (12.75), is just the opposite of the expression in (12.44), obtained when considering the vertical normal stress σ_{zz}. It follows that the final expression for the second integral can be written as

$$\frac{\sigma_2}{q} = \frac{c_p}{a} \{h_2(\xi + 1) - h_2(\xi - 1) + h_3(\xi + 1) - h_3(\xi - 1)\}. \tag{12.79}$$

The function $h_2(\xi)$ is the opposite of the function given in (12.64), i.e.

$$h_2(\xi) = -\frac{1}{\pi} \Re \left\{ \frac{4\beta_s g_p g_s^2}{(1 - 2\beta_s^2)^2 + 4\beta_s^2 g_p g_s} \right\} \frac{H(\tau - \rho)}{\sqrt{\tau^2 - \rho^2}}, \tag{12.80}$$

where the dimensionless complex parameter β_s is defined by (12.65),

$$\beta_s = \left(\tau\xi + i\zeta\sqrt{\tau^2 - \rho^2} \right)/\rho^2, \tag{12.81}$$

and the parameters g_p and g_s are defined by (12.66)

$$g_p = \sqrt{\eta^2 - \beta_s^2}, \qquad g_s = \sqrt{1 - \beta_s^2}. \tag{12.82}$$

The expression for $h_3(\xi)$ is the opposite of the value in (12.67), i.e.

$$h_3(\xi) = \frac{1}{\pi} \frac{4\beta_q(1 - \beta_q^2)(1 - 2\beta_q^2)^2 \sqrt{\beta_q^2 - \eta^2}}{(1 - 2\beta_q^2)^4 + 16\beta_q^4(1 - \beta_q^2)(\beta_3^2 - \eta^2)}$$

$$\times \frac{H(\tau - \tau_q) - H(\tau - \tau_s)}{\sqrt{\tau_s^2 - \tau^2}} H(\xi - \eta\rho), \tag{12.83}$$

where the dimensionless real parameter β_q is defined by (12.68),

$$\beta_q = \left(\xi\tau - \zeta\sqrt{\rho^2 - \tau^2} \right)/\rho^2, \tag{12.84}$$

and the dimensionless parameters τ_q and τ_s are defined by (12.69),

$$\tau_q = \eta\xi + \zeta\sqrt{1 - \eta^2}, \qquad \tau_s = \rho. \tag{12.85}$$

The horizontal normal stress σ_{xx} can be obtained by substituting the results derived above into (12.73), using the further elaborations of $\overline{\sigma}_1$ and $\overline{\sigma}_2$. This gives

$$\frac{\sigma_{xx}}{\sigma_0} = h_1(\xi + 1) - h_1(\xi - 1) + h_2(\xi + 1) - h_2(\xi - 1) + h_3(\xi + 1) - h_3(\xi - 1),$$
$$\tag{12.86}$$

where σ_0 is a reference stress defined as

$$\sigma_0 = \frac{qc_s}{a}, \tag{12.87}$$

and where the functions $h_1(\xi)$, $h_2(\xi)$ and $h_3(\xi)$ are defined in (12.77), (12.80) and (12.83). It should be noted that the functions $h_2(\xi)$ and $h_3(\xi)$ are antisymmetric in ξ.

Computer Program

The horizontal normal stress σ_{xx} can be calculated by the function StripPulse-Sxx shown below. This function uses the functions delta, h1, h2, and h3, which are also shown. The delta function is approximated by a parabolic arc of small width and of unit area.

```
double delta(double t,double z,double e)
{
 double f;
 if ((t<z-e)||(t>z+e)) f=0;else f=3*(e*e-(t-z)*(t-z))/(4*e*e*e);
 return(f);
}
double h1(double x,double z,double t,double nu)
{
 double n,nn,rr,pi,s,tt,tr,tz,xx,zz,eps;complex a,b,bb,b1,gp,gs,d,e;
 pi=4*atan(1.0);nn=(1-2*nu)/(2*(1-nu));n=sqrt(nn);eps=0.01;tt=t*t;
 xx=x*x;zz=z*z;rr=xx+zz;tr=tt-nn*rr;tz=tt-nn*zz;
 if (tr<=0) s=0;else
  {
   a=complex(t/tz,nn*x*z/(tz*sqrt(tr)));b=complex(t*x/rr,(z/rr)*sqrt(tr));
   bb=b*b;b1=1-2*bb;gp=sqrt(nn-bb);gs=sqrt(1-bb);
   d=a*b1*(1-2*nn+2*bb);e=b1*b1+4*bb*gp*gs;s=imag(d/e)/pi;
  }
 if (x<0) s+=(1-2*nn)*delta(t,n*z,eps);
 return(s);
}
double h2(double x,double z,double t,double nu)
{
 double n,nn,rr,pi,s,tt,tr,xx,zz;complex b,bb,b1,gp,gs,d,e;
 pi=4*atan(1.0);nn=(1-2*nu)/(2*(1-nu));n=sqrt(nn);tt=t*t;
 xx=x*x;zz=z*z;rr=xx+zz;tr=tt-rr;
 if (tr<=0) s=0;else
  {
   b=complex(t*x/rr,(z/rr)*sqrt(tr));bb=b*b;b1=1-2*bb;
   gp=sqrt(nn-bb);gs=sqrt(1-bb);d=4*b*gp*(1-bb);
   e=b1*b1+4*bb*gp*gs;s=-real(d/e)/(pi*sqrt(tr));
  }
 return(s);
}
double h3(double x,double z,double t,double nu)
{
 double n,nn,rr,rt,pi,s,b,bb,tt,xx,zz,tq,b2,c,d;
 pi=4*atan(1.0);nn=(1-2*nu)/(2*(1-nu));n=sqrt(nn);tt=t*t;
 xx=x*x;zz=z*z;rr=xx+zz;rt=rr-tt;tq=fabs(n*x)+z*sqrt(1-nn);
 if ((rt<=0)||(t<=tq)||(xx<=nn*rr)) s=0;else
  {
   if (x>0) b=x*t/rr-(z/rr)*sqrt(rt);else b=x*t/rr+(z/rr)*sqrt(rt);
   bb=b*b;b2=(1-2*bb)*(1-2*bb);c=4*b*(1-bb)*b2*sqrt(bb-nn);
   d=b2*b2+16*bb*bb*(1-bb)*(bb-nn);s=c/(pi*d*sqrt(rt));
  }
 return(s);
}
double StripPulseSxx(double x,double z,double t,double nu)
{
 double s;
 s=h1(x+1,z,t,nu)-h1(x-1,z,t,nu)+h2(x+1,z,t,nu)-h2(x-1,z,t,nu);
 s+=h3(x+1,z,t,nu)-h3(x-1,z,t,nu);
 return(s);
}
```

Examples

The horizontal normal stress σ_{xx} in the point $x = 0$, $z = a$, is shown, as a function of time, in the Figs. 12.7 and 12.8, for two values of Poisson's ratio: $\nu = 0$ and $\nu = 0.45$. In the first figure the possible first singularity, indicating the arrival of the compression wave under the load does not appear (its strength appears to be zero), the second singularity indicates the arrival of the (negative) compression waves emanating from the end points of the load (which arrive at time $t = a\sqrt{2}/c_p$), and the third singularity indicates the arrival of the shear waves from these points, at time $t = a\sqrt{2}/c_s$.

12.1.4 The Shear Stress

The last stress component to be evaluated is the shear stress σ_{xz}. It is recalled from (12.9) that the Laplace transform of this quantity is

$$\overline{\sigma}_{xz} = -\frac{i\mu s^2}{2\pi} \int_{-\infty}^{\infty} \{2\alpha\gamma_p C_p \exp(-\gamma_p sz) + (2\alpha^2 + 1/c_s^2)C_s \exp(-\gamma_s sz)\}$$

$$\times \exp(-is\alpha x)\,d\alpha. \tag{12.88}$$

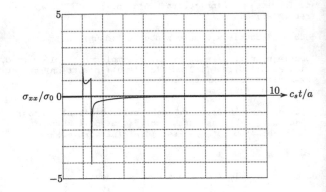

Fig. 12.7 Strip pulse—horizontal normal stress, $\nu = 0$, $x/a = 0$, $z/a = 1$

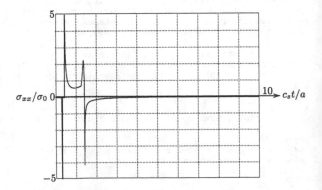

Fig. 12.8 Strip pulse—horizontal normal stress, $\nu = 0.45$, $x/a = 0$, $z/a = 1$

Substitution of the expressions (12.11) and (12.12) for the two constants C_p and C_s gives

$$\overline{\sigma}_{xz} = \overline{\sigma}_1 + \overline{\sigma}_2, \qquad (12.89)$$

where

$$\overline{\sigma}_1 = -\frac{2iq}{\pi} \int_{-\infty}^{\infty} \frac{\sin(s\alpha a)}{\alpha} \frac{(2\alpha^2 + 1/c_s^2)\alpha\gamma_p}{(2\alpha^2 + 1/c_s^2)^2 - 4\alpha^2\gamma_p\gamma_s} \exp[-s(\gamma_p z + i\alpha x)]\, d\alpha, \qquad (12.90)$$

$$\overline{\sigma}_2 = \frac{2iq}{\pi} \int_{-\infty}^{\infty} \frac{\sin(s\alpha a)}{\alpha} \frac{(2\alpha^2 + 1/c_s^2)\alpha\gamma_p}{(2\alpha^2 + 1/c_s^2)^2 - 4\alpha^2\gamma_p\gamma_s} \exp[-s(\gamma_s z + i\alpha x)]\, d\alpha. \qquad (12.91)$$

It may be interesting to note that it is immediately clear from these two expressions that for $z = 0$ the two integrals cancel, so that the boundary condition along the upper surface, that the shear stress vanishes, is indeed satisfied. Inspection also shows that in these expressions the point $\alpha = 0$ is not a singularity.

Using the same methods as for the other stress components leads to the following expression for the shear stress

$$\frac{\sigma_{xz}}{\sigma_0} = h_1(\xi + 1) - h_1(\xi - 1) + h_2(\xi + 1) - h_2(\xi - 1) + h_3(\xi + 1) - h_3(\xi - 1), \qquad (12.92)$$

where σ_0 is a reference stress defined as

$$\sigma_0 = \frac{qc_s}{a}, \qquad (12.93)$$

and where the functions $h_1(\xi)$, $h_2(\xi)$ and $h_3(\xi)$ are defined as follows, using the same dimensionless variables as before.

The function $h_1(\xi)$ is

$$h_1(\xi) = \frac{2}{\pi}\Re\left\{ \frac{(\eta^2 - \beta_p^2)(1 - 2\beta_p^2)}{(1 - 2\beta_p^2)^2 + 4\beta_p^2\sqrt{1 - \beta_p^2}\sqrt{\eta^2 - \beta_p^2}} \right\} \frac{H(\tau - \eta\rho)}{\sqrt{\tau^2 - \eta^2\rho^2}}, \qquad (12.94)$$

where

$$\beta_p = \left(\xi\tau + i\zeta\sqrt{\tau^2 - \eta^2\rho^2}\right)/\rho^2. \qquad (12.95)$$

The function $h_2(\xi)$ is

$$h_2(\xi) = -\frac{2}{\pi}\Re\left\{ \frac{\sqrt{1 - \beta_s^2}\sqrt{\eta^2 - \beta_s^2}(1 - 2\beta_s^2)}{(1 - 2\beta_s^2)^2 + 4\beta_s^2\sqrt{1 - \beta_s^2}\sqrt{\eta^2 - \beta_s^2}} \right\} \frac{H(\tau - \rho)}{\sqrt{\tau^2 - \rho^2}}, \qquad (12.96)$$

where

$$\beta_s = \left(\xi\tau + i\zeta\sqrt{\tau^2 - \rho^2}\right)/\rho^2. \qquad (12.97)$$

The function $h_3(\xi)$ is

$$h_3(\xi) = \frac{2}{\pi} \frac{(1 - 2\beta_q^2)^3 \sqrt{\beta_q^2 - \eta^2}\sqrt{1 - \beta_q^2}}{(1 - 2\beta_q^2)^4 + 16\beta_q^4(1 - \beta_q^2)(\beta_q^2 - \eta^2)}$$

$$\times \frac{H(\tau - \tau_q) - H(\tau - \tau_s)}{\sqrt{\tau_s^2 - \tau^2}} H(\xi - \eta\rho), \qquad (12.98)$$

where

$$\beta_q = \left(\xi\tau - \zeta\sqrt{\rho^2 - \tau^2}\right)/\rho^2. \qquad (12.99)$$

These expressions have been derived assuming that $x > 0$. For $x < 0$ the values can be obtained by noting from the original integrals that the functions must be symmetric in x. The shear stress itself should be antisymmetric.

Computer Program

The shear stress σ_{xz} can be calculated by the function StripPulseSxz shown below. This function uses the functions h1, h2, and h3, which are also shown.

```
double h1(double x,double z,double t,double nu)
{
 double xa,n,nn,rr,pi,s,tt,tr,xx,zz;complex b,bb,b1,gp,gs,d,e;
 xa=fabs(x);pi=4*atan(1.0);nn=(1-2*nu)/(2*(1-nu));n=sqrt(nn);
 tt=t*t;xx=xa*xa;zz=z*z;rr=xx+zz;tr=tt-nn*rr;
 if (tr<=0) s=0;
 else
   {
    b=complex(t*xa/rr,(z/rr)*sqrt(tr));bb=b*b;b1=1-2*bb;
    gp=sqrt(nn-bb);gs=sqrt(1-bb);d=(nn-bb)*b1;e=b1*b1+4*bb*gp*gs;
    s=2*real(d/e)/(pi*sqrt(tr));
   }
 return(s);
}
double h2(double x,double z,double t,double nu)
{
 double xa,n,nn,rr,pi,s,tt,tr,xx,zz;complex b,bb,b1,gp,gs,d,e;
 xa=fabs(x);pi=4*atan(1.0);nn=(1-2*nu)/(2*(1-nu));n=sqrt(nn);
 tt=t*t;xx=xa*xa;zz=z*z;rr=xx+zz;tr=tt-rr;
 if (tr<=0) s=0;
 else
   {
    b=complex(t*xa/rr,(z/rr)*sqrt(tr));bb=b*b;b1=1-2*bb;
    gp=sqrt(nn-bb);gs=sqrt(1-bb);d=gp*gs*b1;e=b1*b1+4*bb*gp*gs;
    s=-2*real(d/e)/(pi*sqrt(tr));
   }
 return(s);
}
double h3(double x,double z,double t,double nu)
{
 double xa,n,nn,rr,pi,s,b,bb,tt,xx,zz,rt,tq,b2,c,d;
 xa=fabs(x);pi=4*atan(1.0);nn=(1-2*nu)/(2*(1-nu));n=sqrt(nn);tt=t*t;
 xx=xa*xa;zz=z*z;rr=xx+zz;rt=rr-tt;tq=fabs(n*x)+z*sqrt(1-nn);
 if ((rt<=0)||(t<=tq)||(xx<=nn*rr)) s=0;
 else
   {
    b=xa*t/rr-(z/rr)*sqrt(rt);bb=b*b;b2=(1-2*bb)*(1-2*bb);
```

```
    c=b2*(1-2*bb)*sqrt(bb-nn)*sqrt(1-bb);d=b2*b2+16*bb*bb*(1-bb)*(bb-nn);
    s=2*c/(pi*d*sqrt(rt));
    }
  return(s);
}
double StripPulseSxz(double x,double z,double t,double nu)
{
  double s;
  s=h1(x+1,z,t,nu)-h1(x-1,z,t,nu);
  s+=h2(x+1,z,t,nu)-h2(x-1,z,t,nu);
  s+=h3(x+1,z,t,nu)-h3(x-1,z,t,nu);
  return(s);
}
```

Examples

The shear stress σ_{xz} in the point $x = a$, $z = a$, is shown, as a function of time, in the Figs. 12.9 and 12.10 for two values of Poisson's ratio: $\nu = 0$ and $\nu = 0.45$.

In these figures the first singularity indicates the arrival of the compression wave under the load, and the further singularities indicate the arrival of the compression waves and shear waves emanating from the end points of the loaded strip.

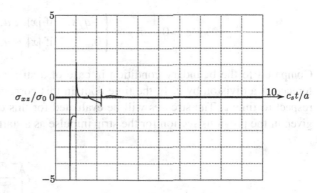

Fig. 12.9 Strip pulse—shear stress, $\nu = 0$, $x/a = 1$, $z/a = 1$

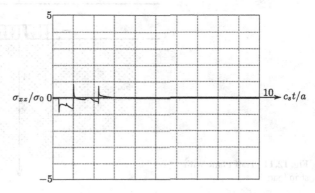

Fig. 12.10 Strip pulse—shear stress, $\nu = 0.45$, $x/a = 1$, $z/a = 1$

12.2 Strip Load

The second problem to be considered in this chapter is the case of a strip load on
an elastic half plane, i.e. a load that is applied at time $t = 0$, and then remains
constant in time, see Fig. 12.11. The solution will be obtained by an integration of
the solution of the problem of a strip pulse, considered in the previous section, over
the time parameter t.

The elastostatic equivalent of this problem is a classical problem of applied mechanics (Timoshenko and Goodier, 1970; Sneddon, 1951). This means that the elastodynamic solutions to be derived in this chapter should reduce to the elastostatic
limits if $t \to \infty$. Also, the solutions should reduce to those obtained for a line load
in the previous chapter, if the width of the loaded strip ($2a$) becomes very small.

In this case the boundary conditions are

$$z = 0 : \sigma_{zx} = 0, \tag{12.100}$$

$$z = 0 : \sigma_{zz} = \begin{cases} -q\,H(t), & \text{if } |x| < a, \\ 0, & \text{if } |x| > a. \end{cases} \tag{12.101}$$

The Laplace transform of the last condition is

$$z = 0 : \overline{\sigma}_{zz} = \begin{cases} -q/s, & \text{if } |x| < a, \\ 0, & \text{if } |x| > a. \end{cases} \tag{12.102}$$

Compared to the boundary condition in case of a strip pulse, see (12.3), the difference is a division by s. In the time domain this corresponds to integration with
respect to time t. The stresses will be evaluated for this case, taking the solutions
given in the previous section for the strip impulse as a start.

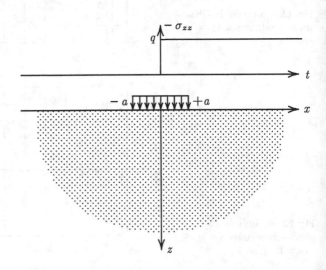

Fig. 12.11 Half plane with
strip load

12.2.1 The Isotropic Stress

The isotropic stress is, on the basis of a time integration of (12.39),

$$\frac{\sigma}{q} = (1 - \eta^2)\{f(\xi + 1, \zeta, \tau) + \Delta f(\xi + 1, \zeta, \tau) - f(\xi - 1, \zeta, \tau) - \Delta f(\xi - 1, \zeta, \tau)\},$$

$$(12.103)$$

where

$$f(\xi, \zeta, \tau) = \int_{\eta\rho}^{\tau} h(\xi, \zeta, \kappa) \, d\kappa, \qquad (12.104)$$

$$\Delta f(\xi, \zeta, \tau) = \int_{\eta\rho}^{\tau} \Delta h(\xi, \zeta, \kappa) \, d\kappa. \qquad (12.105)$$

The factor c_s/a in the reference value of the stress has been omitted, because $dt = (c_s/a) d\kappa$. In both integrals the lower limit of integration has been set equal to $\eta\rho$, because for $\kappa < \eta\rho$ the actual functions contain a factor zero. The two integrals (12.104) and (12.105) will be considered separately.

The First Integral

In the first integral the integrand is, with (12.32),

$$h(\xi, \zeta, \kappa) = \frac{1}{\pi} \Im \left\{ \frac{\kappa\sqrt{\kappa^2 - \eta^2\rho^2} + i\eta^2\xi\zeta}{(\kappa^2 - \eta^2\zeta^2)\sqrt{\kappa^2 - \eta^2\rho^2}} \frac{(1 - 2\beta_p^2)}{(1 - 2\beta_p^2)^2 + 4\beta_p^2 g_p g_s} \right\}, \quad (12.106)$$

where β_p is defined in (12.29),

$$\beta_p = \left(\tau\xi + i\zeta\sqrt{\tau^2 - \eta^2\rho^2}\right)/\rho^2, \qquad (12.107)$$

and the parameters g_p and g_s are defined in (12.30),

$$g_p = \sqrt{\eta^2 - \beta_p^2}, \qquad g_s = \sqrt{1 - \beta_p^2}. \qquad (12.108)$$

Because of the complex character of the expression (12.106) a numerical integration seems to be required. For such a numerical integration a complication is that there is a singularity at the lower limit, for $\kappa = \eta\rho$, of the character $1/\sqrt{\kappa^2 - \eta^2\rho^2}$. Although this is an integrable singularity, the results may be easier to compute, and more accurate, if the function is written as

$$h(\xi, \zeta, \kappa) = \frac{k(\xi, \zeta, \eta\rho)}{(\kappa^2 - \eta^2\zeta^2)\sqrt{\kappa^2 - \eta^2\rho^2}} + h^*(\xi, \zeta, \kappa), \qquad (12.109)$$

where

$$h^*(\xi, \zeta, \kappa) = \frac{k(\xi, \zeta, \kappa) - k(\xi, \zeta, \eta\rho)}{(\kappa^2 - \eta^2\zeta^2)\sqrt{\kappa^2 - \eta^2\rho^2}}, \qquad (12.110)$$

and

$$k(\xi,\zeta,\kappa) = \frac{1}{\pi}\Im\left\{\frac{(\kappa\sqrt{\kappa^2 - \eta^2\rho^2} + i\eta^2\xi\zeta)(1 - 2\beta_p^2)}{(1 - 2\beta_p^2)^2 + 4\beta_p^2\sqrt{1 - \beta_p^2}\sqrt{\eta^2 - \beta_p^2}}\right\}.\tag{12.111}$$

A standard integral is

$$\int_{\eta\rho}^{\tau}\frac{d\kappa}{(\kappa^2 - \eta^2\zeta^2)\sqrt{\kappa^2 - \eta^2\rho^2}} = \frac{1}{\eta^2\zeta\sqrt{\rho^2 - \zeta^2}}\arctan\left(\frac{\zeta\sqrt{\tau^2 - \eta^2\rho^2}}{\tau\sqrt{\rho^2 - \zeta^2}}\right),\tag{12.112}$$

where $\zeta < \rho$. The validity of this integral may be verified by differentiating the right hand side with respect to τ.

Substitution of (12.109) into (12.104) gives, using the integral (12.112),

$$f(\xi,\zeta,\tau) = \frac{k(\xi,\zeta,\eta\rho)}{\eta^2\zeta\sqrt{\rho^2 - \zeta^2}}\arctan\left(\frac{\zeta\sqrt{\tau^2 - \eta^2\rho^2}}{\tau\sqrt{\rho^2 - \zeta^2}}\right) + \int_{\eta\rho}^{\tau}h^*(\xi,\zeta,\kappa)\,d\kappa,\tag{12.113}$$

where the value of the function $h^*(\xi,\zeta,\kappa)$ at the lower limit of integration is zero,

$$h^*(\xi,\zeta,\eta\rho) = 0.\tag{12.114}$$

The integral in (12.113) can be computed accurately by numerical integration.

The Second Integral

The integrand of the second integral is, with (12.36)

$$\Delta h(\xi,\zeta,\kappa) = \delta(\tau - \eta\zeta)\{1 - H(\xi)\}.\tag{12.115}$$

Substitution into (12.105) gives

$$\Delta f(\xi,\zeta,\tau) = H(\tau - \eta\zeta)\{1 - H(\xi)\}.\tag{12.116}$$

This represents a compression wave just below the load.

Computer Program

The isotropic stress can be calculated as a function of ξ, ζ, τ and ν by the function StripLoadS shown below. The functions k, f and df, used by this function, are also shown. The numerical integration is performed using Simspon's rule.

```
double k(double x,double z,double w,double nu)
{
  double n,nn,rr,pi,s,ww,xx,zz,wr;
  complex a,b,bb,b1,gp,gs,d,e;
  pi=4*atan(1.0);nn=(1-2*nu)/(2*(1-nu));n=sqrt(nn);
```

```
ww=w*w;xx=x*x;zz=z*z;rr=xx+zz;wr=ww-nn*rr;
if (wr<=0) s=0;
else
  {
    a=complex(w*sqrt(wr),nn*x*z);b=complex(w*x/rr,(z/rr)*sqrt(wr));bb=b*b;
    b1=1-2*bb;gp=sqrt(nn-bb);gs=sqrt(1-bb);d=a*b1;e=b1*b1+4*bb*gp*gs;
    s=imag(d/e)/pi;
  }
 return(s);
}
double f(double x,double z,double t,double nu)
{
 double n,nn,r,rr,xa,xx,zz,rz,tt,b,p,pp,pr,pz,s,s1,s2,s3,h,hh,eps;
 h=0.001;eps=h*h;xa=x;nn=(1-2*nu)/(2*(1-nu));n=sqrt(nn);
 if (fabs(xa)<eps) {if (x>=0) xa=eps;else xa=-eps;}
 tt=t*t;xx=xa*xa;zz=z*z;rr=xx+zz;r=sqrt(rr);rz=rr-zz;
 p=n*r+h;pp=p*p;b=k(xa,z,p,nu);hh=h*h;if (rz<hh) rz=hh;
 if (tt<=nn*rr) s=0;else
   {
    s=(b/(nn*z*sqrt(rz)))*atan((z*sqrt(tt-nn*rr))/(t*sqrt(rz)));s1=0;
    while (pp<tt)
      {
       p+=h;pp=p*p;pz=pp-nn*zz;pr=pp-nn*rr;s2=(k(xa,z,p,nu)-b)/(pz*sqrt(pr));
       p+=h;pp=p*p;pz=pp-nn*zz;pr=pp-nn*rr;s3=(k(xa,z,p,nu)-b)/(pz*sqrt(pr));
       s+=(s1+4*s2+s3)*h/3;s1=s3;
      }
   }
 return(s);
}
double df(double x,double z,double t,double nu)
{
 double s,n,nn;
 nn=(1-2*nu)/(2*(1-nu));n=sqrt(nn);
 s=1;if ((t<=n*z)||(x>=0)) s=0;return(s);
}
double StripLoadS(double x,double z,double t,double nu)
{
 double s,nn;
 nn=(1-2*nu)/(2*(1-nu));
 s=(1-nn)*(f(x+1,z,t,nu)+df(x+1,z,t,nu)-f(x-1,z,t,nu)-df(x-1,z,t,nu));
 return(s);
}
```

In these functions special care has been taken to avoid discontinuities or singularities at x=1 and x=-1, by introducing a small parameter eps. The magnitude of the step in the numerical integration is denoted by h. Accuracy may be further improved by giving this parameter a smaller value, at the price of computation time, of course.

Examples

Figures 12.12 and 12.13 show the isotropic stress as a function of the lateral coordinate x, for $z/a = 1.0$ and $\nu = 0.4$, and for two values of time: $c_s t/a = 2$ and $c_s t/a = 20$.

In the second figure the elastostatic values, which should obtain for $t \to \infty$, are also shown, by small asterisks. These elastostatic stresses are, see e.g.

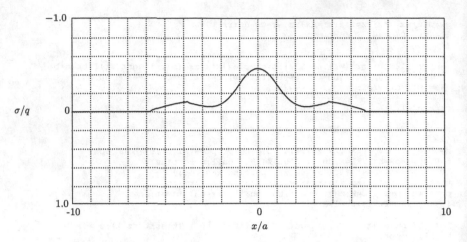

Fig. 12.12 Strip load—isotropic stress, $\nu = 0.4$, $z/a = 1$, $c_s t/a = 2$

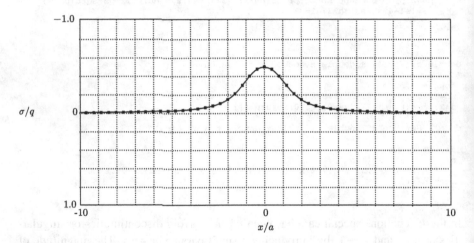

Fig. 12.13 Strip load—isotropic stress, $\nu = 0.4$, $z/a = 1$, $c_s t/a = 20$

Sneddon (1951),

$$\frac{\sigma}{q} = -\frac{1}{\pi}\left\{\arctan\left(\frac{x+a}{z}\right) - \arctan\left(\frac{x-a}{z}\right)\right\}. \qquad (12.117)$$

The agreement appears to be very good. It might be concluded from the figures that the elastodynamic solution presented here is in agreement with the elastostatic limit.

This conclusion is a little too fast, however. By taking into account a wider range of values of the lateral coordinate x/a, it appears that there is a difference between the elastodynamic solution and the elastostatic solution, as is illustrated in Fig. 12.14. This figure shows the isotropic stress for five increasing values of the dimensionless time parameter, namely $c_s t/a = 20, 40, 60, 80, 100$, in a range up to

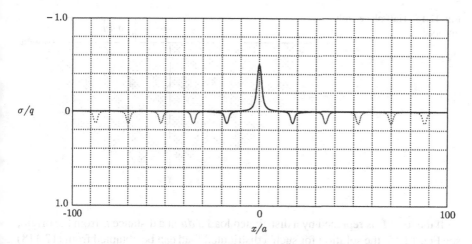

Fig. 12.14 Strip load—isotropic stress, $\nu = 0$, $z/a = 1$, $c_s t/a = 20, 40, 60, 80, 100$

$x/a = 100$, with the resolution of the graphs gradually decreasing, in order to distinguish between the various responses. As in the case of the line load, it appears that the solution consists of the ultimate elastostatic solution, plus a time dependent effect of a local disturbance, moving at the speed of the Rayleigh wave. This disturbance travels at constant speed, and with a constant amplitude and constant shape towards infinity. If time becomes really infinitely large the disturbance vanishes beyond the boundary at infinity, of course, and this confirms the conclusion that the elastodynamics solution is a proper generalization of the elastostatic solution, after all. But is important to realize that the Rayleigh wave disturbance in a two dimensional elastic problem does not exhibit any geometrical damping, and is visible for all finite values of time.

Figure 12.14 has been drawn for the case $\nu = 0$ and $z/a = 1$. For other values of these parameters the Rayleigh wave disturbance also appears, but at somewhat modified intensity or speed. Actually, in each case the velocity corresponds very well with the theoretical velocity of the Rayleigh wave, as determined in Chap. 9 of this book.

Approximation for Large Values of Time

The solution given in this section is exact, but so complex that numerical values can be obtained only by a computer program. It would be useful to have an approximation that would be somewhat easier to handle. Such an approximation can be obtained by extending the approximate solution for a line load derived in the previous chapter, see (11.258),

$$c_s t/z \gg 1 : \quad \frac{\sigma \pi z}{F} \approx -\frac{1}{1 + x^2/z^2} + \frac{m}{1 + (x - c_r t)^2/(w_p z)^2}, \tag{12.118}$$

where F is the magnitude of the line load, c_r is the speed of the Rayleigh wave, and m and w_p are given by (11.256) and (11.257).

Fig. 12.15 Half plane with
line load

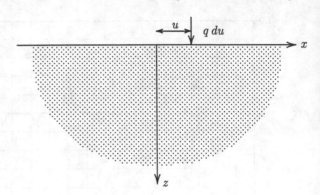

If the load F is replaced by a distributed load $q\,du$ at a distance u from the origin, see Fig. 12.15, the solution for such a distributed load can be obtained from (12.118) through replacing F by $q\,du$, and x by $x - u$. This gives

$$c_s t/z \gg 1 \,:\, \sigma\pi z \approx -\frac{q\,du}{1 + (x - u)^2/z^2} + \frac{mq\,du}{1 + (x - u - c_r t)^2/(w_p z)^2}. \quad (12.119)$$

The isotropic stress due to a distributed load $q(u)$ can now be obtained by integrating over u. For the case of a strip load this involves integrals of the following general form

$$\int_{-a}^{+a} \frac{du}{1 + (u - x)^2/z^2} = \pi z A(x, a, z), \quad (12.120)$$

where $A(x, a, z)$ is an elementary function defined by

$$A(x, a, z) = \frac{1}{\pi}\arctan\left(\frac{x + a}{z}\right) - \frac{1}{\pi}\arctan\left(\frac{x - a}{z}\right). \quad (12.121)$$

Using this notation the approximate expression for the isotropic stress caused by a strip load is

$$c_s t/a \gg 1 \,:\, \frac{\sigma}{q} \approx -A(x, a, z) + m w_p A(x - c_r t, a, w_p z). \quad (12.122)$$

This is indeed a much simpler formula, although it must be noted that it is valid only for large values of the time parameter $c_s t/a$. The approximate formula confirms that for large values of time the solution consists of the elastostatic solution and the Rayleigh wave, where the shape and the magnitude of the Rayleigh wave disturbance remains unchanged as it travels to infinity.

It may be recalled from the previous chapter that the parameters in the approximate solution are

$$m = \frac{(1 - \eta^2)(2\beta^2 - 1)}{2M\beta^2 w}, \quad w_p = \sqrt{1 - \eta^2/\beta^2}, \quad (12.123)$$

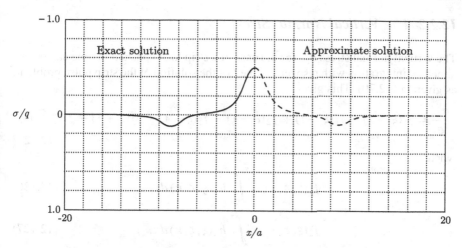

Fig. 12.16 Strip load—isotropic stress, $\nu = 0$, $z/a = 1$, $c_s t/a = 10$

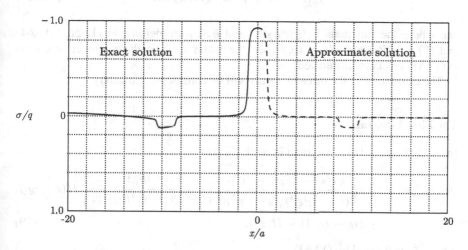

Fig. 12.17 Strip load—isotropic stress, $\nu = 0.5$, $z/a = 0.1$, $c_s t/a = 10$

where η, β and M are defined by

$$\eta = c_s/c_p, \quad \beta = c_s/c_r, \quad M = \frac{1 - 4\beta^2 + 8(1 - \eta^2)\beta^6}{\beta^2(2\beta^2 - 1)^2}. \tag{12.124}$$

Figure 12.16 shows a comparison of the exact analytical solution (in the left half of the figure) and the approximate solution (in the right half, by the dashed curve), for $\nu = 0$ and $z/a = 1$ and $c_s t/a = 10$. The maximum error in the approximate solution at that time is 0.018. For larger values of time the error is further reduced.

Figure 12.17 shows the results for $\nu = 0.5$, $z/a = 0.1$ and $c_s t/z = 10$. In this case the maximum error is 0.026, and again the error becomes smaller with time.

12.2.2 The Vertical Normal Stress

The vertical normal stress σ_{zz} for the case of a strip load constant in time can be obtained by integration with respect to time t of the solution of the strip pulse problem, as given in (12.70). This gives

$$\frac{\sigma_{zz}}{q} = f_1(\xi+1) - f_1(\xi-1) + f_2(\xi+1) - f_2(\xi-1) + f_3(\xi+1) - f_3(\xi-1),$$

(12.125)

where

$$f_1(\xi,\zeta,\tau) = \int_{\eta\rho}^{\tau} h_1(\xi,\zeta,\kappa)\,d\kappa,$$

(12.126)

$$f_2(\xi,\zeta,\tau) = \int_{\rho}^{\tau} h_2(\xi,\zeta,\kappa)\,d\kappa,$$

(12.127)

$$f_3(\xi,\zeta,\tau) = \int_{\tau_q}^{\tau} h_3(\xi,\zeta,\kappa)\,d\kappa,$$

(12.128)

and where the functions $h_1(\xi)$, $h_2(\xi)$ and $h_3(\xi)$ are defined in (12.56), (12.64) and (12.67). Because $d\kappa = (c_s/a)\,dt$ the factor c_s/a in the reference stress σ_0 in (12.58) has been eliminated.

The evaluation of the three integrals will be considered separately.

The First Integral

In the first integral the integrand is, with (12.56),

$$h_1(\xi,\zeta,\kappa) = \frac{1}{\pi}\Im\left\{\frac{\kappa\sqrt{\kappa^2 - \eta^2\rho^2} + i\eta^2\xi\zeta}{(\kappa^2 - \eta^2\zeta^2)\sqrt{\kappa^2 - \eta^2\rho^2}}\frac{(1 - 2\beta_p^2)^2}{(1 - 2\beta_p^2)^2 + 4\beta_p^2 g_p g_s}\right\} H(\kappa - \eta\rho)$$

$$+ \delta(\kappa - \eta\zeta)\{1 - H(\xi)\},$$

(12.129)

where β_p is defined by (12.53),

$$\beta_p = \left(\kappa\xi + i\zeta\sqrt{\kappa^2 - \eta^2\rho^2}\right)/\rho^2,$$

(12.130)

and the parameters g_p and g_s are defined by (12.54),

$$g_p = \sqrt{\eta^2 - \beta_p^2}, \qquad g_s = \sqrt{1 - \beta_p^2}.$$

(12.131)

To avoid the difficulties caused by the singularity at the lower limit of integration, for $\kappa = \eta\rho$, the function $h_1(\xi,\zeta,\kappa)$ is written as

$$h_1(\xi,\zeta,\kappa) = \frac{k_1(\xi,\zeta,\eta\rho)}{(\kappa^2 - \eta^2\zeta^2)\sqrt{\kappa^2 - \eta^2\rho^2}}H(\kappa - \eta\rho) + h_1^*(\xi,\zeta,\kappa)$$

$$+ \delta(\kappa - \eta\zeta)\{1 - H(\xi)\},$$

(12.132)

where

$$h_1^*(\xi, \zeta, \kappa) = \frac{k_1(\xi, \zeta, \kappa) - k_1(\xi, \zeta, \eta\rho)}{(\kappa^2 - \eta^2\zeta^2)\sqrt{\kappa^2 - \eta^2\rho^2}}, \tag{12.133}$$

and

$$k_1(\xi, \zeta, \kappa) = \frac{1}{\pi}\Im\left\{\frac{(\kappa\sqrt{\kappa^2 - \eta^2\rho^2} + i\eta^2\xi\zeta)(1 - 2\beta_p^2)^2}{(1 - 2\beta_p^2)^2 + 4\beta_p^2\sqrt{1 - \beta_p^2}\sqrt{\eta^2 - \beta_p^2}}\right\}. \tag{12.134}$$

Substitution of (12.132) into (12.126) gives, using the integral (12.112),

$$f_1(\xi, \zeta, \tau) = \frac{k_1(\xi, \zeta, \eta\rho)}{\eta^2\zeta\sqrt{\rho^2 - \zeta^2}}\arctan\left(\frac{\zeta\sqrt{\tau^2 - \eta^2\rho^2}}{\tau\sqrt{\rho^2 - \zeta^2}}\right)H(\tau - \eta\rho)$$

$$+ \int_{\eta\rho}^{\tau} h_1^*(\xi, \zeta, \kappa)\,d\kappa + H(\tau - \eta\zeta)\{1 - H(\xi)\}, \tag{12.135}$$

where the value of the function $h_1^*(\xi, \zeta, \kappa)$ at the lower limit of integration is zero,

$$h_1^*(\xi, \zeta, \eta\rho) = 0. \tag{12.136}$$

The integral in (12.135) can be computed numerically.

The Second Integral

In the second integral, equation (12.127), the integrand is, with (12.64),

$$h_2(\xi, \zeta, \kappa) = \frac{1}{\pi}\Re\left\{\frac{4\beta_s g_p g_s^2}{(1 - 2\beta_s^2)^2 + 4\beta_s^2 g_p g_s}\right\}\frac{H(\kappa - \rho)}{\sqrt{\kappa^2 - \rho^2}}, \tag{12.137}$$

where the parameter β_s is defined by equation (12.65),

$$\beta_s = \left(\kappa\xi + i\zeta\sqrt{\kappa^2 - \rho^2}\right)/\rho^2, \tag{12.138}$$

and the parameters g_p and g_s are defined by (12.66),

$$g_p = \sqrt{\eta^2 - \beta_s^2}, \qquad g_s = \sqrt{1 - \beta_s^2}. \tag{12.139}$$

In this case there is again a singularity at the beginning of the integration interval, but this point is now located at $\kappa = \rho$. In order to avoid the numerical difficulties caused by this singularity the function $h_2(\xi, \zeta, \kappa)$ is written as

$$h_2(\xi, \zeta, \kappa) = \frac{k_2(\xi, \zeta, \rho)}{\sqrt{\kappa^2 - \rho^2}}H(\kappa - \rho) + h_2^*(\xi, \zeta, \kappa), \tag{12.140}$$

where

$$h_2^*(\xi, \zeta, \kappa) = \frac{k_2(\xi, \zeta, \kappa) - k_2(\xi, \zeta, \rho)}{\sqrt{\kappa^2 - \rho^2}}, \tag{12.141}$$

and

$$k_2(\xi, \zeta, \kappa) = \frac{1}{\pi} \Re \left\{ \frac{4\beta_s(1 - \beta_s^2)\sqrt{\eta^2 - \beta_s^2}}{(1 - 2\beta_s^2)^2 + 4\beta_s^2\sqrt{1 - \beta_s^2}\sqrt{\eta^2 - \beta_s^2}} \right\}. \tag{12.142}$$

Substitution of (12.140) into (12.127) gives

$$f_2(\xi, \zeta, \tau) = k_2(\xi, \zeta, \rho) \log\left(\frac{\tau + \sqrt{\tau^2 - \rho^2}}{\rho}\right) H(\tau - \rho) + \int_\rho^\tau h_2^*(\xi, \zeta, \kappa) \, d\kappa, \tag{12.143}$$

where the value of the function $h_2^*(\xi, \zeta, \kappa)$ at the lower limit integration is zero,

$$h_2^*(\xi, \zeta, \rho) = 0. \tag{12.144}$$

The integral in (12.143) can be computed numerically.

The Third Integral

In the third integral, (12.128), the integrand is, with (12.67),

$$h_3(\xi) = -\frac{1}{\pi} \frac{4\beta_q(1 - \beta_q^2)(1 - 2\beta_q^2)^2\sqrt{\beta_q^2 - \eta^2}}{(1 - 2\beta_q^2)^4 + 16\beta_q^4(1 - \beta_q^2)(\beta_q^2 - \eta^2)}$$

$$\times \frac{H(\kappa - \tau_q) - H(\kappa - \tau_s)}{\sqrt{\tau_s^2 - \kappa^2}} H(\xi - \eta\rho), \tag{12.145}$$

where the real parameter β_q is defined by (12.68),

$$\beta_q = \left(\xi\kappa - \zeta\sqrt{\rho^2 - \kappa^2}\right)/\rho^2, \tag{12.146}$$

and the parameters τ_q and τ_s are defined by (12.69),

$$\tau_q = \eta\xi + \zeta\sqrt{1 - \eta^2}, \qquad \tau_s = \rho. \tag{12.147}$$

It may be noted that this third integral gives a non-zero contribution only if $\xi > \eta\rho$. The lower limit of integration is τ_q and the upper limit is τ_s, or τ when this is smaller. The integral can be computed numerically.

Computer Program

The vertical normal stress can be calculated as a function of ξ, ζ, τ and ν by the function StripLoadSzz shown below. The auxiliary functions used by this function, in which the three integrals are calculated, are also shown.

```
double k1(double x,double z,double t,double nu)
  {
    double n,nn,rr,pi,s,tt,tr,xx,zz;complex a,b,bb,b1,gp,gs,d,e;
    pi=4*atan(1.0);nn=(1-2*nu)/(2*(1-nu));n=sqrt(nn);
    tt=t*t;xx=x*x;zz=z*z;rr=xx+zz;tr=tt-nn*rr;if (tr<=0) s=0;
    else
    {
    a=complex(t*sqrt(tr),nn*x*z);b=complex(t*x/rr,(z/rr)*sqrt(tr));bb=b*b;
    b1=1-2*bb;gp=sqrt(nn-bb);gs=sqrt(1-bb);d=a*b1*b1;e=b1*b1+4*bb*gp*gs;
    s=imag(d/e)/pi;
    }
  return(s);
}
double f1(double x,double z,double t,double nu)
  {
   double n,nn,r,rr,xx,zz,rz,tt,b,p,pp,pr,pz,s,s1,s2,s3,xa,h,hh,eps;
   h=0.001;eps=h*h;xa=x;nn=(1-2*nu)/(2*(1-nu));n=sqrt(nn);
   if (fabs(xa)<eps) {if (x>=0) xa=eps;else xa=-eps;}
   tt=t*t;xx=xa*xa;zz=z*z;rr=xx+zz;r=sqrt(rr);
   hh=h*h;p=n*r+hh;pp=p*p;b=k1(xa,z,p,nu);rz=rr-zz;if (rz<hh) rz=hh;
   if (tt<=pp) s=0;else
     {
     s=(b/(nn*z*sqrt(rz)))*atan((z*sqrt(tt-nn*rr))/(t*sqrt(rz)));s1=0;
     while (pp<tt)
       {
       p+=h;pp=p*p;pz=pp-nn*zz;pr=pp-nn*rr;s2=(k1(xa,z,p,nu)-b)/(pz*sqrt(pr));
       p+=h;pp=p*p;pz=pp-nn*zz;pr=pp-nn*rr;s3=(k1(xa,z,p,nu)-b)/(pz*sqrt(pr));
       s+=(s1+4*s2+s3)*h/3;s1=s3;
       }
     }
   if ((t>n*z)&&(xa<0)) s+=1;
   return(s);
}
double k2(double x,double z,double t,double nu)
  {
    double n,nn,rr,pi,s,tt,tr,xx,zz;complex b,bb,b1,gp,gs,d,e;
    pi=4*atan(1.0);nn=(1-2*nu)/(2*(1-nu));n=sqrt(nn);
    tt=t*t;xx=x*x;zz=z*z;rr=xx+zz;tr=tt-rr;if (tr<=0) s=0;
    else
    {
    b=complex(t*x/rr,(z/rr)*sqrt(tr));bb=b*b;b1=1-2*bb;
    gp=sqrt(nn-bb);gs=sqrt(1-bb);d=4*b*(1-bb)*gp;e=b1*b1+4*bb*gp*gs;
    s=real(d/e)/pi;
    }
  return(s);
}
double f2(double x,double z,double t,double nu)
  {
   double r,rr,tt,tr,b,p,pp,pr,s,s1,s2,s3,xa,h,hh,eps;
   h=0.001;eps=h*h;xa=x;if (fabs(xa)<eps) {if (x>=0) xa=eps;else xa=-eps;}
   tt=t*t;rr=xa*xa+z*z;r=sqrt(rr);tr=t/r;h=0.001;hh=h*h;
   p=r+hh;pp=p*p;b=k2(xa,z,p,nu);if (tt<=pp) s=0;else
   {
   s=b*log(tr+sqrt(tr*tr-1));s1=0;
   while (pp<tt)
     {
     p+=h;pp=p*p;pr=pp-rr;s2=(k2(xa,z,p,nu)-b)/sqrt(pr);
     p+=h;pp=p*p;pr=pp-rr;s3=(k2(xa,z,p,nu)-b)/sqrt(pr);
     s+=(s1+4*s2+s3)*h/3;s1=s3;
```

```
    }
  }
  return(s);
}
double k3(double x,double z,double t,double nu)
{
  double n,nn,rr,pi,s,b,bb,tt,xx,zz,tq,b2,c,d;
  pi=4*atan(1.0);nn=(1-2*nu)/(2*(1-nu));n=sqrt(nn);
  tt=t*t;xx=x*x;zz=z*z;rr=xx+zz;tq=n*fabs(x)+z*sqrt(1-nn);
  if ((tt>=rr)||(t<=tq)||(xx<=nn*rr)) s=0;
  else
    {
      if (x>0) b=x*t/rr-(z/rr)*sqrt(rr-tt);else b=x*t/rr+(z/rr)*sqrt(rr-tt);
      bb=b*b;b2=(1-2*bb)*(1-2*bb);c=4*b*(1-bb)*b2*sqrt(bb-nn);
      d=b2*b2+16*bb*bb*(1-bb)*(bb-nn);s=-c/(pi*d*sqrt(rr-tt));
    }
  return(s);
}
double f3(double x,double z,double t,double nu)
{
  int m;double n,nn,xx,zz,r,rr,p,s,h,ts,tq;
  nn=(1-2*nu)/(2*(1-nu));n=sqrt(nn);
  xx=x*x;zz=z*z;rr=xx+zz;r=sqrt(rr);ts=r;tq=n*fabs(x)+z*sqrt(1-nn);
  if ((t<=tq)||(xx<=nn*rr)) s=0;else
    {
      if (t<ts) ts=t;m=10000;h=(ts-tq)/m;s=0;p=tq+h/2;
      while (p<ts) {s+=k3(x,z,p,nu)*h;p+=h;}
    }
  return(s);
}
double StripLoadSzz(double x,double z,double t,double nu)
{
  double s;
  s=f1(x+1,z,t,nu)-f1(x-1,z,t,nu)+f2(x+1,z,t,nu)-f2(x-1,z,t,nu);
  s+=f3(x+1,z,t,nu)-f3(x-1,z,t,nu);
  return(s);
}
```

Examples

Figures 12.18 and 12.19 show the vertical normal stress as a function of the lateral coordinate x, for $\nu = 0$ and $z/a = 1.0$, and for two values of time, namely $c_s t/a = 2$ and $c_s t/a = 20$. In the second figure the elastostatic values, which should obtain for $t \to \infty$, are also shown, by small asterisks. These elastostatic stresses are, see e.g. Sneddon (1951),

$$\frac{\sigma_{zz}}{q} = -\frac{1}{\pi}\left\{ \arctan\left(\frac{x+a}{z}\right) - \arctan\left(\frac{x-a}{z}\right) + \frac{(x+a)z}{(x+a)^2+z^2} - \frac{(x-a)z}{(x-a)^2+z^2} \right\}. \tag{12.148}$$

The agreement appears to be very good. This has also been found to be the case for other values of ν, so that it can be concluded that the solution is in agreement with the elastostatic solution.

The agreement in Fig. 12.19 may be somewhat misleading, because the range of values of x/a is considerably smaller than the value of $c_s t/a$, so that an eventual effect at the passing of the shear wave and the Rayleigh wave can not be shown. The behaviour of the solution for large values of time is studied below.

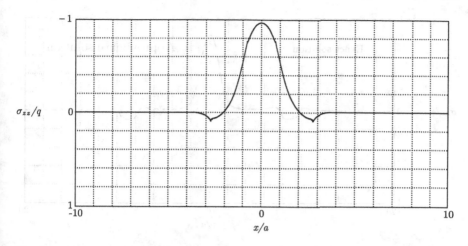

Fig. 12.18 Strip load—vertical normal stress, $v = 0$, $z/a = 1$, $c_s t/a = 2$

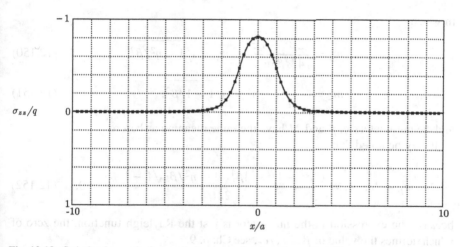

Fig. 12.19 Strip load—vertical normal stress, $v = 0$, $z/a = 1$, $c_s t/a = 20$

Approximation for Large Values of Time

An approximation valid for large values of time can be obtained by integrating the approximation for the case of a line load, as given in (11.287). This gives, using the same type of analysis as used in deriving the expression (12.122) for the isotropic stress, and using the notation $A(x, a, z)$ defined in (12.121),

$$c_s t/a \gg 1 : \frac{\sigma_{zz}}{q} \approx -\frac{1}{\pi} \frac{(x+a)z}{(x+a)^2 + z^2} + \frac{1}{\pi} \frac{(x-a)z}{(x-a)^2 + z^2} - A(x, a, z)$$
$$- 2m_1 w_p A(x - c_r t, a, w_p z) + 2m_2 w_s A(x - c_r t, a, w_s z),$$
$$(12.149)$$

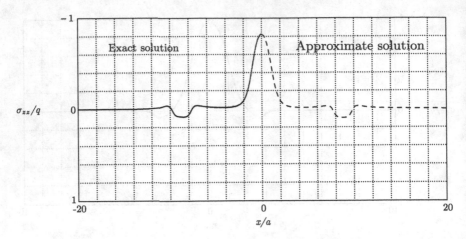

Fig. 12.20 Strip load—vertical normal stress, $\nu = 0$, $z/a = 1$, $c_s t/a = 10$

in which

$$m_1 = \frac{(2\beta^2 - 1)^2}{4M\beta^2 w_p}, \quad w_p = \sqrt{1 - \eta^2/\beta^2}, \tag{12.150}$$

$$m_2 = \frac{\beta^2 w_p}{M}, \quad w_s = \sqrt{1 - 1/\beta^2}, \tag{12.151}$$

and the parameters η, β and M have the same meaning as before.

It may be noted that

$$2m_1 w_p - 2m_2 w_s = \frac{(2\beta^2 - 1)^2 - 4\beta^4\sqrt{1 - \eta^2/\beta^2}\sqrt{1 - 1/\beta^2}}{2M\beta^2} = 0, \tag{12.152}$$

because the expression in the numerator is just the Rayleigh function, the zero of which defines the value of $\beta = c_s/c_r$, see Chap. 9.

To illustrate the accuracy of the approximation (12.149) two examples are shown in Figs. 12.20 and 12.21, for $z/a = 1$ and $c_s t/a = 10$, and two values of Poisson's ratio: $\nu = 0$ and $\nu = 0.5$. The figures show the analytical solution in the left half, and the approximate results in the left half, by a dashed curve. The agreement appears to be very good. The maximum error is 0.023, and this error further decreases for larger values of the time parameter $c_s t/a$.

12.2.3 The Horizontal Normal Stress

The horizontal normal stress σ_{xx} for the case of a strip load constant in time can be obtained by integration with respect to time t of the solution of the strip pulse

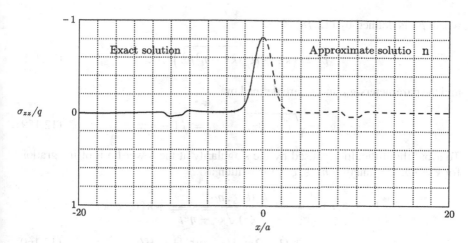

Fig. 12.21 Strip load—vertical normal stress, $\nu = 0.5$, $z/a = 1$, $c_s t/a = 10$

problem, as given in (12.86). This gives

$$\frac{\sigma_{xx}}{q} = f_1(\xi + 1) - f_1(\xi - 1) + f_2(\xi + 1) - f_2(\xi - 1) + f_3(\xi + 1) - f_3(\xi - 1),$$

(12.153)

where

$$f_1(\xi, \zeta, \tau) = \int_{\eta\rho}^{\tau} h_1(\xi, \zeta, \kappa) \, d\kappa,$$

(12.154)

$$f_2(\xi, \zeta, \tau) = \int_{\rho}^{\tau} h_2(\xi, \zeta, \kappa) \, d\kappa,$$

(12.155)

$$f_3(\xi, \zeta, \tau) = \int_{\tau_q}^{\tau} h_3(\xi, \zeta, \kappa) \, d\kappa,$$

(12.156)

and where the functions $h_1(\xi)$, $h_2(\xi)$ and $h_3(\xi)$ are defined in (12.77), (12.80) and (12.83).

Again the evaluation of the three integrals will be considered separately.

The First Integral

In the first integral the integrand is, with (12.77),

$$h_1(\xi, \zeta, \kappa) = \frac{1}{\pi} \Im \left\{ \frac{\kappa\sqrt{\kappa^2 - \eta^2\rho^2} + i\eta^2\xi\zeta}{(\kappa^2 - \eta^2\zeta^2)\sqrt{\kappa^2 - \eta^2\rho^2}} \frac{(1 - 2\beta_p^2)(1 - 2\eta^2 + 2\beta_p^2)}{(1 - 2\beta_p^2)^2 + 4\beta_p^2 g_p g_s} \right\}$$

$$\times H(\kappa - \eta\rho) + (1 - 2\eta^2)\delta(\kappa - \eta\zeta)\{1 - H(\xi)\},$$

(12.157)

where β_p is defined by (12.78),

$$\beta_p = \left(\kappa\xi + i\zeta\sqrt{\kappa^2 - \eta^2\rho^2}\right)/\rho^2, \tag{12.158}$$

and the parameters g_p and g_s are defined by

$$g_p = \sqrt{\eta^2 - \beta_p^2}, \qquad g_s = \sqrt{1 - \beta_p^2}. \tag{12.159}$$

To avoid the difficulties caused by the singularity at the lower limit of integration, for $\kappa = \eta\rho$, the function $h_1(\xi, \zeta, \kappa)$ is written as

$$h_1(\xi, \zeta, \kappa) = \frac{k_1(\xi, \zeta, \eta\rho)}{(\kappa^2 - \eta^2\zeta^2)\sqrt{\kappa^2 - \eta^2\rho^2}} + h_1^*(\xi, \zeta, \kappa)$$
$$+ (1 - 2\eta^2)\delta(\kappa - \eta\zeta)\{1 - H(\xi)\}, \tag{12.160}$$

where

$$h_1^*(\xi, \zeta, \kappa) = \frac{k_1(\xi, \zeta, \kappa) - k_1(\xi, \zeta, \eta\rho)}{(\kappa^2 - \eta^2\zeta^2)\sqrt{\kappa^2 - \eta^2\rho^2}}, \tag{12.161}$$

and

$$k_1(\xi, \zeta, \kappa) = \frac{1}{\pi}\Im\left\{\frac{(\kappa\sqrt{\kappa^2 - \eta^2\rho^2} + i\eta^2\xi\zeta)(1 - 2\beta_p^2)(1 - 2\eta^2 + 2\beta_p^2)}{(1 - 2\beta_p^2)^2 + 4\beta_p^2\sqrt{1 - \beta_p^2}\sqrt{\eta^2 - \beta_p^2}}\right\}. \tag{12.162}$$

Substitution of (12.160) into (12.154) gives, using the integral (12.112),

$$f_1(\xi, \zeta, \tau) = \frac{k_1(\xi, \zeta, \eta\rho)}{\eta^2\zeta\sqrt{\rho^2 - \zeta^2}}\arctan\left(\frac{\zeta\sqrt{\tau^2 - \eta^2\rho^2}}{\tau\sqrt{\rho^2 - \zeta^2}}\right)H(\tau - \eta\rho)$$
$$+ \int_{\eta\rho}^{\tau} h_1^*(\xi, \zeta, \kappa)\, d\kappa + (1 - 2\eta^2)H(\tau - \eta\zeta)\{1 - H(\xi)\}, \tag{12.163}$$

where the value of the function $h_1^*(\xi, \zeta, \kappa)$ at the lower limit integration is zero,

$$h_1^*(\xi, \zeta, \rho) = 0. \tag{12.164}$$

The integral in (12.163) can be computed numerically.

The Second Integral

In the second integral, (12.155), the integrand is, with (12.80),

$$h_2(\xi, \zeta, \kappa) = -\frac{1}{\pi}\Re\left\{\frac{4\beta_2 g_p g_s^2}{(1 - 2\beta_s^2)^2 + 4\beta_s^2 g_p g_s}\right\}\frac{H(\kappa - \rho)}{\sqrt{\kappa^2 - \rho^2}}, \tag{12.165}$$

where β_s is defined by (12.81),

$$\beta_s = \left(\tau\xi + i\zeta\sqrt{\tau^2 - \rho^2}\right)/\rho^2, \tag{12.166}$$

and the parameters g_p and g_s are defined by (12.82),

$$g_p = \sqrt{\eta^2 - \beta_s^2}, \qquad g_s = \sqrt{1 - \beta_s^2}. \tag{12.167}$$

In this case there is again a singularity at the beginning of the integration interval, but this point is now located at $\kappa = \rho$. In order to avoid the numerical difficulties caused by this singularity the function $h_2(\xi, \zeta, \kappa)$ is written as

$$h_2(\xi, \zeta, \kappa) = \frac{k_2(\xi, \zeta, \rho)}{\sqrt{\kappa^2 - \rho^2}} + h_2^*(\xi, \zeta, \kappa), \tag{12.168}$$

where

$$h_2^*(\xi, \zeta, \kappa) = \frac{k_2(\xi, \zeta, \kappa) - k_2(\xi, \zeta, \rho)}{\sqrt{\kappa^2 - \rho^2}}, \tag{12.169}$$

and

$$k_2(\xi, \zeta, \kappa) = -\frac{1}{\pi} \Re\left\{ \frac{4\beta_s(1 - \beta_s^2)\sqrt{\eta^2 - \beta_s^2}}{(1 - 2\beta_s^2)^2 + 4\beta_s^2\sqrt{1 - \beta_s^2}\sqrt{\eta^2 - \beta_s^2}} \right\}. \tag{12.170}$$

Substitution of (12.168) into (12.155) gives

$$f_2(\xi, \zeta, \tau) = k_2(\xi, \zeta, \rho) \log\left(\frac{\tau + \sqrt{\tau^2 - \rho^2}}{\rho}\right) + \int_\rho^\tau h_2^*(\xi, \zeta, \kappa)\,d\kappa, \tag{12.171}$$

where the value of the function $h_2^*(\xi, \zeta, \kappa)$ at the lower limit integration is zero,

$$h_2^*(\xi, \zeta, \rho) = 0. \tag{12.172}$$

The integral in (12.171) can be computed numerically.

The Third Integral

In the third integral, (12.156), the integrand is, with (12.83),

$$h_3(\xi, \zeta, \kappa) = \frac{1}{\pi} \frac{4\beta_q(1 - \beta_q^2)(1 - 2\beta_q^2)^2\sqrt{\beta_q^2 - \eta^2}}{(1 - 2\beta_q^2)^4 + 16\beta_q^4(1 - \beta_q^2)(\beta_q^2 - \eta^2)}$$

$$\times \frac{H(\kappa - \tau_q) - H(\kappa - \tau_s)}{\sqrt{\tau_s^2 - \kappa^2}} H(\xi - \eta\rho), \tag{12.173}$$

where the real parameter β_q is defined by

$$\beta_q = \left(\xi\kappa - \zeta\sqrt{\rho^2 - \kappa^2}\right)/\rho^2, \tag{12.174}$$

and where τ_q and τ_s are defined by (12.69),

$$\tau_q = \eta\xi + \zeta\sqrt{1 - \eta^2}, \qquad \tau_s = \rho. \tag{12.175}$$

It may be noted that this third integral gives a non-zero contribution only if $\xi > \eta\rho$. The lower limit of integration is τ_q and the upper limit is τ_s, or τ when this is smaller. The integral can be computed numerically.

Computer Program

The horizontal normal stress can be calculated as a function of ξ, ζ, τ and ν by the function `StripLoadSxx` shown below. The auxiliary functions used by this function, in which the three integrals are calculated, are also shown.

```
double k1(double x,double z,double t,double nu)
{
  double n,nn,rr,pi,s,tt,tr,xx,zz;complex a,b,bb,b1,gp,gs,d,e;
  pi=4*atan(1.0);nn=(1-2*nu)/(2*(1-nu));n=sqrt(nn);
  tt=t*t;xx=x*x;zz=z*z;rr=xx+zz;tr=tt-nn*rr;if (tr<=0) s=0;
  else
    {
    a=complex(t*sqrt(tr),nn*x*z);b=complex(t*x/rr,(z/rr)*sqrt(tr));bb=b*b;
    b1=1-2*bb;gp=sqrt(nn-bb);gs=sqrt(1-bb);
    d=a*b1*(1+2*bb-2*nn);e=b1*b1+4*bb*gp*gs;s=imag(d/e)/pi;
    }
  return(s);
}
double f1(double x,double z,double t,double nu)
{
  double m,mm,r,rr,xx,zz,rz,tt,b,p,pp,pr,pz,s,s1,s2,s3,xa,h,hh,eps;
  h=0.001;eps=h*h;xa=x;nn=(1-2*nu)/(2*(1-nu));n=sqrt(nn);
  if (fabs(xa)<eps) {if (x>=0) xa=eps;else xa=-eps;}
  tt=t*t;xx=xa*xa;zz=z*z;rr=xx+zz;r=sqrt(rr);
  hh=h*h;p=n*r+hh;pp=p*p;b=k1(xa,z,p,nu);rz=rr-zz;if (rz<hh) rz=hh;
  if (tt<=pp) s=0;else
    {
    s=(b/(nn*z*sqrt(rz)))*atan((z*sqrt(tt-nn*rr))/(t*sqrt(rz)));s1=0;
    while (pp<tt)
      {
      p+=h;pp=p*p;pz=pp-nn*zz;pr=pp-nn*rr;s2=(k1(xa,z,p,nu)-b)/(pz*sqrt(pr));
      p+=h;pp=p*p;pz=pp-nn*zz;pr=pp-nn*rr;s3=(k1(xa,z,p,nu)-b)/(pz*sqrt(pr));
      s+=(s1+4*s2+s3)*h/3;s1=s3;
      }
    }
  if ((t>n*z)&&(xa<0)) s+=1-2*nn;
  return(s);
}
double k2(double x,double z,double t,double nu)
{
  double n,nn,rr,pi,s,tt,tr,xx,zz;complex b,bb,b1,gp,gs,d,e;
  pi=4*atan(1.0);nn=(1-2*nu)/(2*(1-nu));n=sqrt(nn);
  tt=t*t;xx=x*x;zz=z*z;rr=xx+zz;tr=tt-rr;if (tr<=0) s=0;
  else
    {
    b=complex(t*x/rr,(z/rr)*sqrt(tr));bb=b*b;b1=1-2*bb;
```

```
    gp=sqrt(nn-bb);gs=sqrt(1-bb);d=4*b*(1-bb)*gp;e=b1*b1+4*bb*gp*gs;
    s=-real(d/e)/pi;
    }
  return(s);
}
double f2(double x,double z,double t,double nu)
{
  double r,rr,tt,tr,b,p,pp,pr,s,s1,s2,s3,xa,h,hh,eps;
  h=0.001;eps=h*h;xa=x;if (fabs(xa)<eps) {if (x>=0) xa=eps;else xa=-eps;}
  tt=t*t;rr=xa*xa+z*z;r=sqrt(rr);tr=t/r;
  hh=h*h;p=r+hh;pp=p*p;b=k2SxxX(xa,z,p,nu);
  if (tt<=pp) s=0;else
    {
    s=b*log(tr+sqrt(tr*tr-1));s1=0;
    while (pp<tt)
      {
      p+=h;pp=p*p;pr=pp-rr;s2=(k2(xa,z,p,nu)-b)/sqrt(pr);
      p+=h;pp=p*p;pr=pp-rr;s3=(k2(xa,z,p,nu)-b)/sqrt(pr);
      s+=(s1+4*s2+s3)*h/3;s1=s3;
      }
    }
  return(s);
}
double k3(double x,double z,double t,double nu)
{
  double n,nn,rr,pi,s,b,bb,tt,xx,zz,tq,b2,c,d;
  pi=4*atan(1.0);nn=(1-2*nu)/(2*(1-nu));n=sqrt(nn);
  tt=t*t;xx=x*x;zz=z*z;rr=xx+zz;tq=n*fabs(x)+z*sqrt(1-nn);
  if ((tt>=rr)||(t<=tq)||(xx<=nn*rr)) s=0;
  else
    {
    if (x>0) b=x*t/rr-(z/rr)*sqrt(rr-tt);else b=x*t/rr+(z/rr)*sqrt(rr-tt);
    bb=b*b;b2=(1-2*bb)*(1-2*bb);c=4*b*(1-bb)*b2*sqrt(bb-nn);
    d=b2*b2+16*bb*bb*(1-bb)*(bb-nn);s=c/(pi*d*sqrt(rr-tt));
    }
  return(s);
}
double f3(double x,double z,double t,double nu)
{
  int m;double n,nn,xx,zz,r,rr,p,s,h,ts,tq;
  nn=(1-2*nu)/(2*(1-nu));n=sqrt(nn);
  xx=x*x;zz=z*z;rr=xx+zz;r=sqrt(rr);ts=r;tq=n*fabs(x)+z*sqrt(1-nn);
  if ((t<=tq)||(xx<=nn*rr)) s=0;else
    {
    if (t<ts) ts=t;m=10000;h=(ts-tq)/m;s=0;p=tq+h/2;
    while (p<ts) {s+=k3(x,z,p,nu)*h;p+=h;}
    }
  return(s);
}
double StripLoadSxx(double x,double t,double nu)
{
  double s;
  s=f1(x+1,z,t,nu)-f1(x-1,z,t,nu);
  s+=f2(x+1,z,t,nu)-f2(x-1,z,t,nu);
  s+=f3(x+1,z,t,nu)-f3(x-1,z,t,nu);return(s);
}
```

Examples

Figures 12.22 and 12.23 show the horizontal normal stress as a function of the lateral coordinate x, for $\nu = 0$ and $z/a = 1.0$, and for two values of time, namely $c_s t/a = 2$ and $c_s t/a = 20$. In the second figure the elastostatic values, which should obtain for

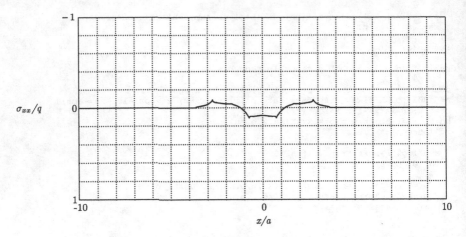

Fig. 12.22 Strip load—horizontal normal stress, $\nu = 0$, $z/a = 1$, $c_s t/a = 2$

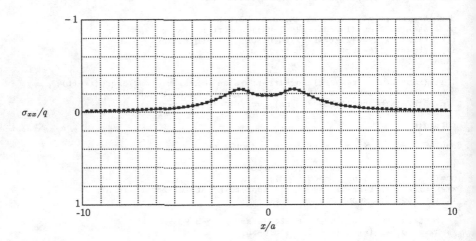

Fig. 12.23 Strip load—horizontal normal stress, $\nu = 0$, $z/a = 1$, $c_s t/a = 20$

$t \to \infty$, are also shown. These elastostatic stresses are, see e.g. Sneddon (1951),

$$\frac{\sigma_{xx}}{q} = -\frac{1}{\pi}\left\{\arctan\left(\frac{x+a}{z}\right) - \arctan\left(\frac{x-a}{z}\right) - \frac{(x+a)z}{(x+a)^2+z^2} + \frac{(x-a)z}{(x-a)^2+z^2}\right\}.$$

(12.176)

The agreement appears to be very good. This has also been found to be the case for other values of ν, so that it can be concluded that the solution is in agreement with the elastostatic solution.

Again, the agreement with the elastostatic solution is so good because the value of time in Fig. 12.23 is large compared to the distance from the loaded area considered. Actually, the elastostatic limit is approached when the distance traveled by the shear wave is large compared to the distance from the loaded area. It may be noted that the compression wave travels faster, and the Rayleigh wave travels only slightly slower than the shear wave, so that the shear wave velocity is indeed a convenient parameter.

Approximation for Large Values of Time

An approximation valid for large values of time can be obtained by integrating the approximation for the case of a line load, as given in (11.301). This gives, using the same type of analysis as used in deriving the expression (12.122) for the isotropic stress, and using the notation $A(x, a, z)$ defined in (12.121),

$$c_s t/a \gg 1 : \quad \frac{\sigma_{xx}}{q} \approx \frac{1}{\pi}\frac{(x+a)z}{(x+a)^2+z^2} - \frac{1}{\pi}\frac{(x-a)z}{(x-a)^2+z^2} - A(x, a, z)$$
$$+ 2m_3 w_p A(x - c_r t, a, w_p z) - 2m_4 w_s A(x - c_r t, a, w_s z),$$
$$(12.177)$$

in which

$$m_3 = \frac{(2\beta^2 - 1)(4\beta^2 + 1 - 2\eta^2)}{2M\beta^2 w_p}, \quad w_p = \sqrt{1 - \eta^2/\beta^2}, \quad (12.178)$$

$$m_4 = \frac{\beta^2 w_p}{M}, \quad w_s = \sqrt{1 - 1/\beta^2}, \quad (12.179)$$

and the parameters η, β and M have the same meaning as before.

To illustrate the accuracy of the approximation (12.177) two examples are shown in Figs. 12.24 and 12.25, for $\nu = 0$ and $c_s t/a = 10$, and for two values of the depth: $z/a = 1$ and $z/a = 0.01$. The figures show the analytical solution in the left half, and the approximate results in the left half, by a dashed curve. The agreement appears

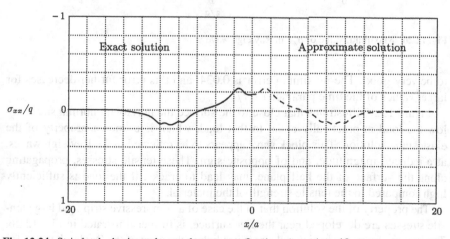

Fig. 12.24 Strip load—horizontal normal stress, $\nu = 0$, $z/a = 1$, $c_s t/a = 10$

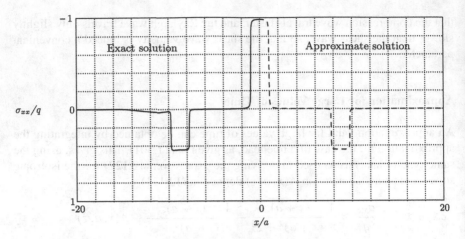

Fig. 12.25 Strip load—horizontal normal stress, $\nu = 0$, $z/a = 0.01$, $c_s t/a = 10$

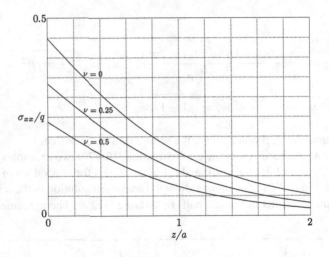

Fig. 12.26 Maximum value of σ_{xx} in Rayleigh wave

to be very good. The maximum error is 0.034, and this error further decreases for larger values of the time parameter $c_s t/a$.

Figure 12.25 illustrates that close to the surface the horizontal normal stress below the load is practically equal to that load, which is a known property of the elastostatic solution. This block wave appears to be reflected in the Rayleigh waves, at a reduced magnitude, and of opposite sign. These tensile stresses, propagating along the surface of the half plane, may lead to cracks, if the load is sufficiently high compared to the tensile strength of the material.

The property of the solution that in the case of a compressive strip load large tensile stresses are developed near the free surface, is further illustrated in Fig. 12.26. This figure shows the horizontal stress just below the crest of the Rayleigh wave, for three values of Poisson's ratio, as a function of depth. The maximum tensile stress occurs if $v = 0$, and its magnitude is about 45% of the magnitude of the compressive load q. This tensile stress rapidly decreases with depth, with the half-width a of the loaded strip as a scaling factor. In a material with a distinct tensile strength cracks may appear to a certain depth, and the depth of these cracks is related to the magnitude of the compressive load q and the tensile strength of the material. Conversely, the tensile strength of the material can be determined from the depth of the cracks and the magnitude of the load.

12.2.4 The Shear Stress

The shear stress σ_{xz} for the case of a strip load constant in time can be obtained by integration with respect to time t of the solution of the strip pulse problem, as given in (12.92). This gives

$$\frac{\sigma_{xz}}{q} = f_1(\xi + 1) - f_1(\xi - 1) + f_2(\xi + 1) - f_2(\xi - 1) + f_3(\xi + 1) - f_3(\xi - 1),$$

(12.180)

where

$$f_1(\xi, \zeta, \tau) = \int_{\eta\rho}^{\tau} h_1(\xi, \zeta, \kappa) d\kappa, \tag{12.181}$$

$$f_2(\xi, \zeta, \tau) = \int_{\rho}^{\tau} h_2(\xi, \zeta, \kappa) d\kappa, \tag{12.182}$$

$$f_3(\xi, \zeta, \tau) = \int_{\tau_q}^{\tau} h_3(\xi, \zeta, \kappa) d\kappa, \tag{12.183}$$

and where the functions $h_1(\xi)$, $h_2(\xi)$ and $h_3(\xi)$ are defined in (12.94), (12.96) and (12.98).

The evaluation of the three integrals will be considered separately.

The First Integral

In the first integral the integrand is, with (12.94),

$$h_1(\xi, \zeta, \kappa) = \frac{2}{\pi} \Re \left\{ \frac{(\eta^2 - \beta_p^2)(1 - 2\beta_p^2)}{(1 - 2\beta_p^2)^2 + 4\beta_p^2 \sqrt{1 - \beta_p^2} \sqrt{\eta^2 - \beta_p^2}} \right\} \frac{H(\kappa - \eta\rho)}{\sqrt{\kappa^2 - \eta^2\rho^2}},$$

(12.184)

where

$$\beta_p = \frac{\xi\kappa + i\zeta\sqrt{\kappa^2 - \eta^2\rho^2}}{\rho^2}. \tag{12.185}$$

To avoid the difficulties caused by the singularity at the lower limit of integration, for $\kappa = \eta \rho$, the function $h_1(\xi, \zeta, \kappa)$ is written as

$$h_1(\xi, \zeta, \kappa) = \frac{k_1(\xi, \zeta, \rho)}{\sqrt{\kappa^2 - \eta^2 \rho^2}} + h_1^*(\xi, \zeta, \kappa), \tag{12.186}$$

where now

$$h_1^*(\xi, \zeta, \kappa) = \frac{k_1(\xi, \zeta, \kappa) - k_1(\xi, \zeta, \eta \rho)}{\sqrt{\kappa^2 - \eta^2 \rho^2}}, \tag{12.187}$$

and

$$k_1(\xi, \zeta, \kappa) = \frac{2}{\pi} \Re \left\{ \frac{(\eta^2 - \beta_p^2)(1 - 2\beta_p^2)}{(1 - 2\beta_p^2)^2 + 4\beta_p^2 \sqrt{1 - \beta_p^2} \sqrt{\eta^2 - \beta_p^2}} \right\} H(\kappa - \eta \rho). \tag{12.188}$$

Substitution of (12.186) into (12.181) gives, using a standard integral,

$$f_1(\xi, \zeta, \tau) = k_1(\xi, \zeta, \eta \rho) \log \left(\frac{\tau + \sqrt{\tau^2 - \eta^2 \rho^2}}{\eta \rho} \right) H(\tau - \eta \rho)$$

$$+ \int_{\eta \rho}^{\tau} h_1^*(\xi, \zeta, \kappa) \, d\kappa, \tag{12.189}$$

where the value of the function $h_1^*(\xi, \zeta, \kappa)$ at the lower limit integration is zero,

$$h_1^*(\xi, \zeta, \eta \rho) = 0. \tag{12.190}$$

The integral in (12.189) can be computed numerically.

The Second Integral

In the second integral, (12.182), the integrand is, with (12.96),

$$h_2(\xi, \zeta, \kappa) = -\frac{2}{\pi} \Re \left\{ \frac{\sqrt{1 - \beta_s^2} \sqrt{\eta^2 - \beta_s^2}(1 - 2\beta_s^2)}{(1 - 2\beta_s^2)^2 + 4\beta_s^2 \sqrt{1 - \beta_s^2} \sqrt{\eta^2 - \beta_s^2}} \right\} \frac{H(\kappa - \rho)}{\sqrt{\kappa^2 - \rho^2}}, \tag{12.191}$$

where

$$\beta_s = \frac{\xi \kappa + i \zeta \sqrt{\kappa^2 - \rho^2}}{\rho^2}. \tag{12.192}$$

In this case there is again a singularity at the beginning of the integration interval, but this point is now located at $\kappa = \rho$. In order to avoid the numerical difficulties caused by this singularity the function $h_2(\xi, \zeta, \kappa)$ is written as

$$h_2(\xi, \zeta, \kappa) = \frac{k_2(\xi, \zeta, \rho)}{\sqrt{\kappa^2 - \rho^2}} + h_2^*(\xi, \zeta, \kappa), \tag{12.193}$$

where

$$h_2^*(\xi, \zeta, \kappa) = \frac{k_2(\xi, \zeta, \kappa) - k_2(\xi, \zeta, \rho)}{\sqrt{\kappa^2 - \rho^2}}, \tag{12.194}$$

and

$$k_2(\xi, \zeta, \kappa) = -\frac{2}{\pi} \Re \left\{ \frac{\sqrt{1 - \beta_s^2} \sqrt{\eta^2 - \beta_s^2} (1 - 2\beta_s^2)}{(1 - 2\beta_s^2)^2 + 4\beta_s^2 \sqrt{1 - \beta_s^2} \sqrt{\eta^2 - \beta_s^2}} \right\} H(\kappa - \rho). \tag{12.195}$$

Substitution of (12.193) into (12.182) gives

$$f_2(\xi, \zeta, \tau) = k_2(\xi, \zeta, \rho) \log \left(\frac{\tau + \sqrt{\tau^2 - \rho^2}}{\rho} \right) H(\tau - \rho) + \int_\rho^\tau h_2^*(\xi, \zeta, \kappa) \, d\kappa, \tag{12.196}$$

where the value of the function $h_2^*(\xi, \zeta, \kappa)$ at the lower limit integration is zero,

$$h_2^*(\xi, \zeta, \rho) = 0. \tag{12.197}$$

The integral in (12.196) can be computed numerically.

The Third Integral

In the third integral, (12.183), the integrand is, with (12.98),

$$h_3(\xi) = \frac{2}{\pi} \frac{(1 - 2\beta_q^2)^3 \sqrt{\beta_q^2 - \eta^2} \sqrt{1 - \beta_q^2}}{(1 - 2\beta_q^2)^4 + 16\beta_q^4 (1 - \beta_q^2)(\beta_q^2 - \eta^2)}$$
$$\times \frac{H(\kappa - \tau_q) - H(\kappa - \tau_s)}{\sqrt{\tau_s^2 - \kappa^2}} H(\xi - \eta\rho), \tag{12.198}$$

where

$$\beta_q = \left(\xi\kappa - \zeta\sqrt{\rho^2 - \kappa^2} \right) / \rho^2. \tag{12.199}$$

and where τ_q and τ_s are defined by (12.69),

$$\tau_q = \eta\xi + \zeta\sqrt{1 - \eta^2}, \qquad \tau_s = \rho. \tag{12.200}$$

It may be noted that this third integral gives a non-zero contribution only if $\xi > \eta\rho$. The lower limit of integration is τ_q and the upper limit is τ_s, or τ when this is smaller. The integral can be computed numerically.

Computer Program

The shear stress can be calculated as a function of ξ, ζ, τ and ν by the function StripLoadSxz shown below. The auxiliary functions used by this function, in which the three integrals are calculated, are also shown.

```
double k1(double x,double z,double t,double nu)
{
 double n,nn,rr,pi,s,tt,tr,xx,zz;complex b,bb,b1,gp,gs,d,e;
 pi=4*atan(1.0);nn=(1-2*nu)/(2*(1-nu));n=sqrt(nn);
 tt=t*t;xx=x*x;zz=z*z;rr=xx+zz;tr=tt-nn*rr;
 if (tr<=0) s=0;
 else
   {
    b=complex(t*x/rr,(z/rr)*sqrt(tr));bb=b*b;
    b1=1-2*bb;gp=sqrt(nn-bb);gs=sqrt(1-bb);
    d=b1*(nn-bb);e=b1*b1+4*bb*gp*gs;s=2*real(d/e)/pi;
   }
 return(s);
}
double f1(double x,double z,double t,double nu)
{
 double n,nn,r,rr,xx,zz,tt,tr,b,p,pp,pr,s,s1,s2,s3,xa,h,hh,eps;
 h=0.001;eps=h*h;xa=x;nn=(1-2*nu)/(2*(1-nu));n=sqrt(nn);
 if (fabs(xa)<eps) {if (x>=0) xa=eps;else xa=-eps;}
 tt=t*t;xx=xa*xa;zz=z*z;rr=xx+zz;r=sqrt(rr);tr=t/(n*r);
 hh=h*h;p=n*r+hh;pp=p*p;b=k1(xa,z,p,nu);
 if (tt<=pp) s=0;else
   {
    s=b*log(tr+sqrt(tr*tr-1));s1=0;
    while (pp<tt)
      {
       p+=h;pp=p*p;pr=pp-nn*rr;s2=(k1(xa,z,p,nu)-b)/sqrt(pr);
       p+=h;pp=p*p;pr=pp-nn*rr;s3=(k1(xa,z,p,nu)-b)/sqrt(pr);
       s+=(s1+4*s2+s3)*h/3;s1=s3;
      }
   }
 return(s);
}
double k2(double x,double z,double t,double nu)
{
 double n,nn,rr,pi,s,tt,tr,xa,xx,zz;complex b,bb,b1,gp,gs,d,e;
 pi=4*atan(1.0);nn=(1-2*nu)/(2*(1-nu));n=sqrt(nn);xa=fabs(x);
 tt=t*t;xx=xa*xa;zz=z*z;rr=xx+zz;tr=tt-rr;
 if (tr<=0) s=0;else
   {
    b=complex(t*xa/rr,(z/rr)*sqrt(tr));bb=b*b;b1=1-2*bb;
    gp=sqrt(nn-bb);gs=sqrt(1-bb);d=gp*gs*b1;e=b1*b1+4*bb*gp*gs;
    s=-2*real(d/e)/pi;
   }
 return(s);
}
double f2(double x,double z,double t,double nu)
{
 double r,rr,tt,tr,b,p,pp,pr,s,s1,s2,s3,xa,h,hh,eps;
 h=0.001;eps=h*h;xa=x;if (fabs(xa)<eps) {if (x>=0) xa=eps;else xa=-eps;}
 tt=t*t;rr=xa*xa+z*z;r=sqrt(rr);tr=t/r;
 hh=h*h;p=r+hh;pp=p*p;b=k2(xa,z,p,nu);
 if (tt<=pp) s=0;else
   {
    s=b*log(tr+sqrt(tr*tr-1));s1=0;
    while (pp<tt)
      {
       p+=h;pp=p*p;pr=pp-rr;s2=(k2(xa,z,p,nu)-b)/sqrt(pr);
       p+=h;pp=p*p;pr=pp-rr;s3=(k2(xa,z,p,nu)-b)/sqrt(pr);
```

```
        s+=(s1+4*s2+s3)*h/3;s1=s3;
        }
    }
    return(s);
}
double k3(double x,double z,double t,double nu)
{
    double n,nn,rr,pi,s,b,bb,tt,xa,xx,zz,tq,b2,c,e;
    pi=4*atan(1.0);nn=(1-2*nu)/(2*(1-nu));n=sqrt(nn);xa=fabs(x);
    tt=t*t;xx=xa*xa;zz=z*z;rr=xx+zz;tq=n*fabs(x)+z*sqrt(1-nn);
    if ((tt>=rr)||(t<=tq)||(xx<=nn*rr)) s=0;else
    {
        b=xa*t/rr-(z/rr)*sqrt(rr-tt);bb=b*b;b2=(1-2*bb)*(1-2*bb);
        c=(1-2*bb)*b2*sqrt(1-bb)*sqrt(bb-nn);
        d=b2*b2+16*bb*bb*(1-bb)*(bb-nn);
        s=2*c/(pi*d*sqrt(rr-tt));
    }
    return(s);
}
double f3(double x,double z,double t,double nu)
{
    int m;double n,nn,xx,zz,r,rr,p,s,h,ts,tq;
    nn=(1-2*nu)/(2*(1-nu));n=sqrt(nn);
    xx=x*x;zz=z*z;rr=xx+zz;r=sqrt(rr);ts=r;tq=n*fabs(x)+z*sqrt(1-nn);
    if ((t<=tq)||(xx<=nn*rr)) s=0;else
    {
        if (t<ts) ts=t;m=10000;h=(ts-tq)/m;s=0;p=tq+h/2;
        while (p<ts) {s+=k3(x,z,p,nu)*h;p+=h;}
    }
    return(s);
}
double StripLoadSxz(double x,double z,double t,double nu)
{
    double s;
    s=f1(x+1,z,t,nu)-f1(x-1,z,t,nu);
    s+=f2(x+1,z,t,nu)-f2(x-1,z,t,nu);
    s+=f3(x+1,z,t,nu)-f3(x-1,z,t,nu);
    if (x<0) s*=-1;
    return(s);
}
```

Examples

Figures 12.27 and 12.28 show the shear stress as a function of the lateral coordinate x, for $v = 0$ and $z/a = 1.0$, and for two values of time: $c_s t/a = 2$ and $c_s t/a = 20$.

In Fig. 12.28 the elastostatic values, which should obtain for $t \to \infty$, are also shown. For $x > 0$ these elastostatic stresses are, see e.g. Sneddon (1951),

$$\frac{\sigma_{xz}}{q} = \frac{1}{\pi}\left\{\frac{z^2}{(x+a)^2+z^2} - \frac{z^2}{(x-a)^2+z^2}\right\}. \tag{12.201}$$

The agreement appears to be very good. This has also been found to be the case for other values of v, so that it can be concluded that the solution is in agreement with the elastostatic solution.

As before, the agreement with the elastostatic solution is so good because the value of time in Fig. 12.28 is large compared to the distance from the loaded area

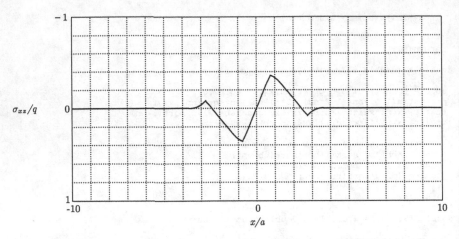

Fig. 12.27 Strip load—shear stress, $\nu = 0$, $z/a = 1$, $c_s t/a = 2$

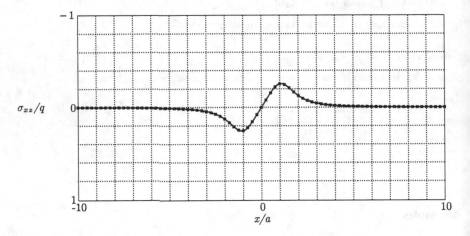

Fig. 12.28 Strip load—shear stress, $\nu = 0$, $z/a = 1$, $c_s t/a = 20$

considered. The actual criterion for the elastostatic limit to be approached is that the distance traveled by the shear wave should be large compared to the distance from the loaded area. As mentioned before the distance traveled by the shear wave is a convenient criterion, because the compression wave travels faster, and the Rayleigh wave travels only slightly slower than the shear wave.

Approximation for Large Values of Time

An approximation valid for large values of time can be obtained by integrating the approximation for the case of a line load, as given in (11.326). This gives, using the same type of analysis as used in deriving the expression (12.122) for the isotropic stress, and using the notation $A(x, a, z)$ defined in (12.121),

$$c_s t/a \gg 1 \; : \; \frac{\sigma_{xz}}{q} \approx \frac{1}{\pi} \frac{z^2}{(x+a)^2 + z^2} - \frac{1}{\pi} \frac{z^2}{(x-a)^2 + z^2}$$

$$- m_5 w_p \log \frac{(x - c_r t + a)^2 + (w_p z)^2}{(x - c_r t - a)^2 + (w_p z)^2}$$

$$+ m_6 w_s \log \frac{(x - c_r t + a)^2 + (w_s z)^2}{(x - c_r t - a)^2 + (w_s z)^2} \qquad (12.202)$$

in which

$$m_5 = \frac{2\beta^2 - 1}{2M}, \quad w_p = \sqrt{1 - \eta^2/\beta^2}, \qquad (12.203)$$

$$m_6 = \frac{(2\beta^2 - 1)w_p}{2M w_s}, \quad w_s = \sqrt{1 - 1/\beta^2}, \qquad (12.204)$$

and the parameters η, β and M have the same meaning as before.

It may be noted that $m_5 w_p = m_6 w_s$, so that the two coefficients of the logarithms in (12.202) are equal.

To illustrate the accuracy of the approximation (12.202) two examples are shown in Figs. 12.29 and 12.30, for $z/a = 1$ and $c_s t/a = 10$, and for two values of Poisson's ratio: $\nu = 0$ and $\nu = 0.5$. The figures show the analytical solution in the left half, and the approximate results in the right half, by a dashed curve. The agreement appears to be very good. The maximum error is 0.015, and this error further decreases for larger values of the time parameter $c_s t/a$.

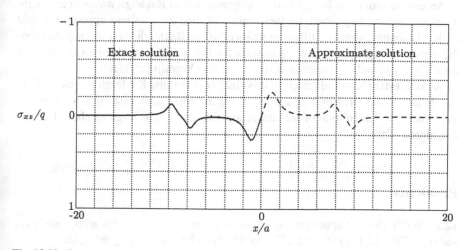

Fig. 12.29 Strip load—shear stress, $\nu = 0$, $z/a = 1$, $c_s t/a = 10$

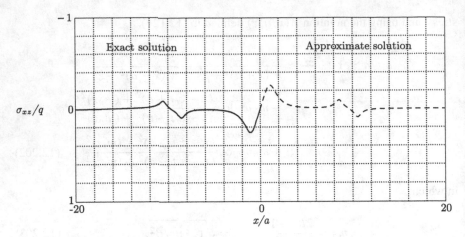

Fig. 12.30 Strip load—shear stress, $\nu = 0.5$, $z/a = 1$, $c_s t/a = 10$

Conclusion

In this chapter the solutions of the problems of a strip pulse and a strip load on an elastic half space have been considered. Using De Hoop's solution method, closed form expressions for the stress components in the half space for the case of a strip pulse have been derived, using a procedure proposed by Stam (1990). The solutions for the case of a strip load, constant in time, have been derived using a numerical integration over time.

The solutions have been validated by verifying that they are proper generalizations of the elastostatic problem, and of the line load problems considered in the previous chapter. They are also in good agreement with results obtained using a numerical (finite element) method (Verruijt et al., 2008).

Special attention has been given to the generation of Rayleigh waves in the vicinity of the surface. In general, for large values of time the solutions appear to consist of the elastostatic stress distribution plus a constant Rayleigh wave disturbance near the surface. Simple analytical expressions for the stresses, valid for large values of time, say $c_s t/a > 10$, have been derived, see also Verruijt (2008a).

An interesting result obtained in this chapter is that the Rayleigh waves result in tensile horizontal stresses propagating along the surface, in case of a compressive strip load. The magnitude of the tensile stresses may be as large as about half the magnitude of the load.

Problems

12.1 Verify that the approximate solutions given in Sect. 12.2 for σ, σ_{zz} and σ_{xx}, see (12.122), (12.149) and (12.177), satisfy the relation $\sigma = (\sigma_{zz} + \sigma_{xx})/2$.

12.2 Verify that these approximate solutions reduce to the approximate solutions given in the previous chapter for a constant line load if the width of the strip $a \rightarrow 0$, using the expression $F = 2aq$ for the total load.

12.3 Consider the values of the approximate solution for the vertical normal stress, (12.177), for $-a < x - c_r t < a$, that is in the vicinity of the passage of the Rayleigh wave. Using (12.152) show that this stress is zero for $z \rightarrow 0$, as required by the boundary condition.

12.4 Similarly, show that the approximate solution for the shear stress, (12.202), satisfies the condition that for $z \rightarrow 0$ this stress is zero, as required by the boundary condition.

Chapter 13
Point Load on an Elastic Half Space

This chapter presents a solution by Pekeris (1955) of the problem of a point load on the surface of an elastic half space. The derivation follows the presentation of the solution in the original paper by Pekeris, but some notations have been modified, and a numerical technique is used to evaluate certain integrals. This enables to generalize the solution by Pekeris for other values of Poisson's ratio than $\nu = 1/4$. The solution for other values of ν was given in closed form by Mooney (1974), and the solution is further analyzed and discussed by Eringen and Suhubi (1975) and by Foinquinos and Roësset (2000).

13.1 Problem

13.1.1 Basic Equations

The problem considered in this paper is a point load on the surface of an elastic half space, applied at time $t = 0$, see Fig. 13.1. The basic differential equations are the equations of motion in the radial and vertical directions r and z, for a linear elastic material, characterized by the Lamé constants λ and μ. These equations are, when expressed in terms of the displacement components u and w in radial and vertical direction, respectively,

$$(\lambda + \mu)\frac{\partial e}{\partial r} + \mu\left(\frac{\partial^2 u}{\partial r^2} + \frac{1}{r}\frac{\partial u}{\partial r} - \frac{u}{r^2} + \frac{\partial^2 u}{\partial z^2}\right) = \rho\frac{\partial^2 u}{\partial t^2}, \qquad (13.1)$$

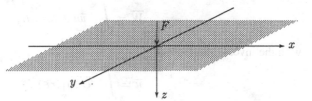

Fig. 13.1 Point load on half space

A. Verruijt, *An Introduction to Soil Dynamics*,
Theory and Applications of Transport in Porous Media 24,
© Springer Science+Business Media B.V. 2010

$$(\lambda + \mu)\frac{\partial e}{\partial z} + \mu\left(\frac{\partial^2 w}{\partial r^2} + \frac{1}{r}\frac{\partial w}{\partial r} + \frac{\partial^2 w}{\partial z^2}\right) = \rho\frac{\partial^2 w}{\partial t^2},\tag{13.2}$$

where e is the volume strain,

$$e = \frac{\partial u}{\partial r} + \frac{u}{r} + \frac{\partial w}{\partial z}.\tag{13.3}$$

The boundary conditions are supposed to describe a vertical point load on a small circular area, applied at time $t = 0$, with the shear stress being zero all along the boundary $z = 0$ for all values of time,

$$z = 0 : \sigma_{zr} = 0,\tag{13.4}$$

$$z = 0 : \sigma_{zz} = \begin{cases} -\dfrac{P}{\pi\epsilon^2} & \text{if } t > 0 \text{ and } r < \epsilon, \\ 0 & \text{if } t < 0 \text{ or } r > \epsilon. \end{cases}\tag{13.5}$$

Here P is the magnitude of the point load and ϵ is the small radius of the loaded area, which tends towards zero.

The Laplace transforms of the displacements are defined by

$$\overline{u} = \int_0^\infty u \exp(-st)\,dt,\tag{13.6}$$

$$\overline{w} = \int_0^\infty w \exp(-st)\,dt.\tag{13.7}$$

Using some elementary properties of the Laplace transform (Churchill, 1972) the basic equations now become, assuming that at time $t = 0$ all displacements and velocities are zero,

$$(\lambda + \mu)\frac{\partial\overline{e}}{\partial r} + \mu\left(\frac{\partial^2\overline{u}}{\partial r^2} + \frac{1}{r}\frac{\partial\overline{u}}{\partial r} - \frac{\overline{u}}{r^2} + \frac{\partial^2\overline{u}}{\partial z^2}\right) = \rho s^2\overline{u},\tag{13.8}$$

$$(\lambda + \mu)\frac{\partial\overline{e}}{\partial z} + \mu\left(\frac{\partial^2\overline{w}}{\partial r^2} + \frac{1}{r}\frac{\partial\overline{w}}{\partial r} + \frac{\partial^2\overline{w}}{\partial z^2}\right) = \rho s^2\overline{w}.\tag{13.9}$$

The axial symmetry of the problem suggests to seek the solution in the form of Hankel integrals (Titchmarsh, 1948; Sneddon, 1951). For this purpose the following Hankel transforms are introduced

$$\overline{U} = \int_0^\infty \overline{u}\,r\,J_1(\xi r)\,dr,\tag{13.10}$$

$$\overline{W} = \int_0^\infty \overline{w}\,r\,J_0(\xi r)\,dr,\tag{13.11}$$

with the inverse transforms

$$\overline{u} = \int_0^\infty \overline{U}\xi\,J_1(r\xi)\,d\xi,\tag{13.12}$$

$$\overline{w} = \int_0^\infty \overline{W} \xi J_0(r\xi) \, d\xi. \tag{13.13}$$

The use of the Bessel functions $J_1(\xi r)$ and $J_0(\xi r)$ in the transforms (13.10) and (13.11) has been suggested by the nature of the radial operators in (13.8) and (13.9), see Sneddon (1951).

Multiplication of (13.8) by $r J_1(\xi r)$, and integration over the interval from $r = 0$ to $r = \infty$ gives, using a transformation of the integrals by partial integration, following the usual Hankel transform methods (Sneddon, 1951),

$$\mu \frac{d^2\overline{U}}{dz^2} - [\rho s^2 + (\lambda + 2\mu)\xi^2]\overline{U} - (\lambda + \mu)\xi \frac{d\overline{W}}{dz} = 0. \tag{13.14}$$

Similarly, multiplication of (13.9) by $r J_0(\xi r)$, and integration over the interval from $r = 0$ to $r = \infty$ gives

$$(\lambda + 2\mu)\frac{d^2\overline{W}}{dz^2} - [\rho s^2 + \mu\xi^2]\overline{W} + (\lambda + \mu)\xi \frac{d\overline{U}}{dz} = 0. \tag{13.15}$$

The form of these differential equations can be somewhat simplified by the introduction of the following parameters,

$$c_s = \sqrt{\mu/\rho}, \tag{13.16}$$

$$k = s/c_s, \tag{13.17}$$

$$\eta^2 = \frac{\mu}{\lambda + 2\mu} = \frac{1 - 2v}{2(1 - v)} = \frac{c_s}{c_p}. \tag{13.18}$$

The quantity c_s is the propagation velocity of shear waves, c_p is the propagation velocity of compression waves, k is a simple scale transformation of the Laplace transform parameter, and η is an auxiliary elastic parameter, completely defined by the value of Poisson's ratio v. Using these parameters the basic equations can be written as

$$\eta^2 \frac{d^2\overline{U}}{dz^2} - (\eta^2 k^2 + \xi^2)\overline{U} - (1 - \eta^2)\xi \frac{d\overline{W}}{dz} = 0, \tag{13.19}$$

$$\frac{d^2\overline{W}}{dz^2} - \eta^2(k^2 + \xi^2)\overline{W} + (1 - \eta^2)\xi \frac{d\overline{U}}{dz} = 0. \tag{13.20}$$

13.2 Solution

The general solution of (13.19) and (13.20) for the half space $z > 0$ is

$$\overline{U} = A\alpha \exp(-\alpha z) + B\xi \exp(-\beta z), \tag{13.21}$$

$$\overline{W} = A\xi \exp(-\alpha z) + B\beta \exp(-\beta z), \tag{13.22}$$

in which

$$\alpha^2 = \xi^2 + k^2, \tag{13.23}$$

$$\beta^2 = \xi^2 + \eta^2 k^2. \tag{13.24}$$

The validity of this solution can easily be verified by substitution into (13.19) and (13.20). The solutions that increase exponentially for $z \to \infty$ have been excluded because of the conditions at infinity.

Inverse transformation of the two equations (13.21) and (13.22) now gives

$$\bar{u} = \int_0^\infty [A\alpha \exp(-\alpha z) + B\xi \exp(-\beta z)]\xi J_1(r\xi)\,d\xi, \tag{13.25}$$

$$\bar{w} = \int_0^\infty [A\xi \exp(-\alpha z) + B\beta \exp(-\beta z)]\xi J_0(r\xi)\,d\xi. \tag{13.26}$$

The boundary conditions are formulated in terms of the stresses σ_{rz} and σ_{zz}. Their Laplace transforms can be related to those of the displacements by the equations

$$\bar{\sigma}_{rz} = \mu\left(\frac{\partial \bar{u}}{\partial z} + \frac{\partial \bar{w}}{\partial x}\right), \tag{13.27}$$

$$\bar{\sigma}_{zz} = (\lambda + 2\mu)\frac{\partial \bar{w}}{\partial z} + \lambda\left(\frac{\partial \bar{u}}{\partial r} + \frac{\bar{u}}{r}\right). \tag{13.28}$$

With (13.25) and (13.26) this gives, using the definitions (13.23) and (13.24),

$$\bar{\sigma}_{rz} = -\mu \int_0^\infty [A(k^2 + 2\xi^2)\exp(-\alpha z) + 2B\beta\xi \exp(-\beta z)]\xi J_1(r\xi)\,d\xi, \tag{13.29}$$

$$\bar{\sigma}_{zz} = -\mu \int_0^\infty [2A\alpha\xi \exp(-\alpha z) + B(k^2 + 2\xi^2)\exp(-\beta z)]\xi J_0(r\xi)\,d\xi, \tag{13.30}$$

The coefficients A and B, which may depend upon the parameters s and ξ, may be determined from the boundary conditions.

The Laplace transforms of the boundary conditions (13.4) and (13.5) are

$$z = 0 : \bar{\sigma}_{zr} = 0, \tag{13.31}$$

$$z = 0 : \bar{\sigma}_{zz} = \begin{cases} -\dfrac{P}{\pi\epsilon^2 s} & \text{if } r < \epsilon, \\ 0 & \text{if } r > \epsilon. \end{cases} \tag{13.32}$$

Using a well known Hankel integral representation, see for instance Erdélyi et al. (1954), formula (8.3.18), the boundary condition (13.32) can also be written as

$$z = 0 : \bar{\sigma}_{zz} = -\int_0^\infty \frac{P}{2\pi s}\xi J_0(r\xi)\,d\xi, \tag{13.33}$$

where the parameter ϵ has been taken infinitely small, so that the load is a point load.

Substituting the general expressions (13.29) and (13.30) into these two boundary conditions leads to the following equations for the determination of the coefficients A and B,

$$A(k^2 + 2\xi^2) + 2B\beta\xi = 0, \tag{13.34}$$

$$2A\alpha\xi + B(k^2 + 2\xi^2) = \frac{P}{2\pi\mu s}. \tag{13.35}$$

Solution of these equations gives

$$A = -\frac{P}{2\pi\mu s}\frac{2\beta\xi}{(k^2 + 2\xi^2)^2 - 4\alpha\beta\xi^2}, \tag{13.36}$$

$$B = \frac{P}{2\pi\mu s}\frac{k^2 + 2\xi^2}{(k^2 + 2\xi^2)^2 - 4\alpha\beta\xi^2}. \tag{13.37}$$

The final expressions for the Laplace transforms of the displacements are

$$\bar{u} = -\frac{P}{2\pi\mu s}\int_0^\infty \frac{2\alpha\beta\exp(-\alpha z) - (k^2 + 2\xi^2)\exp(-\beta z)}{(k^2 + 2\xi^2)^2 - 4\alpha\beta\xi^2}\xi^2 J_1(r\xi)\,d\xi, \tag{13.38}$$

$$\bar{w} = -\frac{P}{2\pi\mu s}\int_0^\infty \frac{2\xi^2\exp(-\alpha z) - (k^2 + 2\xi^2)\exp(-\beta z)}{(k^2 + 2\xi^2)^2 - 4\alpha\beta\xi^2}\beta\xi J_0(r\xi)\,d\xi, \tag{13.39}$$

The remaining mathematical problem is to evaluate the integrals in these expressions, and then to perform the inverse Laplace transform. This is a formidable task. In this paper only the vertical displacements of the surface will be determined.

13.2.1 Vertical Displacement of the Surface

At the surface $z = 0$ of the half space the vertical displacement (13.39) is

$$\bar{w}_0 = \frac{Pk^2}{2\pi\mu s}\int_0^\infty \frac{\beta\xi}{(k^2 + 2\xi^2)^2 - 4\alpha\beta\xi^2}J_0(r\xi)\,d\xi. \tag{13.40}$$

This equation can be somewhat simplified by introducing the following (dimensionless) parameters,

$$x = \xi/k, \tag{13.41}$$

$$a = \alpha/k, \tag{13.42}$$

$$b = \beta/k. \tag{13.43}$$

Noting that $k/s = 1/c_s$, see (13.17), (13.40) can now be written as

$$\bar{w}_0 = \frac{P}{2\pi\mu c_s}\int_0^\infty \frac{bx}{(1 + 2x^2)^2 - 4abx^2}J_0(krx)\,dx. \tag{13.44}$$

In this expression the parameters a and b are defined by

$$a^2 = x^2 + 1, \tag{13.45}$$

$$b^2 = x^2 + \eta^2. \tag{13.46}$$

The integral representation (13.44) can not easily be expressed into elementary functions. Furthermore, the inverse Laplace transform has to be performed. Pekeris (1955) has indicated that a closed form solution can be obtained by first transforming the integration path in (13.44), then performing the inverse Laplace transform, and finally elaborating the remaining integrals. This procedure will be described in some detail below.

13.2.2 The Pekeris Procedure

To modify the integration path the variable x in the solution (13.44) is considered to be the real part of a complex variable $z = x + iy$. The Bateman-Pekeris theorem (see Appendix C) can now be used,

$$\int_0^\infty x f(x) J_0(px) \, dx = -\frac{2}{\pi} \Im \int_0^\infty y f(iy) K_0(py) \, dy, \tag{13.47}$$

which is valid if the function $f(z)$ has no singularities for $\Re(z) > 0$, $\Im(f(z)) = 0$ for $\Im(z) = 0$, and if the function $z^{3/2} f(z)$ tends towards zero for $z \to \infty$. The parameter p must be positive, $p > 0$.

In the case of the integral (13.44) the parameter $p = kr = sr/c_s$, which indeed is always positive. Furthermore in this case the function $f(z)$ is

$$f(z) = \frac{b}{(1 + 2z^2)^2 - 4abz^2}, \tag{13.48}$$

in which, by analytic continuation, the parameters a and b now are defined as

$$a^2 = z^2 + 1, \tag{13.49}$$

$$b^2 = z^2 + \eta^2. \tag{13.50}$$

The function $f(z)$ has singularities on the imaginary axis in the form of branch points, at $z = \pm i$ and $z = \pm \eta i$. It follows from (13.18) that η varies between $\eta = 1/\sqrt{2}$ (for $\nu = 0$) and $\eta = 0$ (for $\nu = \frac{1}{2}$). By assuming appropriate branch cuts in the complex z-plane the function $f(z)$ can be made single-valued in the entire plane, see Fig. 13.2. It is assumed that the arguments of the parameters a and b are chosen such that for $z \to \infty$ they coincide with the argument of z. At infinity the function $f(z)$, as defined by (13.48) behaves as z^{-2} so that the condition that $z^{3/2} f(z) \to 0$ for $z \to \infty$ is certainly satisfied.

The function $f(z)$ may also have poles in the complex z-plane. The location of these poles can be investigated by considering the values for which the denominator

Fig. 13.2 The complex
z-plane

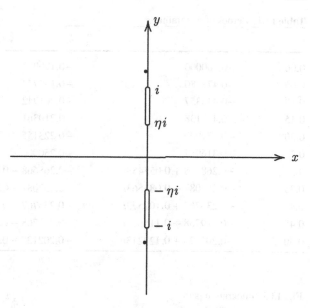

is zero. For this purpose a new variable ζ is introduced, defined as $\zeta = z^2$. The
denominator of $f(z)$ will be zero if

$$(1 + 2\zeta)^2 = 4ab\zeta, \tag{13.51}$$

or, after squaring both sides,

$$(1 + 2\zeta)^4 = 16(\zeta + 1)(\zeta + \eta^2)\zeta^2. \tag{13.52}$$

This leads to an algebraic equation of the third degree in ζ,

$$16(1 - \eta^2)\zeta^3 + 8(3 - 2\eta^2)\zeta^2 + 8\zeta + 1 = 0. \tag{13.53}$$

The zeroes are shown, for various values of Poisson's ratio ν in Table 13.1. Only
the zeroes $\zeta = \zeta_3$ correspond to poles of the function $f(z)$ in the area shown in
Fig. 13.2, the other singularities are located in other blades of the multivalued function. The negative values of ζ_3 indicate that the poles are located along the imaginary
axis in the z-plane. They are indicated in Fig. 13.2 by dots.

It can be concluded that the Bateman-Pekeris theorem can indeed be applied, so
that (13.44) can be transformed into

$$\overline{w}_0 = -\frac{P}{\pi^2 \mu c_s} \Im \int_0^\infty y f(iy) K_0(kry) \, dy, \tag{13.54}$$

where the path of integration should pass the singularities on the imaginary axis on
the right side, see Fig. 13.3.

Table 13.1 Zeroes of denominator

ν	ζ_1	ζ_2	ζ_3
0.00	-0.500000	-0.190983	-1.309017
0.05	-0.473680	-0.195773	-1.280547
0.10	-0.444357	-0.201942	-1.253701
0.15	-0.411138	-0.210361	-1.228500
0.20	-0.371900	-0.223155	-1.204945
0.25	-0.316987	-0.250000	-1.183013
0.30	$-0.268668 + 0.055458i$	$-0.268668 - 0.055458i$	-1.162663
0.35	$-0.253081 + 0.083563i$	$-0.253081 - 0.083563i$	-1.143838
0.40	$-0.236767 + 0.102573i$	$-0.236767 - 0.102573i$	-1.126466
0.45	$-0.219768 + 0.116675i$	$-0.219768 - 0.116675i$	-1.110464
0.50	$-0.202128 + 0.127213i$	$-0.202128 - 0.127213i$	-1.095744

Fig. 13.3 Integration path

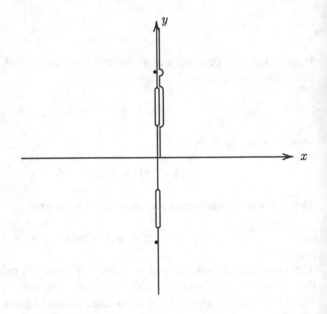

The vertical displacement itself can be obtained from its Laplace transform by application of the complex inversion integral (Churchill, 1972),

$$w_0 = \frac{1}{2\pi i} \int_{\gamma-i\infty}^{\gamma+i\infty} \overline{w}_0 \exp(st)\, ds, \qquad (13.55)$$

where γ should be large enough to ensure that there are no singularities to the right of the integration path. Substitution of (13.54) into (13.55) gives, after interchanging the orders of integration,

$$w_0 = -\frac{P}{\pi^2 \mu c_s} \Im \int_0^\infty y f(iy) \left\{ \frac{1}{2\pi i} \int_{\gamma-i\infty}^{\gamma+i\infty} K_0(sry/c) \exp(st) \, ds \right\} dy. \quad (13.56)$$

The inverse Laplace transform between brackets can be found in standard tables, see for instance Erdélyi et al. (1954), formula (5.15.8),

$$\frac{1}{2\pi i} \int_{\gamma-i\infty}^{\gamma+i\infty} K_0(sry/c) \, ds = \begin{cases} 0, & t < ry/c_s, \\ (t^2 - r^2 y^2/c_s^2)^{-1/2}, & t > ry/c_s. \end{cases} \quad (13.57)$$

It follows that the integrand of the integral (13.56) will contain a factor zero if $y > c_s t/r$. Hence this integral reduces to

$$w_0 = -\frac{P}{\pi^2 \mu r} \Im \int_0^{c_s t/r} \frac{y f(iy)}{\sqrt{c_s^2 t^2/r^2 - y^2}} \, dy. \quad (13.58)$$

In order to further evaluate this integral the behaviour of the function $f(iy)$ along the various parts of the path of integration must be considered. For this purpose it is most convenient to consider the parts between the branch points separately.

The general definition of the function $f(z)$ is, see (13.48),

$$f(z) = \frac{b}{(1 + 2z^2)^2 - 4abz^2}, \quad (13.59)$$

where the parameters a and b are defined by

$$a^2 = z^2 + 1 = (z - i)(z + i), \quad (13.60)$$

$$b^2 = z^2 + \eta^2 = (z - \eta i)(z + \eta i), \quad (13.61)$$

and the arguments of a and b coincide with that of z at infinity. For $z = iy$ the function $f(z)$ becomes

$$f(iy) = \frac{b}{(1 - 2y^2)^2 + 4aby^2}, \quad (13.62)$$

where the values of a and b depend upon the location of the point y on the imaginary axis. In particular

$$0 < y < \eta : a = \sqrt{1 - y^2}, \qquad b = \sqrt{\eta^2 - y^2}, \quad (13.63)$$

$$\eta < y < 1 : a = \sqrt{1 - y^2}, \qquad b = i\sqrt{y^2 - \eta^2}, \quad (13.64)$$

$$1 < y < \infty : a = i\sqrt{y^2 - 1}, \qquad b = i\sqrt{y^2 - \eta^2}. \quad (13.65)$$

It follows from (13.63) that for $0 < y < \eta$ the function $f(iy)$ will be real. This means that any integration along the interval $0 < y < \eta$ will give no contribution to the imaginary part of the integral in (13.58). Now, if the time parameter is so small that $c_s t/r < \eta$ it follows that there will be no contribution at all to the integral, and it can be concluded that then the displacement is zero,

$$c_s t/r < \eta \; : \; w_0 = 0. \tag{13.66}$$

Because η is the ratio of the shear wave velocity and the compression wave velocity, see (13.18), this can also be written as

$$c_p t/r < 1 \; : \; w_0 = 0. \tag{13.67}$$

This result expresses that the displacements are zero until the arrival of the compression wave.

Next consider the behaviour of the integral (13.58) if $\eta < c_s t/r < 1$. Then there will be only contributions to the integral from the range $\eta < y < c_s t/r$, where the largest possible value of $c_s t/r$ is 1. In that range the function $f(iy)$ is, if we write $\tau = c_s t/r$,

$$\eta < y < \tau \; : \; f(iy) = \frac{i\sqrt{y^2 - \eta^2}}{(1 - 2y^2)^2 + 4iy^2\sqrt{1 - y^2}\sqrt{y^2 - \eta^2}}, \tag{13.68}$$

For the evaluation of the integral only the imaginary part is relevant,

$$\eta < y < \tau \; : \; \Im f(iy) = \frac{\sqrt{y^2 - \eta^2}(1 - 2y^2)^2}{(1 - 2y^2)^4 + 16y^4(1 - y^2)(y^2 - \eta^2)}, \tag{13.69}$$

or, after some elaboration,

$$\eta < y < \tau \; : \; \Im f(iy) = \frac{\sqrt{y^2 - \eta^2}(1 - 2y^2)^2}{[1 - 8y^2 + 8(3 - 2\eta^2)y^4 - 16(1 - \eta^2)y^6]}, \tag{13.70}$$

The integral (13.58) now becomes

$$\eta < \tau < 1 \; : \; w_0 = -\frac{P}{\pi^2 \mu r}$$

$$\times \int_\eta^\tau \frac{y\sqrt{y^2 - \eta^2}(1 - 2y^2)^2}{[1 - 8y^2 + 8(3 - 2\eta^2)y^4 - 16(1 - \eta^2)y^6]\sqrt{\tau^2 - y^2}} \, dy, \tag{13.71}$$

where, as before,

$$\tau = c_s t/r. \tag{13.72}$$

This part of the solution can also be written as

$$\eta < \tau < 1 \; : \; w_0 = -\frac{P}{\pi^2 \mu r} G_1(\nu, \tau), \tag{13.73}$$

where now

$$G_1(v, \tau) = \int_\eta^\tau \frac{y\sqrt{y^2 - \eta^2}(1 - 2y^2)^2}{[1 - 8y^2 + 8(3 - 2\eta^2)y^4 - 16(1 - \eta^2)y^6]\sqrt{\tau^2 - y^2}} dy. \quad (13.74)$$

The values of the integral $G_1(v, \tau)$, in the range $\eta < \tau < 1$, can be determined by a numerical integration procedure. The integral has been evaluated in closed form by Pekeris (1955) for the case that $v = \frac{1}{4}$, i.e. $\eta^2 = \frac{1}{3}$. For arbitrary values of v a closed form solution has been given by Mooney (1974).

For values of the dimensionless time variable $\tau > 1$ there will also be a contribution to the integral (13.58) from the interval $1 < y < \tau$. Actually, for large enough values of τ there might also be a contribution to the integral from the small semi-circle around the pole, see Fig. 13.3, but it can be shown that this leads to a completely real value, so that its imaginary part is zero. On the interval $1 < y < \tau$ the function $f(iy)$ is, with (13.65),

$$1 < y < \tau : f(iy) = \frac{-i\sqrt{y^2 - \eta^2}}{(1 - 2y^2)^2 - 4y^2\sqrt{y^2 - 1}\sqrt{y^2 - \eta^2}}, \quad (13.75)$$

which can also be written as

$$1 < y < \tau : f(iy) = \frac{i\sqrt{y^2 - \eta^2}(1 - 2y^2) + 4iy^2(y^2 - \eta^2)\sqrt{y^2 - 1}}{[1 - 8y^2 + 8(3 - 2\eta^2)y^4 - 16(1 - \eta^2)y^6]\sqrt{y^2 - \eta^2}}. \quad (13.76)$$

This is purely imaginary.

The displacement of the surface now can be written as

$$\tau > 1 : w_0 = -\frac{P}{\pi^2 \mu r}\{G_1(v, \tau) + G_2(v, \tau)\}, \quad (13.77)$$

where the function $G_1(v, \tau)$ is the same as before, see (13.74), and the function $G_2(v, \tau)$ is defined by

$$G_2(v, \tau) = \int_1^\tau \frac{4y^3(y^2 - \eta^2)\sqrt{y^2 - 1}}{[1 - 8y^2 + 8(3 - 2\eta^2)y^4 - 16(1 - \eta^2)y^6]\sqrt{\tau^2 - y^2}} dy. \quad (13.78)$$

13.2.3 Numerical Evaluation of the Integrals

The First Integral

The first integral to be evaluated is $G_1(v, \tau)$, see (13.74),

$$G_1(v, \tau) = \int_\eta^\tau \frac{y\sqrt{y^2 - \eta^2}(1 - 2y^2)^2}{[1 - 8y^2 + 8(3 - 2\eta^2)y^4 - 16(1 - \eta^2)y^6]\sqrt{\tau^2 - y^2}} dy. \quad (13.79)$$

The integral can be somewhat simplified, and the singularity at $y = \tau$ can be removed, by the substitution

$$y^2 = \eta^2 + (\tau^2 - \eta^2)\sin^2\theta, \tag{13.80}$$

so that

$$y\,dy = (\tau^2 - \eta^2)\sin\theta\cos\theta\,d\theta, \tag{13.81}$$

$$\sqrt{y^2 - \eta^2} = \sqrt{\tau^2 - \eta^2}\sin\theta, \tag{13.82}$$

and

$$\sqrt{\tau^2 - y^2} = \sqrt{\tau^2 - \eta^2}\cos\theta. \tag{13.83}$$

The integral (13.79) can now be written as

$$G_1 = (\tau^2 - \eta^2)\int_0^{\pi/2} \frac{(1 - 2y^2)^2\sin^2\theta}{1 - 8y^2 + 8(3 - 2\eta^2)y^4 - 16(1 - \eta^2)y^6}\,d\theta, \tag{13.84}$$

where y is defined by (13.80). The integrand of the integral (13.84) has no singularities if $\eta < \tau < 1$. This means that the value of this integral can easily be calculated by a standard numerical algorithm. For larger values of τ the integrand may have a pole, and the integral must be calculated as a Cauchy principal value. The contribution of integration along the semi-circle around the pole can be disregarded because this can be shown to have a zero imaginary part.

The precise value of the location of the pole can be determined by writing $y^2 = p$, and assuming that the denominator is zero if $p = p_R = 1 + d$, where d can be supposed to be a small number, because the values of $|\zeta_3|$ in Table 13.1 are only slightly larger than 1,

$$p = p_R = 1 + d. \tag{13.85}$$

The value $p = p_R$ is a zero of the term between square brackets in the denominator of the integrand of (13.94). It follows that

$$F(p) = 1 - 8p_R + 8(3 - 4\eta^2)p_R^2 - 16(1 - \eta^2)p_R^3 = 0. \tag{13.86}$$

This gives, with (13.85) and using the definition of η^2 in (13.18),

$$d(d + 1)(d + v) = (1 - v)/8, \tag{13.87}$$

from which it follows that

$$d = \frac{(1 - v)/8}{(1 + d)(d + v)}. \tag{13.88}$$

From this equation the value of d can be determined to any desired accuracy by an iterative process, starting with the value $d = (1 - v)/8$. It can be concluded that the value of p_R, which determines the arrival of the Rayleigh wave, can be determined with great accuracy. The corresponding value of y_R then is

$$y_R = \sqrt{p_R} = \sqrt{1+d}. \tag{13.89}$$

and the corresponding value of θ_R is, with (13.80),

$$\theta_R = \arcsin\left(\frac{y_R^2 - \eta^2}{\tau^2 - \eta^2}\right). \tag{13.90}$$

For values of $\tau < y_R$ the pole is not on the path of integration, so that the integral can be evaluated immediately from (13.84), but for values of $\tau > y_R$ the pole is on the path of integration, and the integral must be separated into two parts,

$$\tau > y_R : \ G_1(v, \tau) = G_{11}(v, \tau) + G_{12}(v, \tau), \tag{13.91}$$

where

$$G_{11} = (\tau^2 - \eta^2) \int_0^{\theta_R - \varepsilon} \frac{(1 - 2y^2)^2 \sin^2 \theta}{1 - 8y^2 + 8(3 - 2\eta^2)y^4 - 16(1 - \eta^2)y^6}\, d\theta, \tag{13.92}$$

$$G_{12} = (\tau^2 - \eta^2) \int_{\theta_R + \varepsilon}^{\pi/2} \frac{(1 - 2y^2)^2 \sin^2 \theta}{1 - 8y^2 + 8(3 - 2\eta^2)y^4 - 16(1 - \eta^2)y^6}\, d\theta, \tag{13.93}$$

where ε is a very small number.

The Second Integral

The second integral to be evaluated is $G_2(v, \tau)$, see (13.78),

$$G_2(v, \tau) = \int_1^\tau \frac{4y^3(y^2 - \eta^2)\sqrt{y^2 - 1}}{[1 - 8y^2 + 8(3 - 2\eta^2)y^4 - 16(1 - \eta^2)y^6]\sqrt{\tau^2 - y^2}}\, dy. \tag{13.94}$$

In this case a convenient substitution, which also removes the singularity at $y = \tau$, is

$$y^2 = 1 + (\tau^2 - 1)\sin^2 \theta, \tag{13.95}$$

so that

$$y\, dy = (\tau^2 - 1)\sin \theta \cos \theta\, d\theta, \tag{13.96}$$

$$\sqrt{y^2 - 1} = \sqrt{\tau^2 - 1}\sin \theta, \tag{13.97}$$

and

$$\sqrt{\tau^2 - y^2} = \sqrt{\tau^2 - 1}\cos \theta. \tag{13.98}$$

The integral (13.94) can now be written as

$$G_2(v, \tau) = 4(\tau^2 - 1) \int_0^{\pi/2} \frac{y^2(y^2 - \eta^2)\sin^2\theta}{1 - 8y^2 + 8(3 - 2\eta^2)y^4 - 16(1 - \eta^2)y^6}\, d\theta, \quad (13.99)$$

where the value of y now is determined by (13.95). Again, the integral can be calculated directly if $\tau < y_R$, but for values of $\tau > y_R$ the integration interval must be separated into two parts, to calculate the Cauchy principal value.

13.2.4 Computer Program POINTLOAD

A function that calculates the vertical displacement for given values of v and $c_s t/r$ is reproduced below.

```
double wpekeris(double nu, double t)
{
 int j,k;
 double pi,fac,n,nn,e,f,fa,a,b,g,s,ta,tt,tr,xa,xb,xr,dx,yy,ss,eps,eps1;
 pi=4*atan(1.0);fac=1/(2*pi*pi);eps=0.000001;eps*=eps;eps1=0.001;
 nn=(1-2*nu)/(2*(1-nu));n=sqrt(nn);e=0.000001;e*=e;f=1;b=(1-nu)/8;
 if (nu>0.1) {while(f>e) {a=b;b=(1-nu)/(8*(1+a)*(nu+a));f=fabs(b-a);}}
 else {while (f>e) {a=b;b=sqrt((1-nu)/(8*(1+a)*(1+nu/a)));f=fabs(b-a);}}
 tr=sqrt(1+b); // tr is the arrival time of the Rayleigh wave
 if (t<=n) g=0; // zero displacement before arrival of compression wave
 else if (t<=1)
  {
  tt=t*t;xa=0;xb=pi/2;k=1000*(xb-xa);dx=(xb-xa)/k;fa=0;g=0;
  for (j=0;j<k;j++)
    {
    s=sin(xa+j*dx);ss=s*s;yy=nn+(tt-nn)*ss;a=(1-2*yy)*(1-2*yy)*(tt-nn)*ss;
    b=(1+8*yy*(-1+yy*(3-2*nn-2*yy*(1-nn))));f=a/b;if (j>0) g-=fac*(f+fa)*dx;fa=f;
    }
  }
 else if (t>tr)
  {
  ta=t;if (t<tr-eps) ta=tr+eps1;tt=ta*ta;xr=ArcSin(sqrt((tr*tr-nn)/(tt-nn)));
  xa=0;xb=xr-eps1;k=1000*(xb-xa);dx=(xb-xa)/k;fa=0;g=0;
  for (j=0;j<=k;j++)
    {
    s=sin(xa+j*dx);ss=s*s;yy=nn+(tt-nn)*ss;a=(1-2*yy)*(1-2*yy)*(tt-nn)*ss;
    b=(1+8*yy*(-1+yy*(3-2*nn-2*yy*(1-nn))));f=a/b;if (j>0) g-=fac*(f+fa)*dx;fa=f;
    }
  xa=xr+eps1;xb=pi/2;k=1000*(xb-xa);dx=(xb-xa)/k;
  for (j=0;j<=k;j++)
    {
    s=sin(xa+j*dx);ss=s*s;yy=nn+(tt-nn)*ss;a=(1-2*yy)*(1-2*yy)*(tt-nn)*ss;
    b=(1+8*yy*(-1+yy*(3-2*nn-2*yy*(1-nn))));f=a/b;if (j>0) g-=fac*(f+fa)*dx;fa=f;
    }
  xr=ArcSin(sqrt((tr*tr-1)/(tt-1)));xa=0;xb=xr-eps1;k=1000*(xb-xa);dx=(xb-xa)/k;
  for (j=0;j<=k;j++)
    {
    s=sin(xa+j*dx);ss=s*s;yy=1+(tt-1)*ss;a=4*yy*(tt-1)*(yy-nn)*ss;
    b=(1+8*yy*(-1+yy*(3-2*nn-2*yy*(1-nn))));f=a/b;if (j>0) g-=fac*(f+fa)*dx;fa=f;
    }
  xa=xr+eps1;xb=pi/2;k=1000*(xb-xa);dx=(xb-xa)/k;
  for (j=0;j<=k;j++)
    {
    s=sin(xa+j*dx);ss=s*s;yy=1+(tt-1)*ss;a=4*yy*(tt-1)*(yy-nn)*ss;
    b=(1+8*yy*(-1+yy*(3-2*nn-2*yy*(1-nn))));f=a/b;if (j>0) g-=fac*(f+fa)*dx;fa=f;
```

```
    }
  }
  else
  {
    ta=t;if (t>tr-eps) ta=tr-eps1;tt=ta*ta;xa=0;xb=pi/2;k=1000*(xb-xa);dx=(xb-xa)/k;g=0;
    for (j=0;j<=k;j++)
    {
      s=sin(xa+j*dx);ss=s*s;yy=nn+(tt-nn)*ss;a=(1-2*yy)*(1-2*yy)*(tt-nn)*ss;
      b=(1+8*yy*(-1+yy*(3-2*nn-2*yy*(1-nn))));f=a/b;if (j>0) g-=fac*(f+fa)*dx;fa=f;
    }
    for (j=0;j<=k;j++)
    {
      s=sin(xa+j*dx);ss=s*s;yy=1+(tt-1)*ss;a=4*yy*(tt-1)*(yy-nn)*ss;
      b=(1+8*yy*(-1+yy*(3-2*nn-2*yy*(1-nn))));f=a/b;if (j>0) g-=fac*(f+fa)*dx;fa=f;
    }
  }
  return(g);
}
```

13.2.5 Results

Some results of the computations are shown in Figs. 13.4, 13.5 and 13.6, for the
values $\nu = 0.00$, $\nu = 0.25$ and $\nu = 0.50$. The figures show the vertical displacements
as a function of the variable $c_s t/r$. The case $\nu = 0.25$ is the case for which Pekeris
(1955) obtained a closed form solution. The results of the present computations are
in good agreement with the original results of Pekeris.

The figures show that a first effect occurs when the compression wave arrives, and
somewhat larger displacements occur upon arrival of the shear wave, with a discontinuity in the slope of the curve. A singularity occurs upon arrival of the Rayleigh
wave, but after the passage of this wave the displacements remain constant. The val-

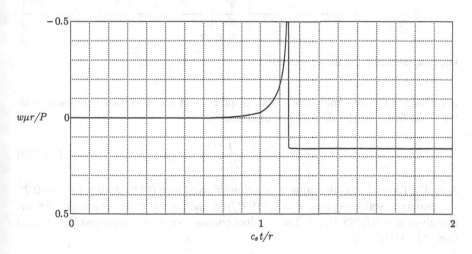

Fig. 13.4 Vertical displacement, $\nu = 0.00$

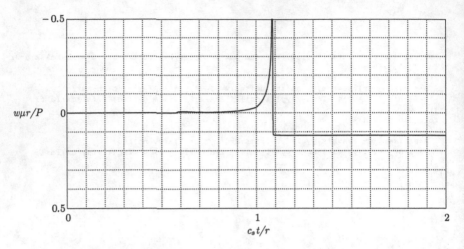

Fig. 13.5 Vertical displacement, $\nu = 0.25$

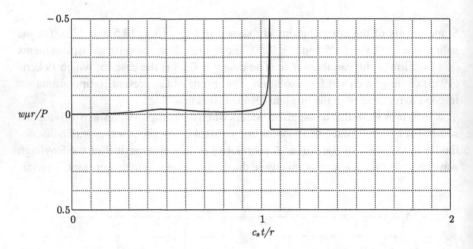

Fig. 13.6 Vertical displacement, $\nu = 0.50$

ues obtained for this final steady state displacement are in perfect agreement with the result from the classical theory of elasticity,

$$w_s = \frac{P(1-\nu)}{2\pi\mu r}.\tag{13.100}$$

Actually, for $\nu = 0.00$ the numerical solution gives $w = 0.159\ P/\mu r$, for $\nu = 0.25$ the numerical solution gives $w = 0.119\ P/\mu r$, and for $\nu = 0.50$ the numerical solution gives $w = 0.0795\ P/\mu r$. These values compare very well to the results obtained using (13.100).

Chapter 14
Moving Loads on an Elastic Half Plane

In this chapter some analytical solutions are derived for problems of vertical loads moving at constant speed over the upper boundary of the half plane $z > 0$. The material is isotropic linear elastic with quasi-viscous damping to represent hysteretic damping. The method used is a Fourier integral method (Sneddon, 1951). The analysis in this chapter is due to Verruijt and Cornejo Córdova (2001). The solution for the undamped case was given by Cole and Huth (1958). A large number of problems for moving loads, on beams, plates and half spaces, has been considered by Fryba (1999).

14.1 Moving Wave

In this section the problem of a moving sinusoidal wave on the half plane $z > 0$ is considered, for an isotropic elastic material with hysteretic damping. This will be used as the basic case for the more general case of a moving strip load or a moving point load. Hysteretic damping is defined as a special type of visco-elastic damping, the special property being that the damping ratio in each full cycle of loading is independent of the frequency of the loading (Hardin, 1965; Verruijt, 1999).

14.1.1 Basic Equations

The basic equations are the equations of motion,

$$\frac{\partial \sigma_{xx}}{\partial x} + \frac{\partial \sigma_{zx}}{\partial z} = \rho \frac{\partial^2 u}{\partial t^2}, \tag{14.1}$$

$$\frac{\partial \sigma_{xz}}{\partial x} + \frac{\partial \sigma_{zz}}{\partial z} = \rho \frac{\partial^2 w}{\partial t^2}. \tag{14.2}$$

A. Verruijt, *An Introduction to Soil Dynamics*,
Theory and Applications of Transport in Porous Media 24,
© Springer Science+Business Media B.V. 2010

For a linear visco-elastic material the stresses are related to the displacements by the following relations.

$$\sigma_{xx} = \lambda\left(\frac{\partial u}{\partial x} + \frac{\partial w}{\partial z}\right) + \lambda t_r \frac{\partial}{\partial t}\left(\frac{\partial u}{\partial x} + \frac{\partial w}{\partial z}\right) + 2\mu\frac{\partial u}{\partial x} + 2\mu t_r \frac{\partial}{\partial t}\frac{\partial u}{\partial x}, \qquad (14.3)$$

$$\sigma_{zz} = \lambda\left(\frac{\partial u}{\partial x} + \frac{\partial w}{\partial z}\right) + \lambda t_r \frac{\partial}{\partial t}\left(\frac{\partial u}{\partial x} + \frac{\partial w}{\partial z}\right) + 2\mu\frac{\partial w}{\partial z} + 2\mu t_r \frac{\partial}{\partial t}\frac{\partial w}{\partial z}, \qquad (14.4)$$

$$\sigma_{xz} = \mu\left(\frac{\partial u}{\partial z} + \frac{\partial w}{\partial x}\right) + \mu t_r \frac{\partial}{\partial t}\left(\frac{\partial u}{\partial z} + \frac{\partial w}{\partial x}\right), \qquad (14.5)$$

where λ and μ are the Lamé constants of the material, and t_r is a relaxation time. In order to describe hysteretic damping the value of t_r should be inversely proportional to the frequency of the loading.

Substitution of (14.3)–(14.5) into (14.1) and (14.2) leads to the basic differential equations

$$(\lambda + \mu)\frac{\partial}{\partial x}\left(\frac{\partial u}{\partial x} + \frac{\partial w}{\partial z}\right) + \mu\left(\frac{\partial^2 u}{\partial x^2} + \frac{\partial^2 u}{\partial z^2}\right)$$

$$+ (\lambda + \mu)t_r \frac{\partial}{\partial t}\frac{\partial}{\partial x}\left(\frac{\partial u}{\partial x} + \frac{\partial w}{\partial z}\right) + \mu t_r \frac{\partial}{\partial t}\left(\frac{\partial^2 u}{\partial x^2} + \frac{\partial^2 u}{\partial z^2}\right) = \rho\frac{\partial^2 u}{\partial t^2}, \qquad (14.6)$$

$$(\lambda + \mu)\frac{\partial}{\partial z}\left(\frac{\partial u}{\partial x} + \frac{\partial w}{\partial z}\right) + \mu\left(\frac{\partial^2 w}{\partial x^2} + \frac{\partial^2 w}{\partial z^2}\right)$$

$$+ (\lambda + \mu)t_r \frac{\partial}{\partial t}\frac{\partial}{\partial z}\left(\frac{\partial u}{\partial x} + \frac{\partial w}{\partial z}\right) + \mu t_r \frac{\partial}{\partial t}\left(\frac{\partial^2 w}{\partial x^2} + \frac{\partial^2 w}{\partial z^2}\right) = \rho\frac{\partial^2 w}{\partial t^2}. \qquad (14.7)$$

These equations can also be written as

$$(\lambda + 2\mu)\frac{\partial^2 u}{\partial x^2} + (\lambda + \mu)\frac{\partial^2 w}{\partial z\partial x} + \mu\frac{\partial^2 u}{\partial z^2}$$

$$+ (\lambda + 2\mu)t_r \frac{\partial^3 u}{\partial x^2\partial t} + (\lambda + \mu)t_r \frac{\partial^3 w}{\partial z\partial x\partial t} + \mu t_r \frac{\partial^3 u}{\partial z^2\partial t} = \rho\frac{\partial^2 u}{\partial t^2}, \qquad (14.8)$$

$$(\lambda + 2\mu)\frac{\partial^2 w}{\partial z^2} + (\lambda + \mu)\frac{\partial^2 u}{\partial z\partial x} + \mu\frac{\partial^2 w}{\partial x^2}$$

$$+ (\lambda + 2\mu)t_r \frac{\partial^3 w}{\partial z^2\partial t} + (\lambda + \mu)t_r \frac{\partial^3 u}{\partial x\partial z\partial t} + \mu t_r \frac{\partial^3 w}{\partial x^2\partial t} = \rho\frac{\partial^2 w}{\partial t^2}. \qquad (14.9)$$

It is assumed that the problem is to determine stresses and displacements in the half plane $z > 0$, subject to the boundary conditions

$$z = 0 : \sigma_{zx} = 0, \qquad (14.10)$$

$$z = 0 : \sigma_{zz} = -p_0 \exp[i\alpha(x - vt)]. \qquad (14.11)$$

These boundary conditions express that the half plane is loaded by a wave load normal to the surface, moving at a speed v, in positive x-direction. Actually only the real part of the boundary condition applies, because the stress σ_{zz} is a real quantity. This means that the real part of all quantities should be taken to obtain physically meaningful results.

14.1.2 Solutions

The solutions are assumed to be of the following form,

$$\alpha u = A \exp[i\alpha(x - vt)] \exp(-a\alpha z), \tag{14.12}$$

$$\alpha w = B \exp[i\alpha(x - vt)] \exp(-a\alpha z), \tag{14.13}$$

where α is a given positive real constant, and the (complex) constant a is unknown. It is assumed that its real part is positive, so that the solution will vanish for $z \to \infty$. It is assumed that the imaginary part of a is negative, so that waves will propagate in positive z-direction (this is Rayleigh's *radiation condition*). If $a = p - iq$ the solution will contain a factor of the form $\exp[i\alpha(qz - vt)]$, where q is a positive number. This ensures that for a fixed value of x the wave is propagated in positive z-direction.

The unknown coefficients A and B in general are complex. A factor α has been added to the variables so that the constants A and B will be dimensionless.

The derivatives needed in the expressions for the stresses are as follows.

$$\frac{\partial u}{\partial x} = i A \exp[i\alpha(x - vt)] \exp(-a\alpha z), \tag{14.14}$$

$$\frac{\partial u}{\partial z} = -Aa \exp[i\alpha(x - vt)] \exp(-a\alpha z), \tag{14.15}$$

$$\frac{\partial w}{\partial x} = i B \exp[i\alpha(x - vt)] \exp(-a\alpha z), \tag{14.16}$$

$$\frac{\partial w}{\partial z} = -Ba \exp[i\alpha(x - vt)] \exp(-a\alpha z), \tag{14.17}$$

$$\frac{\partial^2 u}{\partial x \partial t} = A\alpha v \exp[i\alpha(x - vt)] \exp(-a\alpha z), \tag{14.18}$$

$$\frac{\partial^2 u}{\partial z \partial t} = i A a \alpha v \exp[i\alpha(x - vt)] \exp(-a\alpha z), \tag{14.19}$$

$$\frac{\partial^2 w}{\partial x \partial t} = B\alpha v \exp[i\alpha(x - vt)] \exp(-a\alpha z), \tag{14.20}$$

$$\frac{\partial^2 w}{\partial z \partial t} = i B a \alpha v \exp[i\alpha(x - vt)] \exp(-a\alpha z). \tag{14.21}$$

The second and third order derivatives needed in the basic differential equations (14.6) and (14.7) are as follows.

$$\frac{\partial^2 u}{\partial x^2} = -A\alpha \exp[i\alpha(x - vt)] \exp(-a\alpha z), \qquad (14.22)$$

$$\frac{\partial^2 u}{\partial z^2} = Aa^2\alpha \exp[i\alpha(x - vt)] \exp(-a\alpha z), \qquad (14.23)$$

$$\frac{\partial^2 w}{\partial x^2} = -B\alpha \exp[i\alpha(x - vt)] \exp(-a\alpha z), \qquad (14.24)$$

$$\frac{\partial^2 w}{\partial z^2} = Ba^2\alpha \exp[i\alpha(x - vt)] \exp(-a\alpha z), \qquad (14.25)$$

$$\frac{\partial^2 u}{\partial x \partial z} = -i A a\alpha \exp[i\alpha(x - vt)] \exp(-a\alpha z), \qquad (14.26)$$

$$\frac{\partial^2 w}{\partial x \partial z} = -i B a\alpha \exp[i\alpha(x - vt)] \exp(-a\alpha z), \qquad (14.27)$$

$$\frac{\partial^3 u}{\partial x^2 \partial t} = i A\alpha^2 v \exp[i\alpha(x - vt)] \exp(-a\alpha z), \qquad (14.28)$$

$$\frac{\partial^3 u}{\partial z^2 \partial t} = -i A a^2\alpha^2 v \exp[i\alpha(x - vt)] \exp(-a\alpha z), \qquad (14.29)$$

$$\frac{\partial^3 w}{\partial x^2 \partial t} = i B\alpha^2 v \exp[i\alpha(x - vt)] \exp(-a\alpha z), \qquad (14.30)$$

$$\frac{\partial^3 w}{\partial z^2 \partial t} = -i B a^2\alpha^2 v \exp[i\alpha(x - vt)] \exp(-a\alpha z), \qquad (14.31)$$

$$\frac{\partial^3 u}{\partial x \partial z \partial t} = -A a\alpha^2 v \exp[i\alpha(x - vt)] \exp(-a\alpha z), \qquad (14.32)$$

$$\frac{\partial^3 w}{\partial x \partial z \partial t} = -B a\alpha^2 v \exp[\alpha(x - vt)] \exp(-a\alpha z). \qquad (14.33)$$

Substitution of these expressions into (14.8) and (14.9) now leads to the following system of equations for the determination of the constants A and B,

$$\{[(\lambda + 2\mu) - \mu a^2 - \rho v^2] - 2i\zeta[(\lambda + 2\mu) - \mu a^2]\} A$$
$$+ i(\lambda + \mu)(1 - 2i\zeta)aB = 0, \qquad (14.34)$$

$$i(\lambda + \mu)(1 - 2i\zeta)aA$$
$$+ \{[\mu - (\lambda + 2\mu)a^2 - \rho v^2] - 2i\zeta[\mu - (\lambda + 2\mu)a^2]\} B = 0, \qquad (14.35)$$

where the damping factor ζ is defined by

$$2\zeta = \alpha v t_r. \qquad (14.36)$$

In order to represent hysteretic damping, rather than visco-elastic damping, the product of the parameters $\omega = \alpha v$ and t_r should be considered as a constant. Thus the parameter ζ is an independent material parameter. This means that the relaxation time t_r must be inversely proportional to the frequency $\omega = \alpha v$.

Equations (14.34) and (14.35) can be somewhat simplified by introducing the parameters

$$\eta^2 = \frac{\mu}{\lambda + 2\mu} = \frac{1 - 2v}{2(1 - v)} = \frac{c_s^2}{c_p^2}, \tag{14.37}$$

$$\xi^2 = \frac{\rho v^2}{\mu} = \frac{v^2}{c_s^2}. \tag{14.38}$$

Here c_p and c_s are the propagation velocities of compression waves and shear waves in the elastic material. The system of (14.34) and (14.35) now is

$$\{(1 - \eta^2 a^2)(1 - 2i\zeta) - \eta^2 \xi^2\}A + i(1 - \eta^2)(1 - 2i\zeta)aB = 0, \tag{14.39}$$

$$i(1 - \eta^2)(1 - 2i\zeta)aA + \{(\eta^2 - a^2)(1 - 2i\zeta) - \eta^2 \xi^2\}B = 0. \tag{14.40}$$

It may be noted that the matrix of this system of equations is symmetric.

The system of equations has a non-zero solution only if the determinant Δ is zero. This determinant is defined as

$$\Delta = \begin{vmatrix} (1 - \eta^2 a^2)(1 - 2i\zeta) - \eta^2 \xi^2 & i(1 - \eta^2)(1 - 2i\zeta)a \\ i(1 - \eta^2)(1 - 2i\zeta)a & (\eta^2 - a^2)(1 - 2i\zeta) - \eta^2 \xi^2 \end{vmatrix} \tag{14.41}$$

or, after some elementary elaboration,

$$\Delta = (1 - a^2)^2(1 - 2i\zeta)^2 - (1 + \eta^2)(1 - a^2)(1 - 2i\zeta)\xi^2 + \eta^2 \xi^4. \tag{14.42}$$

The condition that this must be zero leads to the possible roots

$$1 - a_1^2 = \frac{\xi^2}{1 - 2i\zeta}, \qquad 1 - a_2^2 = \frac{\eta^2 \xi^2}{1 - 2i\zeta}. \tag{14.43}$$

For the undamped case ($\zeta = 0$) the roots are real, $1 - a_1^2 = \xi^2$ and $1 - a_2^2 = \eta^2 \xi^2$, in agreement with the known results for this case (Cole and Huth, 1958).

14.1.3 Solution 1

The first solution is determined by the root

$$a_1^2 = 1 - \frac{\xi^2}{1 - 2i\zeta} = \frac{(1 - \xi^2 + 4\zeta^2) - 2i\zeta\xi^2}{1 + 4\zeta^2}. \tag{14.44}$$

This is written as

$$a_1^2 = R_1^2 \exp(-2i\theta_1), \tag{14.45}$$

where R_1 and θ_1 are determined by the equations

$$R_1^4 = \frac{(1 - \xi^2 + 4\zeta^2)^2 + 4\zeta^2\xi^4}{(1 + 4\zeta^2)^2}, \tag{14.46}$$

$$2\theta_1 = \arctan\left(\frac{2\zeta\xi^2}{1 - \xi^2 + 4\zeta^2}\right). \tag{14.47}$$

It is assumed that the angle $2\theta_1$ lies in the interval

$$0 \le 2\theta_1 < \pi. \tag{14.48}$$

It may be noted that this definition of the function $\arctan(x)$ is different from the usual definition of the principal value. The present definition has been used so that the real part of a_1 is always positive and the imaginary part of a_1 is always negative. This ensures that the solution vanishes for $z \to \infty$, and that the waves are going out, from the boundary towards infinity (this is a form of the radiation condition). It may be noted that the present definition of the interval also ensures that $2\theta_1$ is continuous when ξ varies between 0 and ∞. If $2\theta_1$ would be defined in the interval $-\pi/2 < 2\theta_1 < +\pi/2$, the value of $2\theta_1$ would jump from $\pi/2$ to $-\pi/2$ when ξ^2 passes the value $1 + 4\zeta^2$.

The value of a_1 now is

$$a_1 = R_1 \exp(-i\theta_1) = p_1 - iq_1, \tag{14.49}$$

where

$$p_1 = R_1 \cos(\theta_1), \tag{14.50}$$

$$q_1 = R_1 \sin(\theta_1). \tag{14.51}$$

The value of p_1, the real part of a_1, is always positive because of the definition (14.48). This ensures that the solution will vanish for $z \to \infty$, as required. The value of q_1 is also always positive, so that all waves will travel towards infinity.

The values of the real and imaginary parts of a_1 are shown graphically in Fig. 14.1, for relatively small values of ζ. It may be noted that the point $\xi = 0$, $\zeta = 0$ corresponds to $a_1 = 1$. The point $\xi = 1$, $\zeta = 0$ corresponds to $a_1 = 0$, and the point $\xi = 2$, $\zeta = 0$ corresponds to $a_1 = -i\sqrt{3}$. The real part of a_1 is always positive, and the imaginary part is always negative.

Substitution of the value of a_1^2 into either of (14.39) or (14.40) shows that the constants A_1 and B are related by

$$A_1 = -ia_1 B_1 = -i(p_1 - iq_1)B_1 = -(q_1 + ip_1)B_1. \tag{14.52}$$

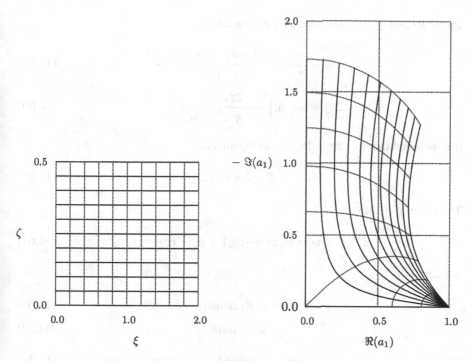

Fig. 14.1 First root a_1 as a function of ξ and ζ, for small values of ζ

Thus the first solution is

$$\alpha u = -ia_1 B_1 \exp[i\alpha(x - vt)] \exp(-a_1\alpha z), \tag{14.53}$$

$$\alpha w = B_1 \exp[i\alpha(x - vt)] \exp(-a_1\alpha z). \tag{14.54}$$

The solution can also be written as

$$\alpha u = -(q_1 + ip_1)B_1 \exp[i\alpha(x + q_1 z - vt)] \exp(-p_1\alpha z), \tag{14.55}$$

$$\alpha w = B_1 \exp[i\alpha(x + q_1 z - vt)] \exp(-p_1\alpha z). \tag{14.56}$$

14.1.4 Solution 2

The second solution is determined by the root

$$a_2^2 = 1 - \frac{\eta^2\xi^2}{1 - 2i\zeta} = \frac{(1 - \eta^2\xi^2 + 4\zeta^2) - 2i\zeta\,\eta^2\xi^2}{1 + 4\zeta^2}. \tag{14.57}$$

This is written as

$$a_2^2 = R_2^2 \exp(-2i\theta_2), \tag{14.58}$$

where R_2 and θ_2 are determined by the equations

$$R_2^4 = \frac{(1 - \eta^2\xi^2 + 4\zeta^2)^2 + 4\zeta^2\eta^4\xi^4}{(1 + 4\zeta^2)^2}, \tag{14.59}$$

$$2\theta_2 = \arctan\left(\frac{2\zeta\eta^2\xi^2}{1 - \eta^2\xi^2 + 4\zeta^2}\right). \tag{14.60}$$

It is assumed that the angle $2\theta_2$ lies in the interval

$$0 \leq 2\theta_2 < \pi. \tag{14.61}$$

The value of a_2 now is

$$a_2 = R_2\exp(-i\theta_2) = p_2 - iq_2, \tag{14.62}$$

where

$$p_2 = R_2\cos(\theta_2), \tag{14.63}$$

$$q_2 = R_2\sin(\theta_2). \tag{14.64}$$

The values of p_2 and q_2 are always positive because of the definition (14.61). This ensures that the solution will vanish for $z \to \infty$, and that the waves will be propagated towards infinity, as required.

The values of the real and imaginary parts of a_2 are shown graphically in Fig. 14.2, for relatively large values of ζ. This is actually the same figure as Fig. 14.1, except that ξ must be replaced by $\eta\xi$. It may be noted that the point $\xi = 0$, $\zeta = 0$ corresponds to $a_2 = 1$. For large values of the damping parameter ζ the values of a_2 all approach the point $a_2 = 1$. The real part of a_2 is always positive, and the imaginary part is always negative.

Substitution of the value of a_2^2 into either of (14.39) or (14.40) shows that the constants A_2 and B_2 are related by

$$B_2 = ia_2A_2 = i(p_2 - iq_2)A_2 = (q_2 + ip_2)A_2. \tag{14.65}$$

Thus the second solution is

$$\alpha u = A_2\exp[i\alpha(x - vt)]\exp(-a_2\alpha z), \tag{14.66}$$

$$\alpha w = ia_2A_2\exp[i\alpha(x - vt)]\exp(-a_2\alpha z). \tag{14.67}$$

This solution can also be written as

$$\alpha u = A_2\exp[i\alpha(x + q_2z - vt)]\exp(-p_2\alpha z), \tag{14.68}$$

$$\alpha w = (q_2 + ip_2)A_2\exp[i\alpha(x + q_2z - vt)]\exp(-p_2\alpha z). \tag{14.69}$$

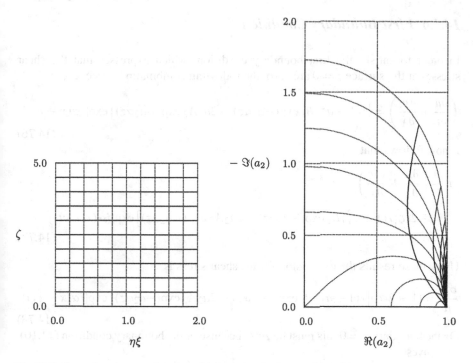

Fig. 14.2 Second root a_2 as a function of ξ/m and ζ, for large values of ζ

14.1.5 Completion of the Solution

By addition of the two possible solutions the general solution is obtained,

$$\alpha u = \left[-ia_1 B_1 \exp(-a_1\alpha z) + A_2 \exp(-a_2\alpha z)\right]\exp[i\alpha(x - vt)], \quad (14.70)$$

$$\alpha w = \left[B_1 \exp(-a_1\alpha z) + ia_2 A_2 \exp(-a_2\alpha z)\right]\exp[i\alpha(x - vt)]. \quad (14.71)$$

The first order derivatives of these functions are

$$\frac{\partial u}{\partial x} = \left[a_1 B_1 \exp(-a_1\alpha z) + i A_2 \exp(-a_2\alpha z)\right]\exp[i\alpha(x - vt)], \quad (14.72)$$

$$\frac{\partial u}{\partial z} = \left[ia_1^2 B_1 \exp(-a_1\alpha z) - a_2 A_2 \exp(-a_2\alpha z)\right]\exp[i\alpha(x - vt)], \quad (14.73)$$

$$\frac{\partial w}{\partial x} = \left[i B_1 \exp(-a_1\alpha z) - a_2 A_2 \exp(-a_2\alpha z)\right]\exp[i\alpha(x - vt)], \quad (14.74)$$

$$\frac{\partial w}{\partial z} = \left[-a_1 B_1 \exp(-a_1\alpha z) - ia_2^2 A_2 \exp(-a_2\alpha z)\right]\exp[i\alpha(x - vt)]. \quad (14.75)$$

It may be noted that the factor α, which was introduced in the general solutions for the displacements, (14.70) and (14.71), does not appear in the derivatives, which are dimensionless.

14.1.6 First Boundary Condition

In order to satisfy the first boundary condition, which expresses that the shear stresses at the surface $z = 0$ are zero, the following combination is needed,

$$\left(\frac{\partial u}{\partial z} + \frac{\partial w}{\partial x}\right) = \left[i(1 + a_1^2)B_1 \exp(-a_1\alpha z) - 2a_2 A_2 \exp(-a_2\alpha z)\right] \exp[i\alpha(x - vt)].$$
(14.76)

It now follows that

$$t_r \frac{\partial}{\partial t}\left(\frac{\partial u}{\partial z} + \frac{\partial w}{\partial x}\right)$$

$$= -2i\zeta\left[i(1 + a_1^2)B_1 \exp(-a_1\alpha z) - 2a_2 A_2 \exp(-a_2\alpha z)\right] \exp[i\alpha(x - vt)].$$
(14.77)

Using these results the expression for the shear stress is

$$\frac{\sigma_{zx}}{\mu} = (1 - 2i\zeta)\left[i(1 + a_1^2)B_1 \exp(-a_1\alpha z) - 2a_2 A_2 \exp(-a_2\alpha z)\right] \exp[i\alpha(x - vt)].$$
(14.78)

On the boundary $z = 0$ this must be zero, because of the boundary condition (14.10). This gives

$$2a_2 A_2 = i(1 + a_1^2)B_1.$$
(14.79)

This enables to write the expression for the shear stress, (14.78), in a slightly simpler form,

$$\frac{\sigma_{zx}}{\mu} = 2a_2 A_2(1 - 2i\zeta)\left[\exp(-a_1\alpha z) - \exp(-a_2\alpha z)\right] \exp[i\alpha(x - vt)]. \quad (14.80)$$

Written in this form it can immediately be seen that the boundary condition of zero shear stress at the upper boundary $z = 0$ is indeed satisfied.

14.1.7 Second Boundary Condition

The second boundary condition refers to the vertical normal stress σ_{zz}, which is prescribed along the boundary $z = 0$, see (14.11),

$$z = 0 : \sigma_{zz} = -p_0 \exp[i\alpha(x - vt)].$$
(14.81)

This stress can be expressed into the displacement components u and w by (14.4), which can also be written as

$$\frac{\sigma_{zz}}{\mu} = \frac{1 - 2\eta^2}{\eta^2}\left(\frac{\partial u}{\partial x} + \frac{\partial w}{\partial z}\right) + \frac{1 - 2\eta^2}{\eta^2} t_r \frac{\partial}{\partial t}\left(\frac{\partial u}{\partial x} + \frac{\partial w}{\partial z}\right) + 2\frac{\partial w}{\partial z} + 2t_r \frac{\partial}{\partial t}\frac{\partial w}{\partial z}.$$
(14.82)

The first term (the volume strain) is, from (14.72) and (14.75),

$$\left(\frac{\partial u}{\partial x} + \frac{\partial w}{\partial z}\right) = i(1 - a_2^2)A_2 \exp(-a_2\alpha z)\exp[i\alpha(x - vt)]. \tag{14.83}$$

It may be noted that the volume strain due to the first solution, with the coefficient B_1, is zero. This suggests that the first solution represents the shear waves.

It follows from (14.83) that

$$t_r \frac{\partial}{\partial t}\left(\frac{\partial u}{\partial x} + \frac{\partial w}{\partial z}\right) = 2\zeta(1 - a_2^2)A_2 \exp(-a_2\alpha z)\exp[i\alpha(x - vt)]. \tag{14.84}$$

The value of $\partial w/\partial z$ has been given in (14.75),

$$\frac{\partial w}{\partial z} = \left[-a_1 B_1 \exp(-a_1\alpha z) - ia_2^2 A_2 \exp(-a_2\alpha z)\right]\exp[i\alpha(x - vt)]. \tag{14.85}$$

Differentiation with respect to t gives

$$t_r \frac{\partial^2 w}{\partial z\partial t} = 2i\zeta\left[a_1 B_1 \exp(-a_1\alpha z) + ia_2^2 A_2 \exp(-a_2\alpha z)\right]\exp[i\alpha(x - vt)]. \tag{14.86}$$

Using these results the expression for the vertical normal stress is

$$\frac{\sigma_{zz}}{\mu} = (1 - 2i\zeta)$$
$$\times \left[i(-2 + 1/\eta^2 - a_2^2/\eta^2)A_2 \exp(-a_2\alpha z) - 2a_1 B_1 \exp(-a_1\alpha z)\right]$$
$$\times \exp[i\alpha(x - vt)], \tag{14.87}$$

or, because it follows from the definition of the roots a_1 and a_2, see (14.43), that $1/\eta^2 - a_2^2/\eta^2 = 1 - a_1^2$,

$$\frac{\sigma_{zz}}{\mu} = -(1 - 2i\zeta)\left[i(1 + a_1^2)A_2 \exp(-a_2\alpha z) + 2a_1 B_1 \exp(-a_1\alpha z)\right]\exp[i\alpha(x - vt)]. \tag{14.88}$$

The boundary condition (14.81) now leads to the equation

$$i(1 + a_1^2)A_2 + 2a_1 B_1 = \frac{p_0}{\mu(1 - 2i\zeta)}, \tag{14.89}$$

which is the second relation between the two constants A_2 and B_1, the first relation being (14.79).

Equations (14.89) makes it possible to write the expression for the vertical normal stress, (14.88), in a slightly simpler form,

$$\frac{\sigma_{zz}}{\mu} = -\frac{p_0}{\mu}\exp(-a_2\alpha z)\exp[i\alpha(x - vt)]$$
$$+ 2a_1(1 - 2i\zeta)B_1\left[\exp(-a_2\alpha z) - \exp(-a_1\alpha z)\right]\exp[i\alpha(x - vt)]. \tag{14.90}$$

Written in this form it can immediately be seen that the boundary condition for the vertical normal stress at the upper boundary $z = 0$ is indeed satisfied.

14.1.8 The Two Constants

The two constants A_2 and B_1 can be determined from (14.79) and (14.89). This gives

$$A_2 = -\frac{p_0}{\mu(1-2i\zeta)} \frac{i(1+a_1^2)}{(1+a_1^2)^2 - 4a_1a_2}, \tag{14.91}$$

$$B_1 = -\frac{p_0}{\mu(1-2i\zeta)} \frac{2a_2}{(1+a_1^2)^2 - 4a_1a_2}. \tag{14.92}$$

This completes the solution of the problem for a traveling wave load.

14.1.9 Final Solution

The final expressions for the displacements are

$$\alpha u = \frac{ip_0}{\mu(1-2i\zeta)} \frac{2a_1a_2 \exp(-a_1\alpha z) - (1+a_1^2)\exp(-a_2\alpha z)}{(1+a_1^2)^2 - 4a_1a_2} \exp[i\alpha(x-vt)], \tag{14.93}$$

$$\alpha w = -\frac{p_0}{\mu(1-2i\zeta)} \frac{2a_2 \exp(-a_1\alpha z) - a_2(1+a_1^2)\exp(-a_2\alpha z)}{(1+a_1^2)^2 - 4a_1a_2} \exp[i\alpha(x-vt)]. \tag{14.94}$$

The final expressions for the stresses are found to be

$$\frac{\sigma_{xx}}{p_0} = \frac{(1+a_1^2)(1-a_1^2+2a_2^2)\exp(-a_2\alpha z) - 4a_1a_2 \exp(-a_1\alpha z)}{(1+a_1^2)^2 - 4a_1a_2} \exp[i\alpha(x-vt)], \tag{14.95}$$

$$\frac{\sigma_{zz}}{p_0} = -\frac{(1+a_1^2)^2 \exp(-a_2\alpha z) - 4a_1a_2 \exp(-a_1\alpha z)}{(1+a_1^2)^2 - 4a_1a_2} \exp[i\alpha(x-vt)], \tag{14.96}$$

$$\frac{\sigma_{xz}}{p_0} = \frac{2ia_2(1+a_1^2)[\exp(-a_2\alpha z) - \exp(-a_1\alpha z)]}{(1+a_1^2)^2 - 4a_1a_2} \exp[i\alpha(x-vt)]. \tag{14.97}$$

The isotropic stress $\sigma_0 = \frac{1}{2}(\sigma_{xx} + \sigma_{zz})$ is of a relatively simple form,

$$\frac{\sigma_0}{p_0} = \frac{(1+a_1^2)(a_2^2 - a_1^2)\exp(-a_2\alpha z)}{(1+a_1^2)^2 - 4a_1a_2} \exp[i\alpha(x-vt)]. \tag{14.98}$$

This equation can also be derived immediately from the volume strain, of course.

14.1.10 The Displacement of the Origin

Of particular interest is the vertical displacement of the surface $z = 0$ below the center of the load, for $x = vt$. With (14.94) this gives

$$\alpha w_d = -\frac{p_0}{\mu(1 - 2i\zeta)} \frac{a_2(1 - a_1^2)}{(1 + a_1^2)^2 - 4a_1 a_2}. \tag{14.99}$$

It follows from the definition of a_1, see (14.44), that $1 - a_1^2 = \xi^2/(1 - 2i\zeta)$, hence

$$\alpha w_d = -\frac{p_0}{\mu(1 - 2i\zeta)^2} \frac{a_2 \xi^2}{(1 + a_1^2)^2 - 4a_1 a_2}. \tag{14.100}$$

The actual value of the displacement is the real part of this expression. This means that the amplitude of the displacement is determined by the absolute value $|w_d|$. This can be expressed in dimensionless form as

$$\frac{\alpha \mu |w_d|}{p_0} = \frac{1}{|1 - 2i\zeta|^2} \frac{|a_2| \xi^2}{|(1 + a_1^2)^2 - 4a_1 a_2|}. \tag{14.101}$$

All quantities in the right hand side of this equation can be expressed in terms of parameters introduced before. In fact,

$$|1 - 2i\zeta|^2 = 1 + 4\zeta^2, \tag{14.102}$$

$$|a_2| = R_2, \tag{14.103}$$

$$|(1 + a_1^2)^2 - 4a_1 a_2| = \sqrt{P^2 + Q^2}, \tag{14.104}$$

where

$$P = (1 + p_1^2 - q_1^2)^2 - 4p_1^2 q_1^2 - 4p_1 p_2 + 4q_1 q_2, \tag{14.105}$$

$$Q = 4(p_1 q_2 + p_2 q_1 - p_1 q_1 (1 + p_1^2 - q_1^2). \tag{14.106}$$

The parameters in these expressions have been defined in Sects. 1.3 and 1.4.

The amplitude of the vertical displacement is shown as a function of the dimensionless velocity of the wave load v/c_s in Fig. 14.3, for $v = 0$ and four values of the damping ratio ζ. The amplitude is shown as a ratio to the static value, obtained for $v/c_s \to 0$. It appears that there is a definite peak in the displacements, which corresponds to the velocity of the Rayleigh wave in the undamped case. Actually, for $v = 0$ the velocity of the Rayleigh wave is $c_r = 0.874c_s$. The magnitude of the peak depends very much upon the value of the damping ratio ζ. A damping ratio $\zeta = 0.1$ reduces the peak value to about three times the static value.

Fig. 14.3 Moving wave load, dynamic amplification factor, $\nu = 0$

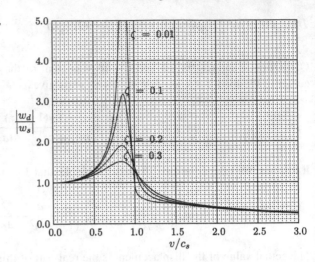

14.2 Moving Strip Load

The solution for a moving strip load can be derived from the previous solution using Fourier transforms. The boundary condition is supposed to be

$$z = 0 \; : \; \sigma_{zz} = \begin{cases} -p_0, & \text{if } |x - vt| < b, \\ 0, & \text{if } |x - vt| > b. \end{cases} \tag{14.107}$$

This boundary condition can also be written as

$$z = 0 \; : \; \sigma_{zz} = -\frac{2p_0}{\pi} \int_0^\infty \frac{\sin(\alpha b) \cos[\alpha(x - vt)]}{\alpha} \, d\alpha. \tag{14.108}$$

Comparing this with the boundary condition for the previous case, see (14.11), and remembering that the real part of this boundary condition and of the solution should be considered only, it follows that the solution of the problem for a moving strip load (say $F(x, z, t)$) can be obtained from the solution of the problem for a moving wave load (say $f(x, z, t, \alpha)$) by multiplication of the real part of the solution by a factor

$$\frac{2}{\pi} \frac{\sin(\alpha b)}{\alpha} \, d\alpha,$$

and then integrating from $\alpha = 0$ to $\alpha = \infty$. Hence

$$F(x, z, t) = \frac{2}{\pi} \int_0^\infty \frac{\sin(\alpha b) \Re\{f(x, z, t, \alpha)\}}{\alpha} \, d\alpha, \tag{14.109}$$

or, because only the function $f(x, y, z, t)$ can be complex,

$$F(x, z, t) = \frac{2}{\pi} \Re \int_0^\infty \frac{\sin(\alpha b) f(x, z, t, \alpha)}{\alpha} \, d\alpha. \tag{14.110}$$

This is simply an application of the Fourier integral, of course. The same result could have been obtained by using the Fourier transform method (Sneddon, 1951). It may be noted that in the formulation used here the parameter α is always positive.

14.2.1 Vertical Displacement of the Surface

Application of the Fourier integral (14.110) to the expression (14.100) for the vertical displacements gives, taking into account that the real part should be taken,

$$w = -\frac{2p_0}{\pi \mu} \Re \int_0^\infty \frac{a_2[2\exp(-a_1\alpha z) - (1 + a_1^2)\exp(-a_2\alpha z)]}{\alpha^2(1 - 2i\zeta)[(1 + a_1^2)^2 - 4a_1a_2]}$$

$$\times \sin(\alpha b)\exp[i\alpha(x - vt)]\,d\alpha. \tag{14.111}$$

Because the coefficients a_1 and a_2 depend upon the velocity factor ξ $(= v/c_s)$ and the damping ratio ζ, but not on α, this can also be written as

$$w = -\frac{p_0}{\mu}\Re\left\{\frac{a_2[2I_1 - (1 + a_1^2)I_2]}{(1 - 2i\zeta)[(1 + a_1^2)^2 - 4a_1a_2]}\right\}, \tag{14.112}$$

where

$$I_1 = \frac{2}{\pi}\int_0^\infty \frac{\sin(\alpha b)\exp\{\alpha[i(x - vt) - a_1z]\}}{\alpha^2}\,d\alpha, \tag{14.113}$$

$$I_2 = \frac{2}{\pi}\int_0^\infty \frac{\sin(\alpha b)\exp\{\alpha[i(x - vt) - a_2z]\}}{\alpha^2}\,d\alpha. \tag{14.114}$$

Although these integrals are of a relatively simple form, they can not be evaluated in analytic form, because of the singularity for $\alpha = 0$. This is a well known property of elastic problems for a half plane with a load on its surface. The displacements can not be determined uniquely, and the solution will have a logarithmic singularity. It is possible, however, to consider some special properties of the solution, by considering some special points, for instance.

As an example one may consider the displacements of the upper surface $z = 0$. Then the two integrals (14.113) and (14.114) are equal,

$$I_0 = I_1 = I_2 = \frac{2}{\pi}\int_0^\infty \frac{\sin(\alpha b)\exp[i\alpha(x - vt)]}{\alpha^2}\,d\alpha, \tag{14.115}$$

and the displacement of the surface is, using the definition of a_1 in (14.44),

$$w_d = -\frac{p_0}{\mu}\Re\left\{\frac{a_2\xi^2}{(1 - 2i\zeta)^2[(1 + a_1^2)^2 - 4a_1a_2]}I_0\right\}. \tag{14.116}$$

Fig. 14.4 Moving strip load, dynamic amplification factor, $\nu = 0.3333$

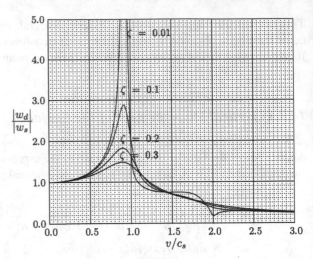

As before, the integral does not converge, but it is independent of ξ and ζ. Of particular interest is the amplitude of the displacement. Because the displacement will always be a sinusoidal function of time, it follows that

$$|w_d| = \frac{p_0}{\mu} \frac{|a_2|\xi^2}{|1 - 2i\zeta|^2|(1 + a_1^2)^2 - 4a_1a_2|} |I_0|. \tag{14.117}$$

The static value is obtained by letting $\xi \to 0$. This gives

$$|w_s| = \frac{p_0}{2\mu|1 - 2i\zeta|} \frac{1}{1 - \eta^2} |I_0| = \frac{p_0(1 - \nu)}{\mu\sqrt{1 + 4\zeta^2}} |I_0|. \tag{14.118}$$

The ratio of the dynamic to the static displacement now is

$$\frac{|w_d|}{|w_s|} = \frac{|a_2|\xi^2}{(1 - \nu)\sqrt{1 + 4\zeta^2}|(1 + a_1^2)^2 - 4a_1a_2|}. \tag{14.119}$$

The dynamic amplification factor is shown in graphical form in Fig. 14.4, for $\nu = \frac{1}{3}$, and four values of the damping factor ζ. Actually, this is the same relationship as for the wave loads, shown in Fig. 14.3. These figures differ only because the value of ν is different.

14.2.2 Vertical Normal Stress

Application of the Fourier integral (14.110) to the expression (14.96) for the vertical normal stress gives, taking into account that the real part should be taken,

$$\frac{\sigma_{zz}}{p_0} = -\Re\left\{\frac{(1 + a_1^2)^2(J_1 + iJ_2) - 4a_1a_2(J_3 + iJ_4)}{(1 + a_1^2)^2 - 4a_1a_2}\right\}, \tag{14.120}$$

where

$$J_1 = \frac{2}{\pi} \int_0^\infty \frac{\sin(\alpha b)\cos[\alpha(x - vt + q_2 z)]}{\alpha} \exp(-p_2\alpha z)\, d\alpha, \quad (14.121)$$

$$J_2 = \frac{2}{\pi} \int_0^\infty \frac{\sin(\alpha b)\sin[\alpha(x - vt + q_2 z)]}{\alpha} \exp(-p_2\alpha z)\, d\alpha, \quad (14.122)$$

$$J_3 = \frac{2}{\pi} \int_0^\infty \frac{\sin(\alpha b)\cos[\alpha(x - vt + q_1 z)]}{\alpha} \exp(-p_1\alpha z)\, d\alpha, \quad (14.123)$$

$$J_4 = \frac{2}{\pi} \int_0^\infty \frac{\sin(\alpha b)\sin[\alpha(x - vt + q_1 z)]}{\alpha} \exp(-p_1\alpha z)\, d\alpha. \quad (14.124)$$

These integrals can be evaluated using the standard integrals

$$2\int_0^\infty \frac{\sin(\alpha x)\cos(\alpha y)}{\alpha} \exp(-\alpha z)\, d\alpha = \arctan\left\{\frac{y + x}{z}\right\} - \arctan\left\{\frac{y - x}{z}\right\},$$
$$(14.125)$$

$$2\int_0^\infty \frac{\sin(\alpha x)\sin(\alpha y)}{\alpha} \exp(-\alpha z)\, d\alpha = \frac{1}{2}\log\left\{\frac{z^2 + (x + y)^2}{z^2 + (x - y)^2}\right\}. \quad (14.126)$$

Using these integrals it follows that

$$J_1 = \frac{1}{\pi}\arctan\left\{\frac{x - vt + q_2 z + b}{p_2 z}\right\} - \frac{1}{\pi}\arctan\left\{\frac{x - vt + q_2 z - b}{p_2 z}\right\}, \quad (14.127)$$

$$J_2 = \frac{1}{2\pi}\log\left\{\frac{p_2^2 z^2 + (x - vt + q_2 z + b)^2}{p_2^2 z^2 + (x - vt + q_2 z - b)^2}\right\}, \quad (14.128)$$

$$J_3 = \frac{1}{\pi}\arctan\left\{\frac{x - vt + q_1 z + b}{p_1 z}\right\} - \frac{1}{\pi}\arctan\left\{\frac{x - vt + q_1 z - b}{p_1 z}\right\}, \quad (14.129)$$

$$J_4 = \frac{1}{2\pi}\log\left\{\frac{p_1^2 z^2 + (x - vt + q_1 z + b)^2}{p_1^2 z^2 + (x - vt + q_1 z - b)^2}\right\}. \quad (14.130)$$

It is now assumed that the width of the strip ($2b$) is very small, and the load p_0 is very large, so that the total load $P = 2p_0 b$ remains finite. The load then is a point load. Then the following approximations can be used,

$$\arctan(x + \varepsilon) - \arctan(x - \varepsilon) \approx \frac{2\varepsilon}{1 + x^2}, \quad (14.131)$$

$$\log\left\{\frac{1 + \varepsilon}{1 - \varepsilon}\right\} \approx 2\varepsilon. \quad (14.132)$$

The expressions (14.127)–(14.130) now become

$$J_1 = \frac{2b}{z}K_1 = \frac{2b}{z}\frac{1}{\pi}\frac{p_2}{p_2^2 + [(x - vt)/z + q_2]^2}, \quad (14.133)$$

$$J_2 = \frac{2b}{z} K_2 = \frac{2b}{z} \frac{1}{\pi} \frac{(x - vt)/z + q_2}{p_2^2 + [(x - vt)/z + q_2]^2}, \tag{14.134}$$

$$J_3 = \frac{2b}{z} K_3 = \frac{2b}{z} \frac{1}{\pi} \frac{p_1}{p_1^2 + [(x - vt)/z + q_1]^2}, \tag{14.135}$$

$$J_4 = \frac{2b}{z} K_4 = \frac{2b}{z} \frac{1}{\pi} \frac{(x - vt)/z + q_1}{p_1^2 + [(x - vt)/z + q_1]^2}. \tag{14.136}$$

We now write, by definition of the factors C and D, which will depend upon ξ and ζ,

$$\frac{(1 + a_1^2)^2}{(1 + a_1^2)^2 - 4a_1a_2} = \frac{1}{2} + C + iD. \tag{14.137}$$

Then

$$\frac{-4a_1a_2}{(1 + a_1^2)^2 - 4a_1a_2} = \frac{1}{2} - C - iD, \tag{14.138}$$

because the sum of these two quantities must be 1. Now taking into account that the stress σ_{zz} is the real part of (14.120), it follows that

$$-\frac{\sigma_{zz} z}{P} = \frac{1}{2}(K_1 + K_3) + C(K_1 - K_3) - D(K_2 - K_4), \tag{14.139}$$

where P is the total load, $P = 2p_0b$, and the functions $K_1 - K_4$ have been defined in (14.133)–(14.136).

Equation (14.139) is the final expression for the vertical normal stress. The functions K_1–K_4 can easily be expressed in terms of the constants defined before, and the variable $(x - vt)/z$. The constants C and D can also be expressed in the constants defined before, as will be shown below.

It follows from (14.137) that

$$C + iD = \frac{(1 + a_1^2)^2 + 4a_1a_2}{2[(1 + a_1^2)^2 - 4a_1a_2]}, \tag{14.140}$$

which enables to determine C and D. Actually, one may write

$$(1 + a_1^2)^2 = f_1 + ig_1 = [(1 + p_1^2 - q_1^2)^2 - 4p_1^2q_1^2] + i[-4p_1q_1(1 + p_1^2 - q_1^2)], \tag{14.141}$$

$$4a_1a_2 = f_2 + ig_2 = [4(p_1p_2 - q_1q_2)] + i[-4(p_1q_2 + p_2q_1)], \tag{14.142}$$

so that

$$C + iD = \frac{f_1 + ig_1 + f_2 + ig_2}{2(f_1 + ig_1 - f_2 - ig_2)} = \frac{(f_1 + f_2) + i(g_1 + g_2)}{2[(f_1 - f_2) + i(g_1 - g_2)]}. \tag{14.143}$$

Fig. 14.5 Moving point load, vertical normal stress, $v = 0.0$, $\zeta = 0.01$

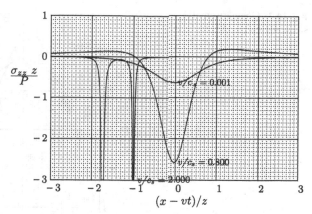

Fig. 14.6 Moving point load, vertical normal stress, $v = 0.0$, $\zeta = 0.1$

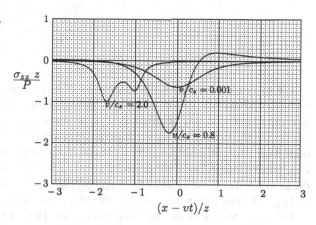

This means that

$$C = \frac{(f_1 + f_2)(f_1 - f_2) + (g_1 + g_2)(g_1 - g_2)}{2[(f_1 - f_2)^2 + (g_1 - g_2)^2]},$$ (14.144)

$$D = \frac{(f_1 - f_2)(g_1 + g_2) - (f_1 + f_2)(g_1 - g_2)}{2[(f_1 - f_2)^2 + (g_1 - g_2)^2]}.$$ (14.145)

These expressions enable to evaluate numerical values of the vertical stress.

The distribution of the vertical normal stress is shown in graphical form in Figs. 14.5 and 14.6 for $v = 0$ and for two values of the damping ratio: $\zeta = 0.01$, $\zeta = 0.1$, respectively.

The two figures show the stress distributions for three values of the velocity, $v/c_s = 0.001$, $v/c_s = 0.8$ and $v/c_s = 2.0$. The case $v/c_s = 0.001$ can be considered to be very close to the static case. The stress distribution for this case is in agreement with the classical Flamant solution (Timoshenko and Goodier, 1970), as is indeed the case in both figures. The maximum value of the stress in this case occurs for $(x - vt)/z = 0$, and its theoretical value is $\sigma_{zz}z/P = 2/\pi = 0.637$. The values that can be read from the two Figs. 14.5 and 14.6 are close to the theoretical value.

Fig. 14.7 Comparison of
solutions, $\nu = 0.333333$

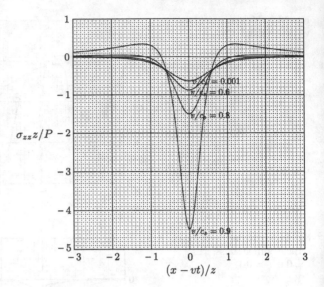

For the case $v/c_s = 0.8$ the results for $\zeta = 0.01$ are in reasonable agreement
with the undamped solution given by Cole and Huth (1958), see also Fryba (1999).
A comparison with this solution will be presented in some more detail later. The
location of the two pulses for the supersonic case $v/c_s = 2.0$ are also in good agreement with the results obtained by Cole and Huth for the undamped problem. Actu-
ally, the two singularities are given by $(x - vt) = -|a_s|z$ and $(x - vt) = -|a_p|z$. In
the undamped case the values of $|a_s|$ and $|a_p|$ are, for $\nu = 0$ and $v/c_s = 2$: $|a_s| = \sqrt{3}$
and $a_p = 1$, which is in agreement with the location of the peaks in Fig. 14.5.

For larger values of the damping ratio ζ, see Fig. 14.6, the results indicate that
the effect of a moderate amount of damping is sufficient to limit the maximum
stresses to the level of the static stresses. This is an important result. It means that the
damping properties of soils eliminate the extreme peaks in the stresses that occur in
an elastic material without internal damping. This does not mean that it is advisable
to construct infrastructure for moving loads on soft soils. Peaks in the stresses may
be avoided by the damping of the material, but the displacements will be very large,
because the material is so soft.

Figure 14.7 shows a comparison of the present solution for a very small value of
the damping ratio with the solution by Cole and Huth (1958) for a purely elastic material, without damping. The left half of the figure indicates the solution of Cole and
Huth, the right half of the figure represents the present solution with $\zeta = 0.000001$.
The two solutions are indistinguishable, confirming that the present solution is a
proper generalization of the earlier solution by Cole and Huth (1958).

14.2.3 Isotropic Stress

Of the other stress components the isotropic stress $\sigma_0 = \frac{1}{2}(\sigma_{xx} + \sigma_{zz})$ is perhaps the
simplest one to evaluate. Application of the Fourier integral (14.110) to the expres-

sion (14.98) gives

$$\frac{\sigma_0}{p_0} = \Re\left\{\frac{(1+a_1^2)(a_2^2 - a_1^2)(J_1 + iJ_2)}{(1+a_1^2)^2 - 4a_1a_2}\right\}, \qquad (14.146)$$

where the integrals J_1 and J_2 have been defined in (14.121) and (14.122).

We now write, by definition of the constants E and F,

$$\frac{(1+a_1^2)(a_2^2 - a_1^2)}{(1+a_1^2)^2 - 4a_1a_2} = E + iF. \qquad (14.147)$$

Furthermore it is again assumed that the width of the loaded strip $(2b)$ is very small. In that case (14.146) reduces to

$$\frac{\sigma_{0z}}{P} = \Re\{(E + iF)(K_1 + iK_2)\} = EK_1 - FK_2. \qquad (14.148)$$

This is the final expression for the isotropic stress. The constants E and F can be calculated by elaborating the definition (14.147). This gives

$$E + iF = \frac{[(1+p_1^2 - q_1^2) - 2ip_1q_1][(p_2^2 - p_1^2 - q_2^2 + q_1^2) + 2i(p_1q_1 - p_2q_2)]}{(f_1 - f_2) + i(g_1 - g_2)}, \qquad (14.149)$$

from which it follows that

$$E = \frac{(us - vt)(f_1 - f_2) + (vs + ut)(g_1 - g_2)}{(f_1 - f_2)^2 + (g_1 - g_2)^2}, \qquad (14.150)$$

$$F = \frac{(vs + ut)(f_1 - f_2) - (us - vt)(g_1 - g_2)}{(f_1 - f_2)^2 + (g_1 - g_2)^2}, \qquad (14.151)$$

where

$$u = 1 + p_1^2 - q_1^2, \qquad (14.152)$$

$$v = -2p_1q_1, \qquad (14.153)$$

$$s = p_2^2 - p_1^2 - q_2^2 + q_1^2, \qquad (14.154)$$

$$t = 2p_1q_1 - 2p_2q_2. \qquad (14.155)$$

This enables to elaborate the isotropic stress.

Some examples are shown in Fig. 14.8, for a practically undamped material, and three values of the velocity. The pseudo-static case $v/c_s = 0.001$ is in agreement with the elastostatic solution (Timoshenko and Goodier, 1970).

14.2.4 Horizontal Normal Stress

The horizontal normal stress σ_{xx} can most simply be evaluated by noting that

$$\sigma_{xx} = 2\sigma_0 - \sigma_{zz}. \qquad (14.156)$$

Fig. 14.8 Moving point load, isotropic stress, $\nu = 0.0$, $\zeta = 0.01$

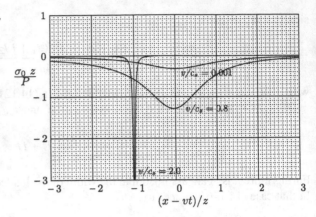

Fig. 14.9 Moving point load, horizontal normal stress, $\nu = 0.0$, $\zeta = 0.01$

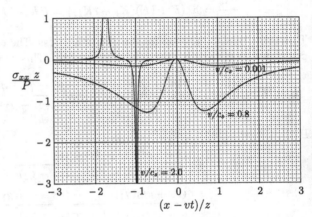

Some examples are shown in Fig. 14.9, for a practically undamped material, and three values of the velocity. As before, the pseudo-static case $v/c_s = 0.001$ is in agreement with the elastostatic solution (Timoshenko and Goodier, 1970).

14.2.5 Shear Stress

Application of the Fourier integral (14.110) to the expression (14.97) for the shear stress gives, taking into account that the real part should be taken,

$$\frac{\sigma_{xz}}{p_0} = \Re\left\{\frac{2ia_2(1 + a_1^2)(J_1 + iJ_2 - J_3 - iJ_4)}{(1 + a_1^2)^2 - 4a_1a_2}\right\}, \tag{14.157}$$

or, in the limiting case $b \to 0$,

$$\frac{\sigma_{xz}z}{P} = \Re\left\{\frac{2ia_2(1 + a_1^2)(K_1 + iK_2 - K_3 - iK_4)}{(1 + a_1^2)^2 - 4a_1a_2}\right\}. \tag{14.158}$$

Fig. 14.10 Moving point load, shear stress, $v = 0.0$, $\zeta = 0.01$

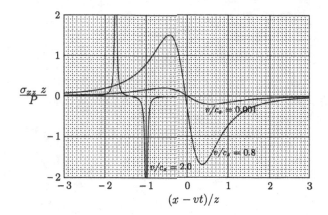

This can be elaborated by writing

$$\frac{2ia_2(1+a_1^2)}{(1+a_1^2)^2 - 4a_1a_2} = G + iH. \tag{14.159}$$

Then the final expression for the shear stress is

$$\frac{\sigma_{xz}z}{P} = G(K_1 - K_3) - H(K_2 - K_4). \tag{14.160}$$

The constants G and H can be calculated by elaborating the definition (14.159),

$$G + iH = \frac{2(q_2 + ip_2)[(1 + p_1^2 - q_1^2) - 2ip_1q_1]}{(f_1 - f_2) + i(g_1 - g_2)}. \tag{14.161}$$

It follows that

$$G = \frac{2(q_2u - p_2u)(f_1 - f_2) + 2(p_2u + q_2v)(g_1 - g_2)}{(f_1 - f_2)^2 + (g_1 - g_2)^2}, \tag{14.162}$$

$$H = \frac{2(p_2u + q_2v)(f_1 - f_2) - 2(q_2u - p_2v)(g_1 - g_2)}{(f_1 - f_2)^2 + (g_1 - g_2)^2}. \tag{14.163}$$

Some examples are shown in Fig. 14.10, for a practically undamped material, and three values of the velocity. As before, the pseudo-static case $v/c_s = 0.001$ is in agreement with the elastostatic solution (Timoshenko and Goodier, 1970).

Chapter 15
Foundation Vibrations

In this chapter the problem of propagation of vibration of waves in soils due to vibrating foundation elements is considered. The type of problem is illustrated in Fig. 15.1. The purpose of the discussions is not to derive rigorous theoretical solutions, but rather to describe practical methods of analysis, based upon theoretical solutions presented in earlier chapters, and in the literature, see e.g. Richart et al. (1970), Gazetas (1991).

15.1 Foundation Response

In this section the response of a footing on an elastic soil will be considered. The mass of the footing (perhaps including the mass of the machine causing the vibrations) is denoted by M. The response of the foundation mass depends upon the applied load and the soil reaction. If the contact pressure between the foundation mass and the soil is denoted by p, the equation of motion of the foundation mass is

$$F - pA = M\frac{d^2w}{dt^2}, \tag{15.1}$$

where A is the area of the footing, and w is the vertical displacement of the footing. It is assumed that the footing is completely rigid, and that it remains in contact with the soil at all times, so that the displacement of the footing w is equal to the displacement of the soil surface immediately beneath it. It is now assumed that the applied force, the soil reaction, and the displacement are all periodic, with a circular

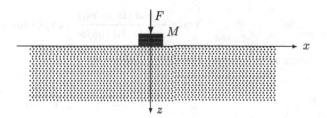

Fig. 15.1 Foundation element on soil

A. Verruijt, *An Introduction to Soil Dynamics*,
Theory and Applications of Transport in Porous Media 24,
© Springer Science+Business Media B.V. 2010

frequency ω,

$$F = \text{Re}[F_0 \exp(i\omega t)], \tag{15.2}$$

$$p = \text{Re}[p_0 \exp(i\omega t)], \tag{15.3}$$

$$w = \text{Re}[w_0 \exp(i\omega t)]. \tag{15.4}$$

Substitution of these equations into (15.1) gives

$$F_0 = p_0 A - \omega^2 M w_0. \tag{15.5}$$

In earlier chapters the response of the soil to a periodic load was generally found to be also periodic, but with a certain phase difference. In general one may therefore write

$$p_0 A = (K + i\omega C) w_0, \tag{15.6}$$

where the dynamic stiffness K and the dynamic damping C may depend upon the frequency ω, and upon parameters such as the shear modulus G, the soil density ρ and the dimensions of the foundation plate, for instance the radius of a circular plate a. Substitution of (15.6) into (15.5) gives

$$F_0 = (K + i\omega C - \omega^2 M) w_0. \tag{15.7}$$

This is the standard form of the differential equation for the response of a single mass system, supported by a spring of stiffness K and a damper having a viscosity C. This system has been considered in great detail in Chap. 1. All results obtained there, such as the occurrence of resonance at certain frequencies, and the influence of damping upon the maximum response, can be immediately applied to the present system, but taking into account that the stiffness K and the damping C may depend upon the frequency ω.

It remains to determine the dynamic parameters K and C for a particular system. This requires the solution of the response problem of the soil for that particular case.

Circular Footing

In Chap. 9 the problem of a circular footing of radius a on an elastic half space has been considered. For this case the relation between the amplitudes was found to be

$$w_0 = \frac{p_0 a \sin(\omega/\omega_0)}{m(\lambda + 2\mu)(\omega/\omega_0)} \exp(-i\omega/\omega_0), \tag{15.8}$$

where m is a material constant, defined by

$$m^2 = \frac{\mu}{\lambda + 2\mu} = \frac{1 - 2v}{2(1 - v)}, \tag{15.9}$$

and where ω_0 is a characteristic frequency, defined by

$$\omega_0^2 = \frac{4\mu}{\rho a^2}. \tag{15.10}$$

It should be noted that this is an approximate solution, derived under the assumption that the horizontal displacements are negligible, compared to the vertical displacements. In (15.6) the inverse relation of (15.8) is needed. This inverse relation is

$$p_0 = \frac{m(\lambda + 2\mu)(\omega/\omega_0)}{a \sin(\omega/\omega_0)} \exp(i\omega/\omega_0)\, w_0. \tag{15.11}$$

Comparing this with (15.6) shows that

$$K + i\omega C = \frac{m(\lambda + 2\mu)(\omega/\omega_0)}{a \sin(\omega/\omega_0)} \exp(i\omega/\omega_0)\, A. \tag{15.12}$$

With $A = \pi a^2$ it now follows that

$$K = m(\lambda + 2\mu)\pi a \frac{(\omega/\omega_0)}{\tan(\omega/\omega_0)}, \tag{15.13}$$

and

$$C = \frac{m(\lambda + 2\mu)\pi a}{\omega_0}. \tag{15.14}$$

These values will be used in the next section.

It is convenient to write the last term of (15.5), which represents the influence of the mass M, in a somewhat different form, which involves the mass of the soil. For this purpose the mass of the foundation M is expressed into a representative mass of the soil by writing

$$M = \frac{4\rho a^3}{1 - \nu} B = \frac{4\rho A a}{\pi(1 - \nu)} B, \tag{15.15}$$

where B is a dimensionless factor, the *mass ratio*. The form of this expression is suggested by the form of a similar factor introduced by Lysmer and Richart (1966). Because ω_0^2 can be expressed as

$$\omega_0^2 = \frac{4\mu}{\rho a^2} = \frac{4m^2(\lambda + 2\mu)}{\rho a^2}, \tag{15.16}$$

see (15.10), the mass M can also be written as

$$M = \frac{16 A m^2(\lambda + 2\mu)}{\pi a(1 - \nu)\omega_0^2} B. \tag{15.17}$$

Substitution of (15.17) and (15.12) into (15.7) finally gives the following relation between the amplitudes of the applied force and the displacement,

$$\frac{F_0}{A} = \frac{m(\lambda + 2\mu)}{a}\left[\frac{\omega}{\omega_0}\frac{\exp(i\omega/\omega_0)}{\sin(\omega/\omega_0)} - \frac{16mB}{\pi(1-\nu)}\left(\frac{\omega}{\omega_0}\right)^2\right]w_0. \qquad (15.18)$$

In the static case, with $\omega \to 0$, the value of $\sin(\omega/\omega_0)$ can be approximated by ω/ω_0, and one obtains

$$\frac{F_0}{A} = \frac{m(\lambda + 2\mu)}{a}w_s. \qquad (15.19)$$

The static displacement w_s is a reference value, which can be used to present the final results in dimensionless form. Using this reference value the dynamic response can be written as

$$\frac{w_0}{w_s} = \left[\frac{\omega}{\omega_0}\frac{\exp(i\omega/\omega_0)}{\sin(\omega/\omega_0)} - \frac{16mB}{\pi(1-\nu)}\left(\frac{\omega}{\omega_0}\right)^2\right]^{-1}. \qquad (15.20)$$

This expression gives the dynamic multiplication factor. For $\nu = 1/3$ the absolute value of the factor w_0/w_s is shown in Fig. 15.2, for various values of the mass ratio B. It appears that for sufficiently large values of the mass ratio B a certain resonance may occur, for frequencies in the order of magnitude $\frac{1}{4}\omega_0$. For frequencies large compared to ω_0 the amplitude of the dynamic vibrations tends towards zero.

It is interesting to also show the force on the soil, as a ratio of the force applied to the foundation mass. From (15.11) and (15.18) it follows that

$$\frac{p_0A}{F_0} = \left[1 - \frac{16mB}{\pi(1-\nu)}\left(\frac{\omega}{\omega_0}\right)\frac{\sin(\omega/\omega_0)}{\exp(i\omega/\omega_0)}\right]^{-1}. \qquad (15.21)$$

This function is shown in Fig. 15.3, for various values of the mass ratio B and for $\nu = 1/3$. When the foundation has no mass ($B = 0$) the entire force is transmitted to the soil, of course. For larger values of the mass ratio the force on the soil may be somewhat smaller than the applied force, because part of the force is used to move

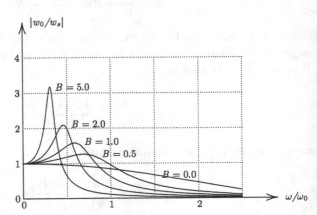

Fig. 15.2 Dynamic response of footing

Fig. 15.3 Force transmitted
to half space

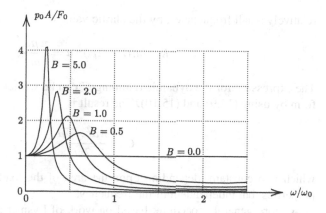

the foundation mass. It can be seen, however, that the force between foundation and
soil may also be considerably larger than the applied force. It may also be mentioned
that for very high frequencies the force transmitted to the soil may be extremely
large, at least in theory.

Although the results shown in Figs. 15.2 and 15.3 should be considered as in-
dicative only, because of the approximate character of the solution used to construct
them, it is interesting to note that the general shape of the functions shown in the
figure is very similar to the behaviour obtained by Lysmer and Richart (1966) for a
circular footing, which is based upon a more rigorous analysis. This gives support
to the results of the approximate analysis presented here.

15.2 Equivalent Spring and Damping

The procedure described in the previous section can easily be generalized, on the
basis of the response function of the soil to a surface load. All that is needed is to
write the relation in the form of (15.6),

$$p_0 A = (K + i\omega C) w_0. \tag{15.22}$$

For one particular case, namely the response of a quasi-elastic half space due to
a uniform load on a circular area, approximate relations have been derived in the
previous section. There it was found, see (15.13) and (15.14) that

$$K = m(\lambda + 2\mu)\pi a \frac{(\omega/\omega_0)}{\tan(\omega/\omega_0)}, \tag{15.23}$$

and

$$C = \frac{m(\lambda + 2\mu)\pi a}{\omega_0}. \tag{15.24}$$

The formula for the spring constant is a function of the frequency ω, but it is a
slowly varying function, so that it can be approximated reasonably well, at least for

relatively small frequencies, by the elastic value

$$K \approx m(\lambda + 2\mu)\pi a = \frac{\pi \mu a}{m}. \tag{15.25}$$

The expression for the dynamic damping C can be written in a somewhat different form by using (15.9) and (15.10). The result is

$$C = \frac{\pi a^2 \sqrt{\rho \mu}}{2m}, \tag{15.26}$$

which is a constant, depending upon the area of the loaded surface, and the soil properties, but independent of the frequency.

A more general procedure, based on work of Lysmer and Richart (1966) is as follows (see also Gazetas, 1991). The spring constant K is defined by the elastic deformation of the footing in static conditions. For a rigid circular plate of radius a, for instance, the relation between the applied load and the displacement is (Timoshenko and Goodier, 1970)

$$w = \frac{\pi(1 - v^2)pa}{2E}, \tag{15.27}$$

so that for this case

$$K = \frac{4\mu a}{1 - v}. \tag{15.28}$$

This closely resembles the expression (15.25), which is of course not very surprising, as they all express the stiffness of an elastic half space. Relations of this form are very common in soil dynamics, see for instance Richart et al. (1970).

The dynamic damping C can be obtained for various cases, at least as a first approximation, by trial and error, and curve fitting. A convenient choice, due to Lysmer, appears to be (Gazetas, 1991)

$$C = \frac{3.4 a^2 \sqrt{\rho \mu}}{1 - v}, \tag{15.29}$$

which closely resembles (15.26). Again, relations of this form are often used in soil dynamics.

The damping ratio ζ, which plays an important role in the analysis of dynamic systems, is usually defined as

$$\zeta = \frac{C}{2\sqrt{KM}}. \tag{15.30}$$

For the case of a rigid circular foundation plate, the mass ratio B is usually defined as follows (Richart et al., 1970, p. 204),

$$B = \frac{(1 - v)M}{4\rho a^3}. \tag{15.31}$$

With (15.13), (15.14) and (15.31), (15.30) can also be written as

$$\zeta = \frac{0.425}{\sqrt{B}}.\tag{15.32}$$

This shows that for a very small mass the damping ratio will be very large, because then the mass ratio is small. For a very large mass the damping ratio is small, because then the mass ratio is large. All this is due to the fact that the dynamic damping of the elastic half space, which is caused by radiation damping, is constant.

Expressions for equivalent dynamic stiffnesses K and for equivalent dynamic dampings C for various cases are given, for instance, by Gazetas (1991). These include foundations with various shapes of the contact area, footings loaded by lateral or rocking loads, and buried foundations.

15.3 Soil Properties

In order to determine the response of a foundation to vibrations the soil parameters needed are the density ρ, the shear modulus G, and the Poisson ratio ν. Of these the density can most easily be determined or estimated. Moreover, its variability is rather restricted: most soils have a density of about 1600 kg/m^3 when completely dry, and about 2000 kg/m^3 when completely saturated.

Poisson's ratio is usually not so easy to measure, or to estimate. Fortunately, its value does not influence the results very much. Common values are in the range from 0.3 (for sand) to 0.5 (for clays, or saturated soils).

The most important parameter is the shear modulus G. Its value may vary between fairly wide limits, and its influence on the results is very large. A complication is that the value of the shear modulus for natural soils depends very much on the magnitude of the shear strains. For very small strains the shear modulus may be a factor 10 or even 100 larger than it is for large strains. A typical example is shown in Fig. 15.4. Thus it is very important to know beforehand the order of magnitude of the shear strains.

The actual value of the shear modulus may be determined from laboratory tests, or from a field test. In field tests it is usually the propagation velocity of shear waves

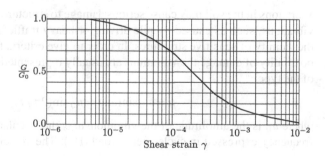

Fig. 15.4 Dynamic shear modulus

c_s that is measured. The shear modulus then follows from the formula

$$G = \rho^2 c_s. \tag{15.33}$$

Again it is of importance to note the dependence of the shear modulus of the shear strain level. A value measured using very small deformations may not be representative for a case in which large deformations can be expected.

15.4 Propagation of Vibrations

In previous chapters the propagation of vibrations in space has been investigated for various cases. These were restricted to linear elastic materials, in which the only damping is due to radiation. In real soils some damping may occur due to irreversible deformations (*material damping*). This is very difficult to estimate theoretically, but it may be quite significant, especially for large vibrations and soft soils. Theoretical solutions, which are available only for a few cases of linear elastic bodies, usually indicate that the amplitude of the vibrations decays with the radius r of the distance in the form r_0/r (for a wave from a cavity in an infinite body), or $\sqrt{r_0/r}$ (for Rayleigh waves on the surface of a half space).

In order to include the effect of material damping as well, in addition to the radiation damping, Bornitz (1931) has suggested to use a formula of the type

$$\frac{w}{w_0} = \sqrt{\frac{r_0}{r}} \exp(-\alpha(r - r_0)). \tag{15.34}$$

The value of the parameter α may vary between $1/(30 \text{ m})$ for very stiff soils to $1/(6 \text{ m})$ for very soft soils (Barkan, 1962).

The best procedure to determine the value of α in engineering practice is to perform a field test, in which the attenuation of the response to the action of a vibrator (or, more simply, to a dropped weight) is measured at various distances from the source of the disturbance.

15.5 Design Criteria

Vibrations in the soil may cause serious damage to structures. This means that strong vibrations, such as caused by pile driving or heavy traffic, may not be allowable in the vicinity of sensitive structures. In order to give criteria for the assessment of the possibility of damage the vibrations are usually represented by a harmonic vibration of the type

$$u = u_0 \sin(\omega t) = u_0 \sin(2\pi f t), \tag{15.35}$$

where u_0 is the amplitude of the vibration, ω is the circular frequency, and f is the frequency expressed in cycles per second (Hz). The velocity corresponding to this vibration is

$$v = \omega u_0 \cos(\omega t) = 2\pi f u_0 \cos(2\pi f t), \tag{15.36}$$

and the acceleration is

$$a = -\omega^2 u_0 \sin(\omega t) = -4\pi^2 f^2 u_0 \sin(2\pi f t). \tag{15.37}$$

If the amplitude of the velocity is denoted by v_0 and the amplitude of the acceleration is denoted by a_0, it follows that

$$v_0 = 2\pi f u_0, \tag{15.38}$$

and

$$a_0 = 4\pi^2 f^2 u_0. \tag{15.39}$$

Design criteria are often expressed in terms of allowable velocities or allowable accelerations, as a function of the frequency. Such criteria can conveniently be represented graphically in a diagram such as shown in Fig. 15.5. In this diagram the relation between frequency, displacement and acceleration is given. The basic variables in the diagram are the frequency f, and the amplitude of the acceleration a_0. The frequency f is constant along vertical lines, and the acceleration is constant along horizontal lines. A point in which the acceleration is a^* corresponds to a displacement $a^*/(4\pi^2 f^2)$. This means that the displacement is constant along lines of constant a/f^2. These lines are shown in the diagram as lines with a slope 2:1. They give the displacement corresponding to a certain acceleration and a certain frequency. Similarly, lines of constant velocity would be lines with a slope 1:1. They are not shown in the figure. As an example one may consider the point on the lower horizontal axis along which the acceleration is $a_0 = 10^{-2}$ m/s, and the displacement is $u_0 = 10^{-2}$ m. With (15.39) the frequency is then found to be $f = 0.159$.

An alternative, even more convenient representation is shown in Fig. 15.6. In this figure the basic parameters are the frequency f and the amplitude of the velocity v_0.

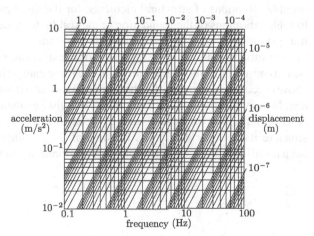

Fig. 15.5 Frequency, displacement and acceleration

Fig. 15.6 Frequency, displacement, velocity and acceleration

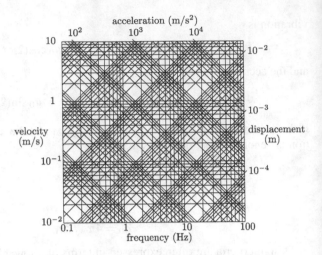

Lines of constant displacement $u_0 = c$ can now be represented by lines with a slope 1:1 because, with (15.38), $v_0 = 2\pi f u_0$. Similarly, lines of constant acceleration, say $a_0 = d$ can be represented by lines having a slope 1:1 in downward direction because, with (15.38) and (15.39), $v_0 = a_0/2\pi f$.

As an example, or a check, one may consider the point for which $f = 1$ and $v_0 = 10^{-1}$ m/s. In this case the corresponding displacement is, with (15.38), $u_0 = 0.0159$ m. This value is indeed indicated by the scale of displacements on the right, together with the upward sloping lines of constant displacement. The acceleration is found to be $a_0 = 0.628$ m/s^2, which is indicated by the scale of accelerations at the top of the figure, together with the downward sloping lines of constant acceleration.

Several countries have established design criteria in standards. Such a standard may, for instance, give a maximum velocity for various categories of buildings, distinguishing between newly constructed general purpose buildings, houses and masonry structures, and monuments or sensitive buildings. Normal values for such allowable velocities are 16 mm/s, 6 mm/s, and 3 mm/s. These values are the allowable vibrations of structural elements, for the three categories of buildings. Allowable vibrations of the foundation or the soil in its immediate vicinity, are usually much smaller, of the order of magnitude of 4 mm/s or 2 mm/s. For tall buildings the allowable vibrations at the top levels of the structure may be considerably larger (say 40 mm/s, 15 mm/s or 8 mm/s for the three categories of buildings mentioned above) because swaying of the structure may occur without causing structural damage. In very tall buildings (skyscrapers) the displacements and velocities may become so large that they cause severe discomfort for the residents, even though the structure itself is perfectly safe. In some of these buildings movable masses have been installed which counteract the natural vibrations, to reduce the discomfort.

Appendix A
Integral Transforms

In this appendix a brief review is given of some integral transform methods. These are techniques used to reduce a differential equation to an algebraic equation. The main transforms are the Laplace transform, the Fourier transform and the Hankel transform. These will be presented here, together with some of their main properties. Derivations of the theorems will be given in condensed form, or not at all. Complete derivations are given by Titchmarsh (1948), Sneddon (1951) and Churchill (1972). Extensive tables of transforms have been published by the staff of the Bateman project (Erdélyi et al., 1954).

Short tables of Laplace transforms, Fourier transforms and Hankel transforms are presented, with references to their derivation, and some numerical illustrations and verifications.

Finally, an elegant and effective method is described for the determination of the inverse Fourier-Laplace transform for certain problems, in particular problems of elastodynamics (De Hoop, 1960).

A.1 Laplace Transforms

A.1.1 Definitions

The Laplace transform is particularly useful for problems in which the variables are defined in a semi-infinite domain $0 < t < \infty$, where t may, for instance, be the time, and $t = 0$ indicates the initial value of time. The Laplace transform of a function $f(t)$ is defined as

$$F(s) = \int_0^\infty f(t) \exp(-st) \, dt, \tag{A.1}$$

where s is a parameter, which is assumed to be sufficiently large for the integral to exist. By the integration over the time domain, for various values of s, the function $f(t)$ is transformed into a function $F(s)$. For various functions the Laplace transform can be calculated, sometimes very easily, sometimes with considerable

Table A.1 Some Laplace transforms

No.	$f(t)$	$F(s) = \int_0^\infty f(t)\exp(-st)\,dt$
1	1	$\frac{1}{s}$
2	t	$\frac{1}{s^2}$
3	t^n	$\frac{n!}{s^{n+1}}$
4	$\exp(at)$	$\frac{1}{s-a}$
5	$\sin(at)$	$\frac{a}{s^2+a^2}$
6	$\cos(at)$	$\frac{s}{s^2+a^2}$
7	$\frac{\sin(at)}{t}$	$\arctan(\frac{a}{s})$

effort. Tables of such transforms are widely available (Churchill, 1972; Erdélyi et al., 1954). A short table is given in Table A.1. The integrals in this table can all be evaluated with little effort, using techniques such as partial integration.

The fundamental property of the Laplace transform appears when considering the transform of the time derivative. Using partial integration this is found to be

$$\int_0^\infty \frac{df(t)}{dt}\exp(-st)\,dt = sF(s) - f(0). \tag{A.2}$$

Thus differentiation with respect to time is transformed into multiplication by s, and subtraction of the initial value $f(0)$.

A.1.2 Example

In order to illustrate the application of the Laplace transform technique consider the differential equation

$$\frac{df(t)}{dt} + 2f = 0, \tag{A.3}$$

with the initial condition $f(0) = 5$. Using the property (A.2) the differential equation (A.3) is transformed into the algebraic equation

$$(s+2)F(s) - 5 = 0, \tag{A.4}$$

the solution of which is

$$F(s) = \frac{5}{s+2}. \tag{A.5}$$

Inverse transformation now gives, using transform no. 4 from Table A.1,

$$f(t) = 5\exp(-2t). \tag{A.6}$$

Substitution into the original differential equation (A.3) will show that this is indeed the correct solution, satisfying the given initial condition.

This example shows that the solution of the problem can be performed in a straightforward way. The main problem is the inverse transformation of the solution (A.5), which depends upon the availability of a sufficiently wide range of Laplace transforms. If the inverse transformation can not be found in a table of transforms it may be possible to use the general inverse transformation theorem (Churchill, 1972), but this requires considerable mathematical skill.

A.1.3 Heaviside's Expansion Theorem

A powerful inversion method is provided by the expansion theorem developed by Heaviside, one of the pioneers of the Laplace transform method. This applies to functions that can be written as a quotient of two polynomials,

$$F(s) = \frac{p(s)}{q(s)}, \tag{A.7}$$

where $q(s)$ must be a polynomial of higher order than $p(s)$. It is assumed that the function $q(s)$ possesses single zeroes only, so that it may be written as

$$q(s) = (s - s_1)(s - s_2) \cdots (s - s_n). \tag{A.8}$$

One may now write

$$F(s) = \frac{p(s)}{q(s)} = \frac{a_1}{s - s_1} + \frac{a_2}{s - s_2} + \cdots + \frac{a_n}{s - s_n}. \tag{A.9}$$

The coefficient a_i can be determined by multiplication of both sides of (A.9) by $(s - s_i)$, and then passing into the limit $s \to s_i$. This gives

$$a_i = \lim_{s \to s_i} \frac{(s - s_i)\, p(s)}{q(s)}. \tag{A.10}$$

Because $q(s_i) = 0$ the limit may be evaluated using L'Hôpital's rule, giving

$$a_i = \frac{p(s_i)}{q'(s_i)}. \tag{A.11}$$

Inverse transformation of the expression (A.9) now gives, using formula no. 4 from Table A.1,

$$f(t) = \sum_{i=1}^{n} \frac{p(s_i)}{q'(s_i)} \exp(s_i t). \tag{A.12}$$

This is Heaviside's expansion theorem. It provides a useful method to determine the inverse Laplace transform of functions of the form (A.7). It can also be used to determine the inverse transform of functions of a more general form, although such inverse transforms can usually be found in a more general way by application of the complex inversion integral (Churchill, 1972).

A.2 Fourier Transforms

A.2.1 Fourier Series

For certain partial differential equations the Fourier transform method can be used to derive solutions. These include problems of potential flow, and elasticity problems, especially in the case of problems for infinite regions, semi-infinite regions, or infinite strips. The main principles of the method will be presented in this section.

The main property of the Fourier transform can most easily be derived by first considering a Fourier series expansion. For this purpose let there be given a function $g(\theta)$, which is periodic with a period 2π, such that $g(\theta + 2\pi) = g(\theta)$. This function can be written as

$$g(\theta) = \frac{1}{2}A_0 + \sum_{k=1}^{\infty}\{A_k\cos(k\theta) + B_k\sin(k\theta)\},\qquad (A.13)$$

where

$$A_k = \frac{1}{\pi}\int_{-\pi}^{+\pi} g(t)\cos(kt)\,dt,\qquad (A.14)$$

and

$$B_k = \frac{1}{\pi}\int_{-\pi}^{+\pi} g(t)\sin(kt)\,dt,\qquad (A.15)$$

These formulas can be derived by multiplication of (A.13) by $\cos(j\theta)$ or $\sin(j\theta)$, and then integrating the result from $\theta = -\pi$ to $\theta = +\pi$. It will then appear that from the infinite series only one term is unequal to zero, namely for $k = j$. This leads to (A.14) and (A.15).

For a function with period $2\pi l$ the Fourier expansion can be obtained from (A.13) by replacing θ with x/l, t by t/l and then renaming $g(x/l)$ as $f(x)$. The result is

$$f(x) = \frac{1}{2}A_0 + \sum_{k=1}^{\infty}\{A_k\cos(kx/l) + B_k\sin(kx/l)\},\qquad (A.16)$$

where now

$$A_k = \frac{1}{\pi l}\int_{-\pi l}^{+\pi l} f(t)\cos(kt/l)\,dt,\qquad (A.17)$$

and

$$B_k = \frac{1}{\pi l} \int_{-\pi l}^{+\pi l} f(t) \sin(kt/l) \, dt. \qquad (A.18)$$

Example

As an example consider the block function defined by

$$f(x) = \begin{cases} 0, & |x| > \pi l/2, \\ 1, & |x| < \pi l/2. \end{cases} \qquad (A.19)$$

For this case the coefficients A_k and B_k can easily be calculated, using the expressions (A.17) and (A.18). The factors B_k are all zero, which is a consequence of the fact that the function $f(x)$ is even, $f(-x) = f(x)$. The factors A_k are equal to zero when k is even, and the uneven terms are proportional to $1/k$. The series (A.16) finally can be written as

$$f(x) = \frac{1}{2} + \frac{2}{\pi} \left\{ \cos\left(\frac{x}{l}\right) - \frac{1}{3} \cos\left(\frac{3x}{l}\right) + \frac{1}{5} \cos\left(\frac{5x}{l}\right) - \frac{1}{7} \cos\left(\frac{7x}{l}\right) + \cdots \right\}.$$

$$(A.20)$$

The first term of this series represents the average value of the function, the second term causes the main fluctuation, and the remaining terms together modify this first sinusoidal fluctuation into the block function.

Figure A.1 shows the approximation of the series (A.20) by its first 40 terms. It appears that the approximation is reasonably good, except very close to the discontinuities. This is a well known effect, often referred to as the *Gibbs* phenomenon (Weisstein, 1999). The approximation becomes better, of course, when more terms are taken into account, but the overshoot near the discontinuities remains.

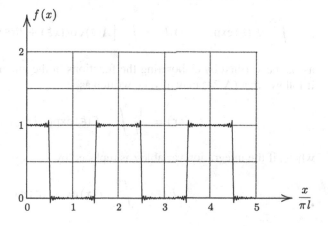

Fig. A.1 Fourier series, 40 terms

A.2.2 From Fourier Series to Fourier Integral

Substitution of (A.17) and (A.18) into (A.16) gives

$$f(x) = \frac{1}{2\pi l} \int_{-\pi l}^{+\pi l} f(t)\, dt + \sum_{k=1}^{\infty} f(t) \frac{1}{\pi l} \int_{-\pi l}^{+\pi l} f(t) \cos[k(t-x)/l]\, dt. \quad (A.21)$$

The interval can be made very large by writing $1/l = \Delta\xi$. Then (A.21) becomes

$$f(x) = \frac{\Delta\xi}{2\pi} \int_{-\pi/\Delta\xi}^{+\pi/\Delta\xi} f(t)\, dt + \sum_{k=1}^{\infty} f(t) \frac{\Delta\xi}{\pi} \int_{-\pi/\Delta\xi}^{+\pi/\Delta\xi} f(t) \cos[k\Delta\xi(t-x)]\, dt. $$
$$(A.22)$$

Writing $k\Delta\xi = \xi$ and letting $\Delta\xi \to 0$ this reduces to

$$f(x) = \frac{1}{\pi} \int_0^{\infty} d\xi \int_{-\infty}^{+\infty} f(t) \cos[\xi(x-t)]\, dt. \quad (A.23)$$

This can also be written as

$$f(x) = \frac{1}{2\pi} \int_0^{\infty} \big[A(\xi) \cos(x\xi) + B(\xi) \sin(x\xi) \big] d\xi, \quad (A.24)$$

where

$$A(\xi) = 2 \int_{-\infty}^{\infty} f(t) \cos(\xi t)\, dt, \quad (A.25)$$

and

$$B(\xi) = 2 \int_{-\infty}^{\infty} f(t) \sin(\xi t)\, dt. \quad (A.26)$$

It can be seen from (A.25) that $A(\xi)$ is an even function, $A(-\xi) = A(\xi)$, and from (A.26) it can be seen that $B(\xi)$ is uneven, $B(-\xi) = -B(\xi)$. Therefore, if $F(\xi) = \frac{1}{2}[A(\xi) + iB(\xi)]$, it follows that

$$\int_{-\infty}^{\infty} F(\xi) \exp(-ix\xi)\, d\xi = \int_0^{\infty} \big[A(\xi) \cos(x\xi) + B(\xi) \sin(x\xi) \big] d\xi, \quad (A.27)$$

as can be verified by elaborating the functions in the integral on the left hand side. It follows that (A.24) may also be written as

$$f(x) = \frac{1}{2\pi} \int_{-\infty}^{\infty} F(\xi) \exp(-ix\xi)\, d\xi, \quad (A.28)$$

where, if the integration variable t is replaced by x,

$$F(\xi) = \int_{-\infty}^{\infty} f(x) \exp(ix\xi)\, dx. \quad (A.29)$$

This is the basic formula of the Fourier transform method. The function $F(\xi)$ is called the Fourier transform of $f(x)$. It may be mentioned that the asymmetry of the formulas is often eliminated by writing a factor $1/\sqrt{2\pi}$ in each of the two integrals.

The main property of the Fourier transform appears when considering the Fourier transform of the second derivative $d^2 f/dx^2$. This is found to be, using partial integration,

$$\int_{-\infty}^{\infty} \frac{d^2 f}{dx^2} \exp(ix\xi)\, dx = -\xi^2\, F(y), \tag{A.30}$$

if it is assumed that $f(x)$ and its derivative df/dx tend towards zero for $\xi \to -\infty$ and $\xi \to \infty$. Thus, under these conditions, the second derivative is transformed into multiplication by $-\xi^2$.

When it is known that the function $f(x)$ is even, $f(-x) = f(x)$, one may write

$$f(x) = \frac{2}{\pi} \int_0^{\infty} F_c(\xi) \cos(x\xi)\, d\xi, \tag{A.31}$$

where now

$$F_c(\xi) = \int_0^{\infty} f(x) \cos(x\xi)\, dx. \tag{A.32}$$

The function $F_c(\xi)$ is called the Fourier cosine-transform of $f(x)$.

For uneven functions, $f(-x) = -f(x)$, the Fourier sine-transform may be used,

$$f(x) = \frac{2}{\pi} \int_0^{\infty} F_s(\xi) \sin(x\xi)\, d\xi, \tag{A.33}$$

where

$$F_s(\xi) = \int_0^{\infty} f(x) \sin(x\xi)\, dx. \tag{A.34}$$

Both for the Fourier cosine transform and for the Fourier sine transform various examples are given in the tables published by Churchill (1972) and Erdélyi et al. (1954).

A.2.3 Application

As an example consider the problem of potential flow in a half plane $y > 0$, see Fig. A.2. Along the upper boundary $y = 0$ the potential is given to be a step function. The differential equation is

$$\frac{\partial^2 f}{\partial x^2} + \frac{\partial^2 f}{\partial y^2} = 0, \tag{A.35}$$

Fig. A.2 Half plane

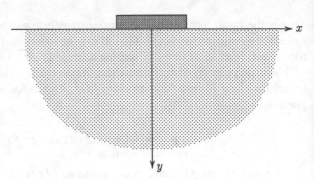

and the boundary condition is supposed to be

$$y = 0 : f = \begin{cases} 0, & |x| > a, \\ p, & |x| < a. \end{cases} \tag{A.36}$$

Because the boundary condition (A.36) is symmetric with respect to the y-axis, it can be expected that the solution will be even, and therefore the Fourier cosine transform (A.32) may be used. The transformed problem is, using (A.30),

$$-\xi^2 F_c + \frac{d F_c}{d y^2} = 0. \tag{A.37}$$

The solution of this ordinary differential equation that vanishes at infinity is

$$F_c = A(\xi) \exp(-\xi y). \tag{A.38}$$

From this it follows that the value at the surface $y = 0$ is

$$y = 0 : F_c = A(\xi). \tag{A.39}$$

The transformed boundary condition is, with (A.36) and (A.32),

$$y = 0 : F_c = \frac{2p}{\pi} \frac{\sin(\xi a)}{\xi}. \tag{A.40}$$

From (A.39) and (A.40) the integration factor $A(\xi)$ can be determined,

$$A(\xi) = \frac{2p}{\pi} \frac{\sin(\xi a)}{\xi}. \tag{A.41}$$

The final solution of the transformed problem is

$$F_c = \frac{2p}{\pi} \frac{\sin(\xi a)}{\xi} \exp(-\xi y). \tag{A.42}$$

The solution of the original problem can now be obtained by the inverse transform, (A.31),

$$f = \frac{2p}{\pi} \int_0^\infty \frac{\sin(\xi a)\cos(\xi x)}{\xi} \exp(-\xi y)\, d\xi. \tag{A.43}$$

Although this integral has been obtained as a Fourier integral, it can actually most easily be found in a table of Laplace transforms, because of the function $\exp(-\xi y)$ in the integral. In such tables the following integral may be found

$$\int_0^\infty \frac{\sin(at)}{t} \exp(-st)\, dt = \arctan\left(\frac{a}{s}\right). \tag{A.44}$$

Using this result, and some trigonometric relations to bring the integrand of (A.43) into the correct form to apply (A.44), the final solution of the problem considered here is found to be

$$f = \frac{p}{\pi} \arctan\left(\frac{a+x}{y}\right) + \frac{p}{\pi} \arctan\left(\frac{a-x}{y}\right). \tag{A.45}$$

It can easily be verified that this solution satisfies the differential equation (A.35) and the boundary condition (A.36). Thus the expression (A.45) is indeed the solution of the problem.

A.2.4 List of Fourier Transforms

In this section a number of Fourier transforms is listed, together with references or indications for their derivation. For some integrals a numerical verification is shown, using Simpson's numerical integration scheme. The numerical results confirm the analytical formulas.

In this section the Fourier cosine transform, see (A.32), is defined as

$$F_c(y) = \int_0^\infty f(x)\cos(xy)\, dx. \tag{A.46}$$

The inverse transform is, with (A.31),

$$f(x) = \frac{2}{\pi} \int_0^\infty F_c(y)\cos(xy)\, dy. \tag{A.47}$$

The Fourier sine transform, see (A.34), is defined as

$$F_s(y) = \int_0^\infty f(x)\sin(xy)\, dx. \tag{A.48}$$

The inverse transform is, with (A.33),

$$f(x) = \frac{2}{\pi} \int_0^\infty F_s(y)\sin(xy)\, dy. \tag{A.49}$$

It may be noted that in some publications the definitions (A.46) and (A.48) contain a factor $\sqrt{2/\pi}$. The inverse transforms then also contain this factor, so that the pair of transforms becomes symmetric. This is especially valuable when constructing a table of transforms.

A well known integral of the Laplace transform type (Churchill, 1972) is, when formulated as a Fourier cosine transform,

$$\int_0^\infty \exp(-xt)\cos(xy)\,dx = \frac{t}{y^2+t^2}. \tag{A.50}$$

Using the inversion theorem (A.47) it follows that

$$\int_0^\infty \frac{1}{x^2+t^2}\cos(xy)\,dx = \frac{\pi}{2t}\exp(-yt). \tag{A.51}$$

Differentiation of (A.51) with respect to t gives

$$\int_0^\infty \frac{1}{(x^2+t^2)^2}\cos(xy)\,dx = \frac{\pi}{4t^3}(1+yt)\exp(-yt). \tag{A.52}$$

Another well known integral of the Laplace transform type (Churchill, 1972) is, when formulated as a Fourier sine transform,

$$\int_0^\infty \exp(-xt)\sin(xy)\,dx = \frac{y}{y^2+t^2}. \tag{A.53}$$

The inverse form of this integral is, with (A.49),

$$\int_0^\infty \frac{x}{x^2+t^2}\sin(xy)\,dx = \frac{\pi}{2}\exp(-yt). \tag{A.54}$$

Differentiation of (A.54) with respect to t gives

$$\int_0^\infty \frac{x}{(x^2+t^2)^2}\sin(xy)\,dx = \frac{\pi y}{4t}\exp(-yt). \tag{A.55}$$

Differentiation of (A.53) with respect to y gives

$$\int_0^\infty x\exp(-xt)\cos(xy)\,dx = \frac{t^2-y^2}{(t^2+y^2)^2}. \tag{A.56}$$

Differentiation of (A.53) with respect to t gives

$$\int_0^\infty x\exp(-xt)\sin(xy)\,dx = \frac{2yt}{(y^2+t^2)^2}. \tag{A.57}$$

A well known discontinuous integral is (Titchmarsh, 1948, p. 177)

$$\int_0^\infty \frac{\sin(xt)}{x}\cos(xy)\,dx = \begin{cases} \frac{\pi}{2}, & y<t, \\ 0, & y>t. \end{cases} \tag{A.58}$$

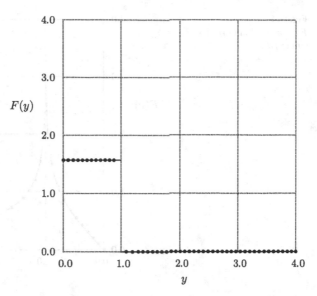

Fig. A.3
$F(y) = \int_0^\infty [\sin(x)/x] \times \cos(xy)\, dx$

A numerical verification of the integral (A.58) is shown in Fig. A.3. The analytical results, as given by (A.58), are indicated by the fully drawn lines. Some numerical results, calculated using Simpson's integration formula, are indicated by the dots. An integral similar to (A.58) is (Titchmarsh, 1948, p. 179)

$$\int_0^\infty \frac{\sin(xt)}{x} \sin(xy)\, dx = \frac{1}{2} \log \left| \frac{t+y}{t-y} \right|. \tag{A.59}$$

A comparison with the results of a numerical computation of this integral is shown in Fig. A.4. Again the analytical results are indicated by the fully drawn lines, and the numerical data are indicated by the dots.

Some integrals of the Weber-Schafheitlin type (Watson, 1944, p. 405) are

$$\int_0^\infty J_0(xt) \sin(xy)\, dx = \begin{cases} 0, & y < t, \\ (y^2 - t^2)^{-1/2}, & y > t. \end{cases} \tag{A.60}$$

$$\int_0^\infty J_0(xt) \cos(xy)\, dx = \begin{cases} (t^2 - y^2)^{-1/2}, & y < t, \\ 0, & y > t. \end{cases} \tag{A.61}$$

$$\int_0^\infty \frac{J_0(xt)}{x} \sin(xy)\, dx = \begin{cases} \arcsin(y/t), & y < t, \\ \frac{1}{2}\pi, & y > t. \end{cases} \tag{A.62}$$

It may be noted that (A.61) may be derived from (A.62) by differentiation with respect to the parameter y.

Fig. A.4
$F(y) = \int_0^\infty [\sin(x)/x] \times$
$\sin(xy)\,dx$

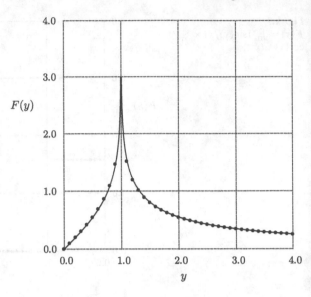

Differentiating (A.62) with respect to t gives

$$\int_0^\infty J_1(xt)\sin(xy)\,dx = \begin{cases} (y/t)(t^2 - y^2)^{-1/2}, & y < t, \\ 0, & y > t. \end{cases} \tag{A.63}$$

Finally, another useful Weber-Schafheitlin integral is (Watson, 1944, p. 405)

$$\int_0^\infty J_1(xt)\cos(xy)\,dx = \begin{cases} \frac{1}{t}, & y < t, \\ -\frac{t}{\sqrt{y^2 - t^2}(y + \sqrt{y^2 - t^2})}, & y > t. \end{cases} \tag{A.64}$$

A.3 Hankel Transforms

A.3.1 Definitions

For problems with radial symmetry a useful solution method is provided by the Hankel transform. This transform is defined by

$$F(y) = \int_0^\infty xf(x)J_0(xy)\,dx. \tag{A.65}$$

The inverse transform is

$$f(x) = \int_0^\infty yF(y)J_0(xy)\,dy. \tag{A.66}$$

For a derivation of this relation the reader is referred to the literature, see e.g. Sneddon (1951).

The main property of the Hankel transform is that it transforms the operator often appearing in radially symmetric problems into a simple multiplication,

$$\int_0^\infty \left[\frac{d^2 f}{dx^2} + \frac{1}{x}\frac{df}{dx}\right] x J_0(xy)\, dx = -y^2 F(y). \tag{A.67}$$

This property can be derived by using partial integration, and noting that the Bessel function $w = J_0(xy)$ satisfies the differential equation

$$\frac{d^2 w}{dx^2} + \frac{1}{x}\frac{dw}{dx} + y^2 w = 0. \tag{A.68}$$

Thus the combination $d^2 f/dx^2 + (1/x)\, df/dx$ is transformed into multiplication of the Hankel transform $F(y)$ by $-y^2$. This means that a differential equation in which this combination of derivatives appears may be transformed into an algebraic equation. In many cases this algebraic equation is relatively simple to solve, but the problem then remains to find the inverse transform. For the inverse transformation tables of transforms may be consulted, but if the tables do not give the inverse transform, it may be a formidable mathematical problem to derive it.

A.3.2 List of Hankel Transforms

In this section a number of Hankel transforms is listed, together with references or indications for their derivation. For some integrals a numerical verification is shown, using Simpson's numerical integration scheme. The numerical results confirm the analytical formulas.

A pair of integrals of the Weber-Schafheitlin type is (Abramowitz and Stegun, 1964, 11.4.33, 11.4.34)

$$\int_0^\infty \frac{J_1(xt)}{x} J_0(xy)\, dx = \begin{cases} \frac{2}{\pi} E(\frac{y^2}{t^2}), & y < t, \\ \frac{2y}{\pi t}\{E(\frac{t^2}{y^2}) - (1 - \frac{t^2}{y^2})K(\frac{t^2}{y^2})\}, & y > t. \end{cases} \tag{A.69}$$

In these equations the functions $K(x)$ and $E(x)$ are complete elliptic integrals of the first and second kind, respectively. A short list of values of these functions, adapted from Abramowitz and Stegun (1964) is given in Table A.2.

Two well known integrals of the Hankel transform type are (Sneddon, 1951, p. 528)

$$\int_0^\infty \frac{x}{(x^2 + t^2)^{1/2}} J_0(xy)\, dx = \frac{1}{y}\exp(-yt). \tag{A.70}$$

$$\int_0^\infty \frac{x}{(x^2 + t^2)^{3/2}} J_0(xy)\, dx = \frac{1}{t}\exp(-yt). \tag{A.71}$$

Table A.2 Complete elliptic integrals

x	$K(x)$	$E(x)$
0.0	1.57079	1.57079
0.1	1.61244	1.53076
0.2	1.65962	1.48903
0.3	1.71389	1.44536
0.4	1.77751	1.39939
0.5	1.85407	1.35064
0.6	1.94956	1.29842
0.7	2.07536	1.24167
0.8	2.25720	1.17848
0.9	2.57809	1.10477
1.0	∞	1.00000

Differentiation of (A.71) with respect to t gives

$$\int_0^\infty \frac{x}{(x^2+t^2)^{5/2}} J_0(xy)\,dx = \frac{1+yt}{3t^3}\exp(-yt). \tag{A.72}$$

Differentiating this again with respect to t gives

$$\int_0^\infty \frac{x}{(x^2+t^2)^{7/2}} J_0(xy)\,dx = \frac{3+3yt+(yt)^2}{15t^5}\exp(-yt). \tag{A.73}$$

The inverse form of (A.71) is

$$\int_0^\infty \exp(-xt)\,x\,J_0(xy)\,dx = \frac{t}{(y^2+t^2)^{3/2}}. \tag{A.74}$$

This integral can also be considered as a Laplace transform.
Integration of the integral (A.74) with respect to t gives

$$\int_0^\infty \exp(-xt)\,J_0(xy)\,dx = \frac{1}{(y^2+t^2)^{1/2}}. \tag{A.75}$$

Equation (A.75) is the inverse form of (A.70). It is a well known integral, which can also be considered as a Laplace transform, and can be found in many tables (see e.g. Churchill, 1972, p. 327).

A comparison of analytical and numerical computations of the integral (A.75) is shown in Fig. A.5.

The Fourier transforms (A.60)–(A.64) can also be considered as Hankel transforms. Written in the form of Hankel transforms these integrals are as follows.

$$\int_0^\infty \sin(xt)\,J_0(xy)\,dx = \begin{cases} (t^2-y^2)^{-1/2}, & y < t, \\ 0, & y > t. \end{cases} \tag{A.76}$$

Fig. A.5 $F(y) = \int_0^\infty \exp(-x) J_0(xy) \, dx$

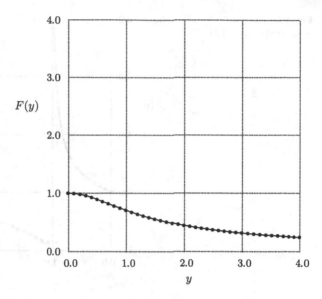

$$\int_0^\infty \cos(xt) \, J_0(xy) \, dx = \begin{cases} 0, & y < t, \\ (y^2 - t^2)^{-1/2}, & y > t. \end{cases} \quad \text{(A.77)}$$

$$\int_0^\infty \frac{\sin(xt)}{x} J_0(xy) \, dx = \begin{cases} \frac{1}{2}\pi, & y < t, \\ \arcsin(t/y), & y > t. \end{cases} \quad \text{(A.78)}$$

$$\int_0^\infty \sin(xt) \, J_1(xy) \, dx = \begin{cases} 0, & y < t, \\ (t/y)(y^2 - t^2)^{-1/2}, & y > t. \end{cases} \quad \text{(A.79)}$$

$$\int_0^\infty \cos(xt) \, J_1(xy) \, dx = \begin{cases} -\dfrac{y}{\sqrt{t^2 - y^2}\,(t + \sqrt{t^2 - y^2})}, & y < t, \\ \frac{1}{y}, & y > t. \end{cases} \quad \text{(A.80)}$$

A numerical verification of the integral (A.80) is shown in Fig. A.6.

A.4 De Hoop's Inversion Method

A.4.1 Introduction

An elegant method to determine the inverse Laplace-Fourier transform for certain problems was developed by De Hoop (1970). In this method the Laplace transform and the Fourier transform are used, and the integration path for the inverse Fourier transform is modified in such a way that the two inverse transforms can be traded off. Actually, the inverse Fourier integral is transformed so that it obtains the form of a Laplace transform. The inverse Laplace transform then simply is the function itself.

Fig. A.6 $F(y) =$
$-\int_0^\infty \cos(x) J_1(xy)\, dx$

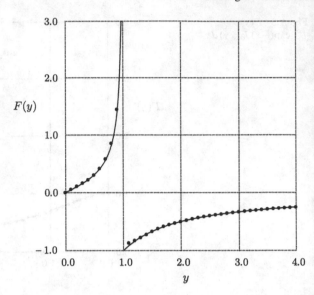

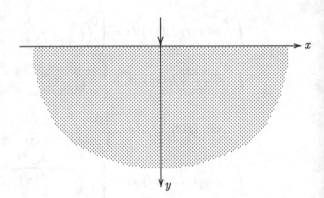

Fig. A.7 Half plane with
impulse load

The method is particularly useful for the solution of problems of elastodynamics.
It will be illustrated here by an example of general dynamics. Many references to
other applications are given by Duffy (1994).

A.4.2 Example

As an example consider a problem of dynamics for a half plane, see Fig. A.7, defined
by the partial differential equation

$$\frac{\partial^2 w}{\partial t^2} = c^2 \left\{ \frac{\partial^2 w}{\partial x^2} + \frac{\partial^2 w}{\partial y^2} \right\}, \tag{A.81}$$

where c is a given constant (the wave velocity), defined by $c^2 = \mu/\rho$, where μ is the elasticity of the material, and ρ its density. The differential equation (A.81) is a basic equation of acoustics (Morse and Ingard, 1968).

It is assumed that the boundary condition at the upper boundary $y = 0$ describes a line impulse,

$$y = 0 : \mu \frac{\partial w}{\partial y} = -P\delta(x)\delta(t), \tag{A.82}$$

where P denotes the strength of the impulse, and μ is the elasticity of the medium.

Solution by Laplace and Fourier Transforms

The Laplace transform of the variable w is defined as

$$\overline{w} = \int_0^\infty w \exp(-st) \, dt. \tag{A.83}$$

If it is assumed that at time $t = 0$ the system is at rest, the differential equation (A.81) is transformed into

$$\frac{\partial^2 \overline{w}}{\partial x^2} + \frac{\partial^2 \overline{w}}{\partial y^2} = (s^2/c^2)\overline{w}. \tag{A.84}$$

The Fourier transform of the function $\overline{w}$ is defined, using (A.29), as

$$\overline{W} = \int_{-\infty}^\infty \overline{w} \exp(ix\xi) \, dx. \tag{A.85}$$

Using the property (A.30) the differential equation (A.84) is further transformed into

$$\frac{\partial^2 \overline{W}}{\partial y^2} = (s^2/c^2 + \xi^2)\overline{W}. \tag{A.86}$$

For mathematical convenience the parameter ξ is written as $\xi = s\alpha$. This gives

$$\frac{\partial^2 \overline{W}}{\partial y^2} = s^2 k^2 \overline{W}, \tag{A.87}$$

where

$$k^2 = 1/c^2 + \alpha^2. \tag{A.88}$$

It may be noted that the inverse Fourier transform is, with (A.28),

$$\overline{w} = \frac{1}{2\pi} \int_{-\infty}^\infty \overline{W} \exp(-ix\xi) \, d\xi, \tag{A.89}$$

or, with $\xi = s\alpha$,

$$\overline{w} = \frac{s}{2\pi} \int_{-\infty}^{\infty} \overline{W} \exp(-is\alpha x)\, d\alpha. \qquad (A.90)$$

The general solution of the ordinary differential equation (A.87) is, assuming that the solution should vanish for $y \to \infty$,

$$\overline{W} = A \exp(-sky), \qquad (A.91)$$

The integration constant A must be determined from the boundary condition at the surface $y = 0$.

The Laplace transform of the boundary condition (A.82) is

$$y = 0 : \mu \frac{\partial \overline{w}}{\partial y} = -P\delta(x), \qquad (A.92)$$

and the Fourier transform of this condition is, using (A.85),

$$y = 0 : \mu \frac{\partial \overline{W}}{\partial y} = -P. \qquad (A.93)$$

It follows from (A.91) and (A.93) that

$$A = \frac{P}{\mu sk}. \qquad (A.94)$$

Substitution of this result in the general solution (A.91) gives

$$\overline{W} = \frac{P}{\mu sk} \exp(-sky). \qquad (A.95)$$

Inverse Fourier transformation using (A.90) gives

$$\overline{w} = \frac{P}{2\pi\mu} \int_{-\infty}^{\infty} \frac{\exp[-s(i\alpha x + ky)]}{k}\, d\alpha. \qquad (A.96)$$

where k is defined by (A.88), i.e. $k^2 = 1/c^2 + \alpha^2$.

Inverse Transform

The remaining mathematical problem is to evaluate the integral (A.96), and then to determine the inverse Laplace transform. In general this may be a formidable problem.

An elegant way to determine the inverse transforms was developed by De Hoop (1960). In this method the integration variable α is first replaced by $p = i\alpha$. The integral (A.96) then becomes

$$\overline{w} = \frac{P}{2\pi i\mu} \int_{-i\infty}^{i\infty} \frac{\exp[-s(px + ky)]}{k}\, dp, \qquad (A.97)$$

Fig. A.8 Integration path in the complex p-plane

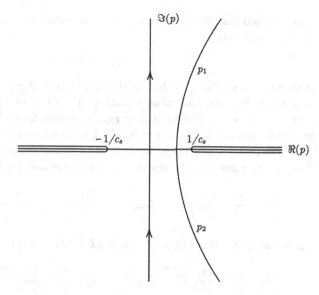

where now the parameter k can be expressed, with (A.88), into p as

$$k^2 = 1/c^2 - p^2. \tag{A.98}$$

The integrand of the integral is continued analytically in the complex p-plane, see Fig. A.8.

Two branch cuts are necessary, from the branch points at $p = \pm 1/c$ to infinity. It is most convenient to let the branch cuts follow the real axis, as shown in the figure.

The integration path in the complex p-plane now is modified to a curved path, also indicated in Fig. A.8. It is assumed that along this curved path a parameter t can be defined so that

$$t = px + ky. \tag{A.99}$$

It is assumed that t is a real and positive parameter. It will later be identified with the time.

It follows from (A.98) and (A.99) that

$$k^2 = 1/c^2 - p^2 = (t^2 - 2tpx + p^2x^2)/y^2, \tag{A.100}$$

or

$$r^2p^2 - 2txp + t^2 - y^2/c^2 = 0, \tag{A.101}$$

where

$$r^2 = x^2 + y^2. \tag{A.102}$$

Equation (A.101) is a quadratic equation in p, with two solutions,

$$p_1 = \frac{tx}{r^2} + \frac{iy}{r^2}\sqrt{t^2 - r^2/c^2}, \qquad p_2 = \frac{tx}{r^2} - \frac{iy}{r^2}\sqrt{t^2 - r^2/c^2}. \tag{A.103}$$

It is assumed that along the two parts of the integration path the variable t varies between the limits

$$r/c < t < \infty. \tag{A.104}$$

It follows that on the upper half of the curve in Fig. A.8 (indicated by p_1) the value of p varies from the real value $p = x/rc$, if $t = r/c$, to a complex value $p = (x + iy)t/r^2$, if $t \to \infty$. The point $p = x/rc$ is always located between the origin and the branch point $p = 1/c$, if $x > 0$, which will be assumed here.

It now remains to express the integral (A.97) in terms of the new variable t. For this purpose it may first be noted from the expressions (A.103) that

$$\frac{dp_1}{dt} = \frac{x}{r^2} + \frac{iy}{r^2}\frac{t}{\sqrt{t^2 - r^2/c^2}}, \qquad \frac{dp_2}{dt} = \frac{x}{r^2} - \frac{iy}{r^2}\frac{t}{\sqrt{t^2 - r^2/c^2}}. \tag{A.105}$$

Furthermore, it follows from (A.99) that $k = t/y - px/y$. This gives, with (A.103),

$$k_1 = \frac{ty}{r^2} - \frac{ix}{r^2}\sqrt{t^2 - r^2/c^2}, \qquad k_2 = \frac{ty}{r^2} + \frac{ix}{r^2}\sqrt{t^2 - r^2/c^2}, \tag{A.106}$$

or

$$\frac{ik_1}{\sqrt{t^2 - r^2/c^2}} = \frac{x}{r^2} + \frac{iy}{r^2}\frac{t}{\sqrt{t^2 - r^2/c^2}}, \qquad \frac{ik_2}{\sqrt{t^2 - r^2/c^2}} = \frac{x}{r^2} - \frac{iy}{r^2}\frac{t}{\sqrt{t^2 - r^2/c^2}}. \tag{A.107}$$

It now follows from (A.105) and (A.107) that

$$\frac{1}{k_1}\frac{dp_1}{dt} = \frac{i}{\sqrt{t^2 - r^2/c^2}}, \qquad \frac{1}{k_2}\frac{dp_2}{dt} = -\frac{i}{\sqrt{t^2 - r^2/c^2}}. \tag{A.108}$$

Substitution into the integral (A.97) now gives, noting that this consists of two branches p_1 and p_2, with the integration path on p_1 from $t = r/c$ to $t = \infty$, and on p_2 from $t = \infty$ to $t = r/c$,

$$\overline{w} = \frac{P}{\pi\mu}\int_{r/c}^{\infty}\frac{\exp(-st)}{\sqrt{t^2 - r^2/c^2}}\,dt. \tag{A.109}$$

This can also be written as

$$\overline{w} = \frac{P}{\pi\mu}\int_{0}^{\infty}\frac{H(t - r/c)}{\sqrt{t^2 - r^2/c^2}}\exp(-st)\,dt, \tag{A.110}$$

where $H(t - r/c)$ is Heaviside's unit step function.

Equation (A.110) has precisely the form of a Laplace transform. It can be concluded that the original function w is

$$w = \frac{P}{\pi\mu}\frac{H(t - r/c)}{\sqrt{t^2 - r^2/c^2}}. \tag{A.111}$$

This completes the solution of the problem. It may be noted that it has been determined without using any explicit form of an inverse Fourier transform formula, or an inverse Laplace transform formula. Actually, the inverse Fourier transform has been modified into a Laplace transform, and the inverse Laplace transform of a Laplace transform is just the original function itself.

Appendix B
Dual Integral Equations

This appendix gives a general procedure for the solution of a system of dual integral equations, as developed by Sneddon (1966).

The system of equations is supposed to be

$$\int_0^\infty F(\xi)A(\xi)J_0(r\xi)\,d\xi = f(r), \quad 0 \leq r < a, \tag{B.1}$$

$$\int_0^\infty \xi A(\xi)J_0(r\xi)\,d\xi = 0, \quad r > a, \tag{B.2}$$

where the functions $F(\xi)$ and $f(r)$ are given in their domain of definition, and $A(\xi)$ is unknown.

The function $A(\xi)$ is represented by a finite Fourier transform

$$A(\xi) = \int_0^a \phi(t)\cos(\xi t)\,dt. \tag{B.3}$$

Using partial integration this can also be written as

$$A(\xi) = \phi(a)\frac{\sin(\xi a)}{\xi} - \int_0^a \phi'(t)\frac{\sin(\xi t)}{\xi}\,dt. \tag{B.4}$$

Using the integral (A.76),

$$\int_0^\infty \sin(\xi t)J_0(\xi r)\,d\xi = 0, \quad r > t, \tag{B.5}$$

it follows that (B.2) is automatically satisfied.

We now use the integral

$$\int_0^s \frac{r}{(s^2 - r^2)^{1/2}}J_0(\xi r)\,dr = \frac{\sin(\xi s)}{\xi}, \tag{B.6}$$

which is the inverse form of the integral (A.76), when this is considered as the Hankel transform of the function $\sin(\xi t)/\xi$.

A. Verruijt, *An Introduction to Soil Dynamics*,
Theory and Applications of Transport in Porous Media 24,
© Springer Science+Business Media B.V. 2010

Application of the operation defined by (B.6) to (B.1) gives

$$\int_0^\infty F(\xi)A(\xi)\frac{\sin(\xi s)}{\xi}\,d\xi = h(s), \quad 0 \le s < a, \tag{B.7}$$

where

$$h(s) = \int_0^s \frac{rf(r)}{(s^2 - r^2)^{1/2}}\,dr. \tag{B.8}$$

Substitution of (B.3) into (B.7) gives

$$\int_0^a \phi(t)\left\{\int_0^\infty F(\xi)\frac{\sin(\xi s)\cos(\xi t)}{\xi}\,d\xi\right\}dt = h(s), \quad 0 \le s < a, \tag{B.9}$$

A well known Fourier cosine transform is (Titchmarsh, 1948, p. 177)

$$\int_0^\infty \frac{\sin(\xi s)\cos(\xi t)}{\xi}\,d\xi = \begin{cases} \frac{\pi}{2}, & t < s, \\ 0, & t > s. \end{cases} \tag{B.10}$$

It now follows that (B.9) can be written as

$$\frac{\pi}{2}\int_0^s \phi(t)\,dt + \int_0^a \phi(t)\left\{\int_0^\infty [F(\xi) - 1]\frac{\sin(\xi s)\cos(\xi t)}{\xi}\,d\xi\right\}dt = h(s),$$
$$0 \le s < a. \tag{B.11}$$

Differentiation with respect to s leads to the Fredholm integral equation

$$\phi(s) + \int_0^a K(t, s)\phi(t)dt = H(s), \quad 0 \le s < a. \tag{B.12}$$

where $K(t, s)$ is the kernel function

$$K(t, s) = \frac{2}{\pi}\int_0^\infty [F(\xi) - 1]\cos(\xi s)\cos(\xi t)\,d\xi, \tag{B.13}$$

and $H(s)$ is the given function

$$H(s) = \frac{2}{\pi}h'(s) = \frac{2}{\pi}\frac{d}{ds}\int_0^s \frac{rf(r)}{(s^2 - r^2)^{1/2}}\,dr. \tag{B.14}$$

The problem has now been reduced to the solution of the Fredholm integral equation (B.12). If this can be solved, the unknown function $A(\xi)$ can be determined using (B.3). It may be noted that in the special case that $F(\xi) = 1$, the kernel function vanishes, and the integral equation (B.12) reduces to the explicit solution $\phi(s) = H(s)$.

Appendix C
Bateman-Pekeris Theorem

This appendix presents the Bateman-Pekeris theorem (Bateman and Pekeris, 1945), which is used in Chap. 13.
The theorem is

$$\int_0^\infty x f(x) J_0(px) \, dx = -\frac{2}{\pi} \Im \int_0^\infty y f(iy) K_0(py) \, dy, \qquad (C.1)$$

where $p > 0$, and $f(z)$ is an analytic function of z in the half plane $\Re(z) > 0$, such that $f(z)$ is real if z is real, and satisfies the condition

$$\lim_{z \to \infty} z^{3/2} f(z) = 0 \quad (\Re(z) > 0). \qquad (C.2)$$

In order to prove this theorem the basic integral is written as

$$N(p) = \int_0^\infty x f(x) J_0(px) \, dx. \qquad (C.3)$$

The Bessel function $J_0(z)$ can be written as the sum of two Hankel functions (Abramowitz and Stegun, 1964, 9.1.3 and 9.1.4),

$$J_0(z) = \frac{1}{2} H_0^{(1)}(z) + \frac{1}{2} H_0^{(2)}(z), \qquad (C.4)$$

so that the integral $N(p)$ can be decomposed into two parts

$$N(p) = N_1(p) + N_2(p), \qquad (C.5)$$

where

$$N_1(p) = \frac{1}{2} \int_0^\infty x f(x) H_0^{(1)}(px) \, dx, \qquad (C.6)$$

and

$$N_2(p) = \frac{1}{2} \int_0^\infty x f(x) H_0^{(2)}(px) \, dx. \qquad (C.7)$$

A. Verruijt, *An Introduction to Soil Dynamics*,
Theory and Applications of Transport in Porous Media 24,
© Springer Science+Business Media B.V. 2010

Fig. C.1 Quarter plane
$\Re(z) > 0, \Im(z) > 0$

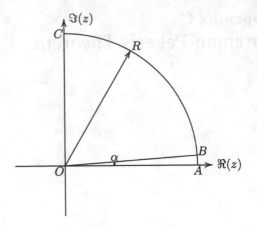

The integral $N_1(p)$ will be considered first. Because the function $f(z)$, by assumption, is analytic in the half plane $\Re(z) > 0$, and $H_0^{(1)}(z)$ is an analytic function of z in the entire plane except at infinity, the function $z f(z) H_0^{(1)}(pz)$ is analytic in the quarter plane $\Re(z) > 0, \Im(z) > 0$. This means that the integral along the contour OABCO in Fig. C.1 is zero,

$$\oint z f(z) H_0^{(1)}(pz) \, dz = 0, \tag{C.8}$$

for every value of the radius R.

It can be shown that the integral over the arc AB tends towards zero for large values of the radius R. For this purpose it may be noted that the asymptotic behaviour of the function $H_0^{(1)}(z)$ is (Abramowitz and Stegun, 1964, 9.2.3)

$$H_0^{(1)}(z) \approx \left(\frac{2}{\pi z}\right)^{1/2} \exp\left(ix - y - \frac{1}{4}i\pi\right) \quad (-\pi < \arg(z) < 2\pi). \tag{C.9}$$

Along the arc AB the Bessel function is $H_0^{(1)}(pz) \approx (2/\pi Rp)^{1/2}$. The integral of the function $z f(z) H_0^{(1)}(pz)$ along the arc AB is, approximately,

$$I_{AB} = \int_{AB} z f(z) H_0^{(1)}(pz) \, dz \approx R f(A) \left(\frac{2}{\pi Rp}\right)^{1/2} R\alpha. \tag{C.10}$$

Because of the condition (C.2) this will tend towards zero if $R \to \infty$.

Along the arc BC the integral will also tend towards zero, because there the integral can be overestimated by

$$\frac{1}{2}\pi R^2 f(R\exp(i\varphi)) \left(\frac{2}{\pi Rp}\right)^{1/2} \exp(-\alpha Rp). \tag{C.11}$$

For $R \to \infty$ this will certainly tend towards zero, because of the exponential factor,

$$\lim_{R \to \infty} I_{BC} = 0. \tag{C.12}$$

Because the contour integral is zero, see (C.8), and the integral along the arc AC is zero, it follows, with (C.6), that

$$N_1(p) = -\frac{1}{2} \int_0^\infty y f(iy) H_0^{(1)}(ipy) \, dy. \tag{C.13}$$

The Hankel function of imaginary argument can be expressed into the modified Bessel function $K_0(z)$ (Abramowitz and Stegun, 1964, 9.6.4),

$$K_0(y) = \frac{1}{2} i \pi H_0^{(1)}(iy), \tag{C.14}$$

so that the final expression for the integral $N_1(p)$ is

$$N_1(p) = \frac{i}{\pi} \int_0^\infty y f(iy) K_0(py) \, dy. \tag{C.15}$$

In a similar way the integral $N_2(p)$ can be transformed into an integral along the imaginary axis. Because at infinity the behaviour of the function $H_0^{(2)}(z)$ is different from that of $H_0^{(1)}(z)$, however,

$$H_0^{(2)}(z) \approx \left(\frac{2}{\pi z}\right)^{1/2} \exp\left(-ix + y + \frac{1}{4}i\pi\right) \quad (-2\pi < \arg(z) < \pi), \tag{C.16}$$

the contour must now be closed by a quarter circle in the lower right half plane. This will give

$$N_2(p) = -\frac{1}{2} \int_0^\infty y f(-iy) H_0^{(2)}(-ipy) \, dy. \tag{C.17}$$

Again this can be expressed into the modified Bessel function $K_0(z)$, using the formula (Abramowitz and Stegun, 1964, 9.6.4),

$$K_0(y) = -\frac{1}{2} i \pi H_0^{(2)}(-iy). \tag{C.18}$$

The final expression for the integral $N_2(p)$ is

$$N_2(p) = -\frac{i}{\pi} \int_0^\infty y f(-iy) K_0(py) \, dy. \tag{C.19}$$

Because the function $f(z)$ is real along the real axis, it follows from the reflection principle (Titchmarsh, 1948, p. 155) that

$$f(-iy) = \overline{f(iy)}. \tag{C.20}$$

Thus it follows that

$$N_2(p) = \overline{N_1(p)}. \tag{C.21}$$

With (C.5) and (C.15) it now follows that

$$N(p) = -\frac{2}{\pi} \Im \int_0^\infty y\, f(iy) K_0(py)\, dy. \tag{C.22}$$

This proves the theorem.

References

M. Abramowitz and I.A. Stegun. *Handbook of Mathematical Functions*. National Bureau of Standards, Washington, 1964.

J.D. Achenbach. *Wave Propagation in Elastic Solids*. Elsevier, Amsterdam, 1975.

K. Aki and P.G. Richards. *Quantitative Seismology*, 2nd edition. University Science Books, Sausalito, 2002.

F.B.J. Barends. Dynamics of elastic plates on a flexible subsoil. *LGM-Mededelingen*, **21**:127–134, 1980.

D.D. Barkan. *Dynamics of Bases and Foundations*. McGraw-Hill, New York, 1962.

H. Bateman. *Tables of Integral Transforms*. McGraw-Hill, New York, 1954. 2 vols.

H. Bateman and C.L. Pekeris. Transmission of light from a point source in a medium bounded by diffusely reflecting parallel plane surfaces. *J. Opt. Soc. Am.*, **35**:651–657, 1945.

W.G. Bickley and A. Talbot. *Vibrating Systems*. Clarendon Press, Oxford, 1961.

M.A. Biot. General theory of three-dimensional consolidation. *J. Appl. Phys.*, **12**:155–164, 1941.

M.A. Biot. Theory of propagation of elastic waves in a fluid-saturated porous solid. *J. Acoust. Soc. Am.*, **28**:168–191, 1956.

M.A. Biot and D.G. Willis. The elastic coefficients of the theory of consolidation. *J. Appl. Mech.*, **24**:594–601, 1957.

A.W. Bishop. The influence of an undrained change in stress on the pore pressure in porous media of low compressibility. *Géotechnique*, **23**:435–442, 1973.

G. Bornitz. *Über die Ausbreitung der von Großkolbenmaschinen erzeugten Bodenschwingungen in die Tiefe*. Springer, Berlin, 1931.

J.E. Bowles. *Analytical and Computer Methods in Foundation Engineering*. McGraw-Hill, New York, 1974.

R.B.J. Brinkgreve and P.A. Vermeer. *PLAXIS, Finite Element Code for Soil and Rock Analysis*. Swets & Seitlinger, Lisse, 2002.

L.P.E. Cagniard, E.A. Flinn and C.H. Dix. *Reflection and Refraction of Progressive Seismic Waves*. McGraw-Hill, New York, 1962.

H.S. Carslaw and J.C. Jaeger. *Operational Methods in Applied Mathematics*, 2nd edition. Oxford University Press, London, 1948.

C.H. Chapman. *Fundamentals of Seismic Wave Propagation*. Cambridge University Press, Cambridge, 2004.

R.V. Churchill. *Operational Mathematics*, 3rd edition. McGraw-Hill, New York, 1972.

J. Cole and J. Huth. Stresses produced in a half plane by moving loads. *ASME J. Appl. Mech.*, **25**:433–436, 1958.

O. Coussy. *Poromechanics*. Wiley, Chichester, 2004.

C.W. Cryer. A comparison of the three-dimensional consolidation theories of Biot and Terzaghi. *Q. J. Mech. Appl. Math.*, **16**:401–412, 1963.

B.M. Das. *Principles of Soil Dynamics*. PWS-Kent, Boston, 1993.

A. Verruijt, *An Introduction to Soil Dynamics*,
Theory and Applications of Transport in Porous Media 24,
© Springer Science+Business Media B.V. 2010

R. De Boer. *Theory of Porous Media*. Springer, Berlin, 2000.

A.T. De Hoop. A modification of Cagniard's method for solving seismic pulse problems. *Appl. Sci. Res. B*, **8**:349–356, 1960.

A.T. De Hoop. The surface line source problem in elastodynamics. *De Ingenieur*, **82**:ET19-21, 1970.

G. De Josselin de Jong. Wat gebeurt er in de grond tijdens het heien? *De Ingenieur*, **68**:B77–B88, 1956.

G. De Josselin de Jong. Consolidatie in drie dimensies. *LGM-Mededelingen*, **7**:25–73, 1963.

E.H. De Leeuw. The theory of three-dimensional consolidation applied to cylindrical bodies. *Proc. 6th Int. Conf. Soil Mech. Found. Eng.*, **1**:287–290, 1965.

D.G. Duffy. *Transform Methods for Solving Partial Differential Equations*. CRC Press, Boca Raton, 1994.

A. Erdélyi, W. Magnus, F. Oberhettinger and F.G. Tricomi. *Tables of Integral Transforms*. McGraw-Hill, New York, 1954. 2 vols.

A.C. Eringen and E.S. Suhubi. *Elastodynamics*, volume 2 of Linear Theory. Academic Press, New York, 1975.

M. Ewing, W. Jardetzky and F. Press. *Elastic Waves in Layered Media*. McGraw-Hill, New York, 1957.

R. Foinquinos and J.M. Roësset. Elastic layered half-spaces subjected to dynamic surface loads. In E. Kausel, G. Manolis, editors, *Wave Motion in Earthquake Engineering*, pages 141–191. WIT Press, Southampton, 2000.

L. Fryba. *Vibration of Solids and Structures under Moving Loads*, 3rd edition. Telford, London, 1999.

Y.C. Fung. *Foundations of Solid Mechanics*. Prentice-Hall, Englewood Cliffs, 1965.

F. Gassmann. Über die Elastizität poröser Medien. *Vierteljahrsschr. Naturforsch. Ges. Zurich*, **96**:1–23, 1951.

G. Gazetas. Foundation vibrations. In H.Y. Fang, editor, *Foundation Engineering Handbook*. Van Nostrand Reinhold, New York, 1991.

J. Geertsma. The effect of fluid-pressure decline on volumetric changes of porous rocks. *Trans. AIME*, **210**:331–343, 1957.

R.E. Gibson. Some results concerning displacements and stresses in a non-homogeneous elastic half-space. *Géotechnique*, **17**:58–67, 1967.

R.E. Gibson, K. Knight and P.W. Taylor. A critical experiment to examine theories of three-dimensional consolidation. *Proc. Eur. Conf. Soil Mech. Wiesbaden*, **1**:69–76, 1963.

K.F. Graff. *Wave Motion in Elastic Solids*. Oxford University Press, London, 1975.

A.E. Green and W. Zerna. *Theoretical Elasticity*. Clarendon Press, Oxford, 1954.

W. Gröbner and N. Hofreiter. *Integraltafel*. Springer, Wien, 1961.

B.O. Hardin. The nature of damping in soils. *J. Soil Mech. Found. Div., Proc. ASCE*, **91**(SM1):63–97, 1965.

M.E. Harr. *Foundations of Theoretical Soil Mechanics*. McGraw-Hill, New York, 1966.

H.G. Hopkins. Dynamic expansion of spherical cavities in metals. *Prog. Solid Mech.*, **1**:83–164, 1960.

I.M. Idriss and H.B. Seed. An analysis of ground motions during the 1957 San Francisco Earthquake. *Bull. Seismol. Soc. Am.*, **58**:2013–2032, 1968.

C.E. Jacob. The flow of water in an elastic artesian aquifer. *Trans. Am. Geophys. Union*, **21**:574–586, 1940.

M.K. Kassir and G.C. Sih. *Three-Dimensional Crack Problems*. Noordhoff, Leyden, 1975.

E. Kausel. *Fundamental Solutions in Elastodynamics*. Cambridge University Press, Cambridge, 2006.

D.E. Knuth. *The TeXbook*. Addison Wesley, Reading, 1986.

H. Kolsky. *Stress Waves in Solids*. Dover, New York, 1963.

S.L. Kramer. *Geotechnical Earthquake Engineering*. Prentice-Hall, Upper Saddle River, 1996.

H. Lamb. *Hydrodynamics*, 6th edition. Cambridge University Press, Cambridge, 1932.

H. Lamb. On the propagation of tremors over the surface of an elastic solid. *Philos. Trans. R. Soc., Ser. A*, **203**:1–42, 1904.

T.W. Lambe and R.V. Whitman. *Soil Mechanics*. Wiley, New York, 1969.

L. Lamport. *LaTeX*, 2nd edition. Addison Wesley, Reading, 1994.

J. Lysmer and F.E. Richart Jr. Dynamic response of footings to vertical loading. *J. Soil Mech. Found. Div., Proc. ASCE*, **92**(SM1):65–91, 1966.

J. Mandel. Consolidation des Sols. *Géotechnique*, **7**:287–299, 1953.

J. Miklowitz. *Elastic Waves and Waveguides*. North-Holland, Amsterdam, 1978.

M.M. Mooney. Some numerical solutions of Lamb's problem. *Bull. Seismol. Soc. Am.*, **64**:473–491, 1974.

P.M. Morse and H. Feshbach. *Methods of Theoretical Physics*. McGraw-Hill, New York, 1953 (reprinted by Feshbach Publishing, Minneapolis, 1981).

P.M. Morse and K.U. Ingard. *Theoretical Acoustics*. McGraw-Hill, New York, 1968.

C.L. Pekeris. The seismic surface pulse. *Proc. Natl. Acad. Sci.*, **41**:469–480, 1955.

W.L. Pilant. *Elastic Waves in the Earth*. Elsevier, Amsterdam, 1979.

W.H. Press, B.P. Flannery, S.A. Teukolsky and W.T. Vetterling. *Numerical Recipes in C*. Cambridge University Press, Cambridge, 1988.

Lord Rayleigh. On waves propagated along the plane surface of an elastic solid. *Proc. Lond. Math. Soc.*, **17**:4–11, 1885.

Lord Rayleigh. *The Theory of Sound*. Macmillan, London, 1894 (reprinted by Dover, New York, 1945).

L. Rendulic. Porenziffer und Porenwasserdruck in Tonen. *Der Bauingenieur*, **17**:559–564, 1936.

F.E. Richart, J.R. Hall and R.D. Woods. *Vibrations of Soils and Foundations*. Prentice-Hall, Englewood Cliffs, 1970.

R.L. Schiffman. A bibliography of consolidation. In J. Bear, M.Y. Corapcioglu, editors, *Fundamentals of Transport Phenomena in Porous Media*, pages 617–669. Martinus Nijhoff, Dordrecht, 1984.

H.B. Seed and I.M. Idriss. *Ground Motions and Soil Liquefaction During Earthquakes*, Earthquake Engineering Research Institute, 1982.

A.P.S. Selvadurai. *Elastic Analysis of Soil-Foundation Interaction*. Elsevier, Amsterdam, 1979.

A.W. Skempton. The pore pressure coefficients A and B. *Géotechnique*, **4**:143–147, 1954.

D.M.J. Smeulders. On Wave Propagation in Saturated and Partially Saturated Porous Media, Ph.D. Thesis, Eindhoven, 1992.

E.A.L. Smith. Pile driving analysis by the wave equation. *Trans. ASCE*, **127**:1145–1193, 1962.

I.N. Sneddon. *Fourier Transforms*. McGraw-Hill, New York, 1951.

I.N. Sneddon. *Mixed Boundary Value Problems in Potential Theory*. North-Holland, Amsterdam, 1966.

I.S. Sokolnikoff. *Mathematical Theory of Elasticity*, 2nd edition. McGraw-Hill, New York, 1956.

A. Sommerfeld. *Partial Differential Equations in Physics*. Academic Press, New York, 1949.

H.J. Stam. The two-dimensional elastodynamic distributed surface load problem. *Geophysics*, **55**(8):1047–1056, 1990.

K. Terzaghi. *Erdbaumechanik auf bodenphysikalischer Grundlage*. Deuticke, Wien, 1925.

K. Terzaghi. *Theoretical Soil Mechanics*. Wiley, New York, 1943.

S.P. Timoshenko and J.N. Goodier. *Theory of Elasticity*, 3rd edition. McGraw-Hill, New York, 1970.

E.C. Titchmarsh. *Theory of Fourier Integrals*, 2nd edition. Clarendon Press, Oxford, 1948.

J.G.M. Van der Grinten. An Experimental Study of Shock-Induced Wave Propagation in Dry, Water-Saturated and Partially Saturated Porous Media, Ph.D. Thesis, Eindhoven, 1987.

A. Verruijt. Discussion on consolidation of a massive sphere. *Proc. 6th Int. Conf. Soil Mech. Montreal*, **3**:401–402, 1965.

A. Verruijt. Elastic storage of aquifers. In R.J.M. De Wiest, editor, *Flow through Porous Media*. Academic Press, New York, 1969.

A. Verruijt. Dynamics of soils with hysteretic damping. In F.B.J. Barends, et al., editors, *Proc. 12th Eur. Conf. Soil Mech. and Geotechnical Engineering*, volume 1, pages 3–14. Balkema, Rotterdam, 1999.

A. Verruijt and C. Cornejo Córdova. Moving loads on an elastic half-plane with hysteretic damping. *ASME J. Appl. Mech.*, **68**:915–922, 2001.

A. Verruijt. An approximation of the Rayleigh stress waves generated in an elastic half plane. *Soil Dyn. Earthq. Eng.*, **28**:159–168, 2008a.

A. Verruijt. Consolidation of soils. In *Encyclopedia of Hydrological Sciences*. Wiley, Chichester, 2008b. doi:10.1002/0470848944.hsa303

A. Verruijt, R.B.J. Brinkgreve and S. Li. Analytical and numerical solution of the elastodynamic strip load problem. *Int. J. Numer. Anal. Methods Geomech.*, **32**:65–80, 2008.

H.F. Wang. *Theory of Linear Poroelasticity*. Princeton University Press, Princeton, 2000.

G.N. Watson. *Theory of Bessel Functions*, 2nd edition. Cambridge University Press, Cambridge, 1944.

E.W. Weisstein. *The CRC Concise Encyclopedia of Mathematics*. CRC Press, Boca Raton, 1999.

H.M. Westergaard. A problem of elasticity suggested by a problem in soil mechanics: soft material reinforced by numerous strong horizontal sheets. In *Contributions to the Mechanics of Solids, Stephen Timoshenko 60th Anniversary Volume*. Macmillan, New York, 1938.

M.J. Wichura. *The PiCTeX Manual*. TeX Users Group, Providence, 1987.

C.R. Wylie. *Advanced Engineering Mathematics*, 2nd edition. McGraw-Hill, New York, 1960.

Author Index

Index

A. Verruijt, *An Introduction to Soil Dynamics*,
Theory and Applications of Transport in Porous Media 24,
© Springer Science+Business Media B.V. 2010

Printed in the United States
By Bookmasters